Conscious States:

The AIM Model of Waking, Sleeping, and Dreaming

About the Author

Allan Hobson (1933–) has always been an experimentalist. In addition to the science described in this book, he has dabbled in a wide variety of activities, especially the design and execution of illustration. Some of these works are the subject of companion volumes written by him and edited by Nicholas Tranquillo. They include the titles: Vermont Professor, London Bridges, The Dream Asylum and five volumes of recent essays. This book was written to celebrate curiosity, originality and collaboration. I hope it will inspire others to follow their quest for new truth wherever it leads them. Copies of the book are available from Amazon.com. For further information, contact: allan_hobson@hms.harvard.edu.

Dedication

The ideas developed in this book were originally elaborated by my former student and collaborator, Robert W. McCarley. Bob's interest in mathematics, his sensitive reading of history, his wry sense of humor, and his toleration of me in the years between 1965 and 2000 are fondly remembered. His subsequent independent career in neuroscientific psychiatry has allowed him to flourish while letting me do my best to imitate his fine example.

Robert W. McCarley (1937–). A talent for mathematics and the history of psychology led to the formal approach to the study of consciousness, which I champion in this book. McCarley appreciated the need for rigor in the definition and quantification of cellular and molecular neurophysiology if one wished to explain the subjective data of consciousness in terms of brain phenomena. He realized that this was precisely what Freud had failed to do in his failed *Project for a Scientific Psychology* (1895) and the transliteration of his erroneous brain science into his speculative dream theory (1900). McCarley's genius helped to create the conceptual and empirical breakthrough that now constitutes the virtual reality model of human consciousness. McCarley has also developed novel approaches to mental illness through his application of evoked potential study of the temporal lobe of schizophrenic patients.

Foreword

Consciousness has been called the last scientific frontier. Neurobiology considers this frontier to have been reached and has begun to study it via the selective analysis of consciousness components. This book uses neurobiology to examine how all of these psychological components are tied together physiologically as the states of waking, sleeping, and dreaming.

Having established the scientific validity of identification and differentiation of the physiological states, their psychological features are defined and measured. This establishes the correlation between physiology and psychology. In this book, emphasis is placed upon the correlation of dream psychology with REM sleep physiology compared with the correlation of psychology with the physiology of waking.

Because this is a very novel and very preliminary effort, the correlations sought and demonstrated are necessarily *formal*. This contrasts with the *content analytic* approach, which has heretofore failed to produce a science of conscious states, and especially a science of dreaming. By formal, I mean, for example, the identification and measurement of such coarse-grained global and universal psychological features of conscious states as sensation, attention, perception, thought, memory, and volition.

Once identified and measured, such formal psychological features of any given conscious state can then be mapped onto the measured formal properties of the physiological state with which they are correlated. Based on the goodness of fit of the correlations, the hypothesis of the unity of psychology and physiology may then be considered. In fact, the fit of dream psychology (e.g. bizarreness) and REM sleep physiology (e.g. aminergic demodulation) is so good as to suggest at least a formal isomorphism between the two levels of analysis. The fit is so good as to suggest a causal identity between psychology and physiology.

Among the scientific implications of this dual-aspect line of reasoning is the elaboration of a three-dimensional model, called AIM, which illustrates the concept of brain/mind isomorphism and shows how that concept can be used to understand not only the vicissitudes of normal consciousness but also the panoply of unusual and abnormal states such as lucid dreaming, hypnosis, and psychosis. The AIM model is presented as a preview of coming attractions at the hands of mathematically sophisticated consciousness scientists like Karl Friston.

The most significant philosophical implication of the dual-aspect monism philosophy elaborated in this book is the *reciprocal causality* principle. Not only are the psychological and physiological levels of analysis linked phenomenologically, but they are also linked causally: mind is causal upon brain and brain is causal upon mind. This is not surprising if they are one and the same thing and only observed and measured separately.

Conscious States:
The AIM Model of Waking, Sleeping, and Dreaming

* First drafts of these chapters were written by Trevor Harley of the University of Dundee.

PART I. THE AIM MODEL

Introduction and Summary, Part I.

In Chapter 1, I define consciousness and elaborate the conscious states paradigm. I explain why I regard sleep and dream research to be an indispensable and solid foundation for consciousness science. The historical and philosophical context of the conscious states paradigm is detailed in Chapters 2 and 3. I first demonstrate the integrative power of the conscious states paradigm to unite the strands of early twentieth century research into the brain (neurobiology), the mind (psychoanalysis) and behavior (operant conditioning and ethology). By seeing the mind as a physical force, the conscious states paradigm moves beyond dualism to dual aspect monism. As a physical force, the mind has causal impact on the brain from which it derives. To a first approximation, the conscious states paradigm thus solves the mind-body problem.

The brain anatomy of conscious state control is the subject of Chapter 4. In this chapter, I emphasize the brain structures, which unite the mental faculties so as to produce a reliable sequence of the waking, sleeping, and dreaming states of consciousness. I assert that consciousness itself will never be anatomically identified. Instead it is seen by me to be a state of matter whose structural basis is beginning to be understood. The failure of early attempts to reduce consciousness to brain structure via lesion and stimulation studies is the subject of Chapter 5. I use the erroneous theory of conscious awareness as a response to external stimulation as a heuristic device to help the reader understand the significance of our alternative theory of intrinsic and spontaneous conscious state generation.

The development of conscious state science has relied heavily on the technical capacity to record the electrical activity of individual neurons in experimental animals and to compare those data with subjective and objective findings in conscious humans. Three basic factors have emerged from this work: Factor A, measuring brain-mind activation, is treated in Chapter 6; Factor I, covered in Chapter 7, assesses information source measured as input-output gating and self stimulation; Factor M, referring to mode mediated by modulation, is discussed in Chapter 8. The three factors are integrated in the AIM model of the conscious states of the brain-mind in Chapter 9.

I regard subjectivity as an essential element of our enterprise. However misleading it may be (because subjects are unaware of their brains), subjectivity, objectively measured, must now be considered an open window on the brain. Consciousness is, I insist, a brain function, and the state dependency of its formal properties makes subjectivity not only a phenomenon to be explained by physiology but a physiological function in and of itself.

Chapter 1. Definitions

A science of consciousness is now emerging via the integrative work of psychologists, physiologists, and philosophers. This book will emphasize the success of this multidisciplinary effort with a selective focus on the alterations in the state of consciousness that occur in waking, sleeping, and dreaming. For our purposes, consciousness may be defined and characterized as shown in Figures 1.1 and 1.2.

Figure 1.1

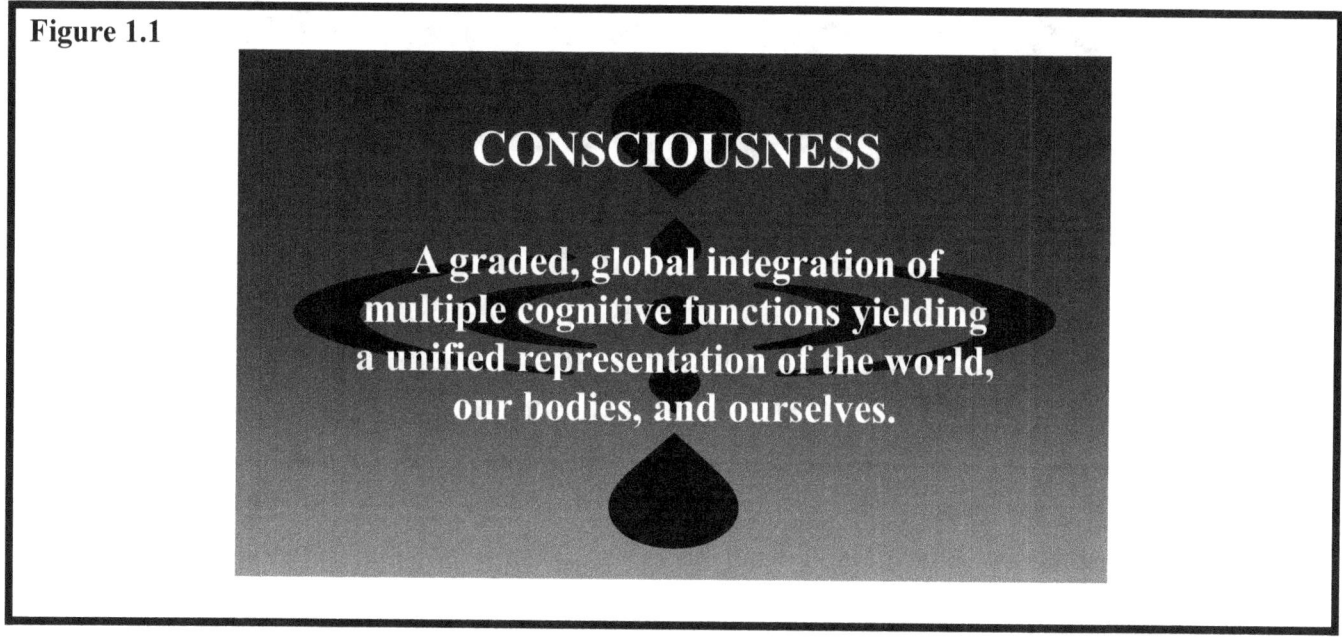

Psychologists are both conscious individuals and professionals who study the mind subjectively and objectively. Their scientific status is gained by mapping back and forth from the subjective to the objective domains and from the self to the other. As shown in Figures 1.3 and 1.4, hypothesis testing is carried out horizontally within and between domains and vertically from the mind to the body and from the body to the mind by physiologists.

It is the explosion of methods in the brain sciences (often called here simply physiology) which gives the modern study of consciousness its new power. Emergent theoretical models are reviewed by philosophers for their mathematical and logical merit. Philosophers may develop novel models and test them using the psychological and physiological approaches outlined in this book.

We say we lose consciousness when we fall asleep but this statement has lost its force owing to the discovery that dreaming, a vivid state of consciousness, occurs in REM sleep. REM sleep was discovered by physiologists who appreciated the psychological significance of their discovery. Thus, there are at least two states of consciousness, waking and dreaming, and one of these is associated with sleep. Dreaming, defined as any mental activity occurring in sleep, is not restricted to REM, indicating further that consciousness is not entirely lost when we are no longer awake.

Enter philosophy. It stands to reason that we do not really "lose" consciousness when we fall asleep. All we lose is waking with its particular brand of consciousness. By falling asleep, we gain access to a quite different state, dream consciousness. This line of thought shows that we have been misled by subjective experience to equate waking with consciousness. We have made the two words synonymous when they are in fact interactive and simultaneous, scientifically distinct and dissociable.

It was first thought that dreaming was an unconscious mental activity, whereas we now recognize that dreaming and waking are two quite different forms of consciousness that do not have more than fleeting access to each other. This means that we need an entirely new way of looking at things. This book will attempt to outline a new approach.

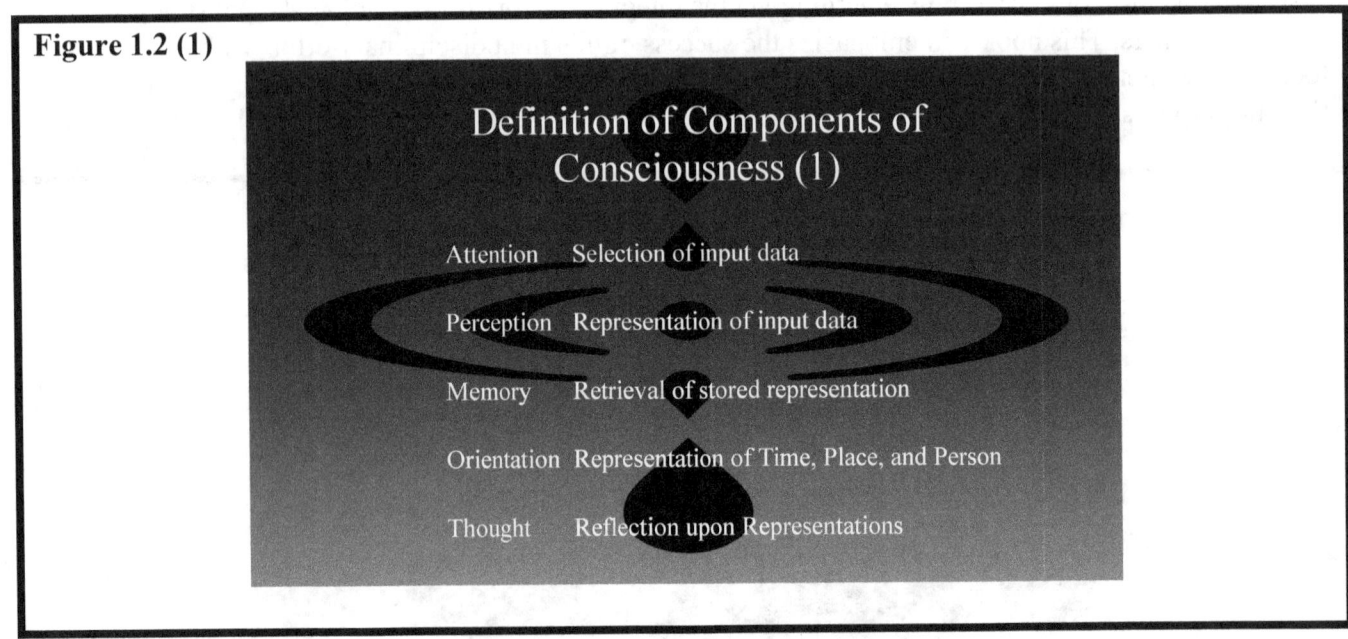

Figure 1.2 (1)

We do not even entirely lose consciousness when we become deeply asleep early in the night. At this time, we may be difficult to rouse and our recall may be non-existent or minimal, but physiology and psychology again conspire to indicate that some very low degree of consciousness may persist even then. Perseverative thought is commonly experienced subjectively in deep sleep when the brain is not inactive, but running at about 50 per cent of waking activation level.

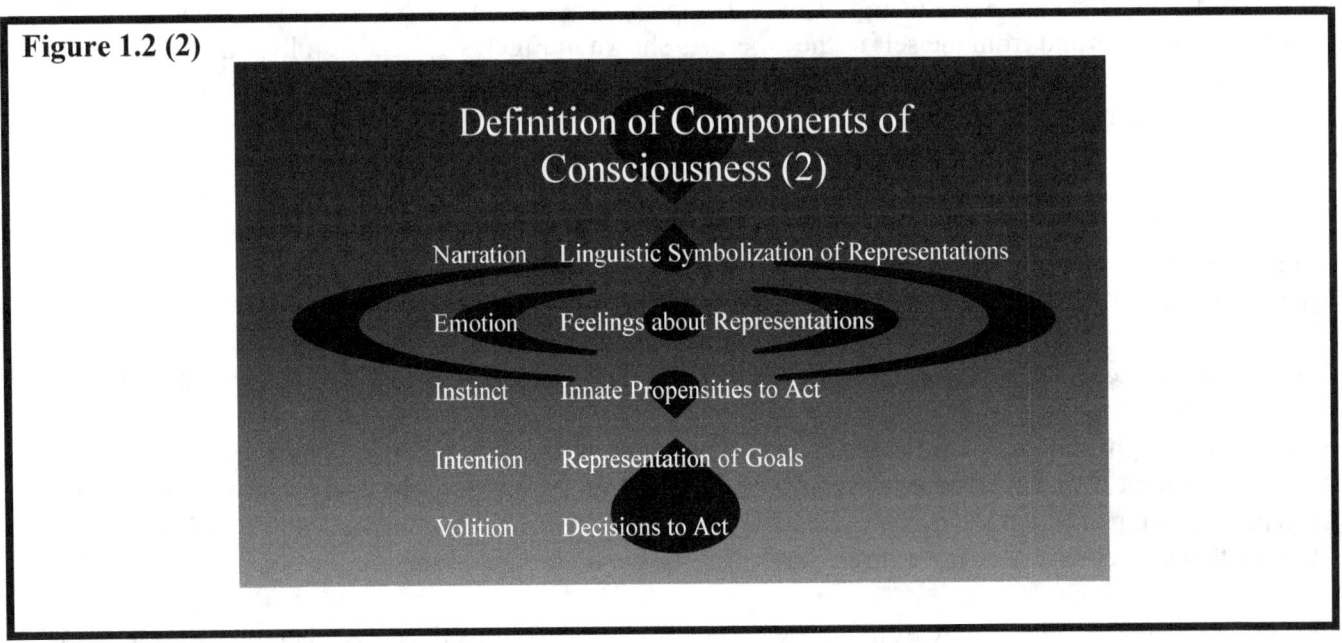

Figure 1.2 (2)

Consciousness is thus not an exclusive concomitant of waking. Nor is consciousness qualitatively or quantitatively fixed. Instead, it is differentiated and graded. As we will spell out in what follows, consciousness is a many-splendored thing which can be looked at subjectively by everyone (in the first person) and measured objectively by psychologists (in the third person). Its brain substrates can furthermore be tracked physiologically. The philosophical implications of all these findings can be used in constructing models of the mind.

The strategy and methods of the conscious states paradigm are depicted in Figures 1.3 and 1.4.

Figure 1.3

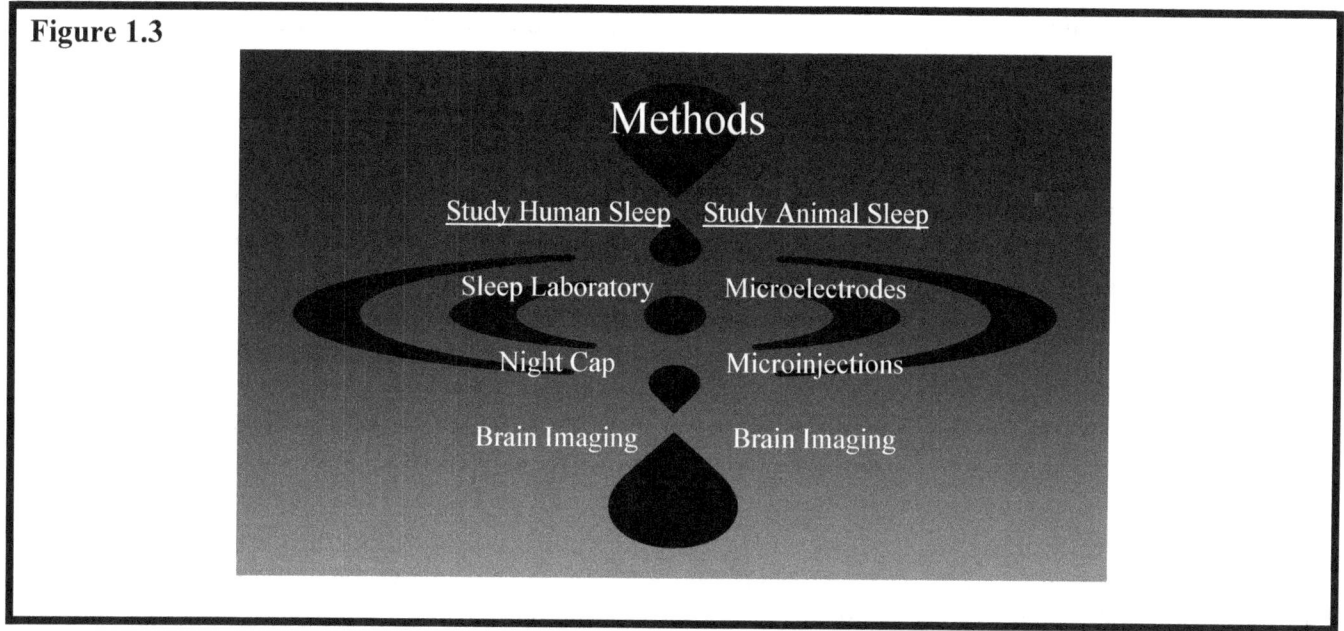

These facts clearly indicate the importance of collaboration between psychologists, physiologists, and philosophers, and the importance of all three disciplines in the minds of all scholars of consciousness. It is difficult, but desirable, to cultivate such a three-way integration, and this book is dedicated to that effort. All serious students of consciousness need to be psychologists, physiologists, and philosophers. No one can be as good at all three disciplines as professionals in any one of them but we all must try, if only to consult collegial experts with active intelligence.

Figure 1.4

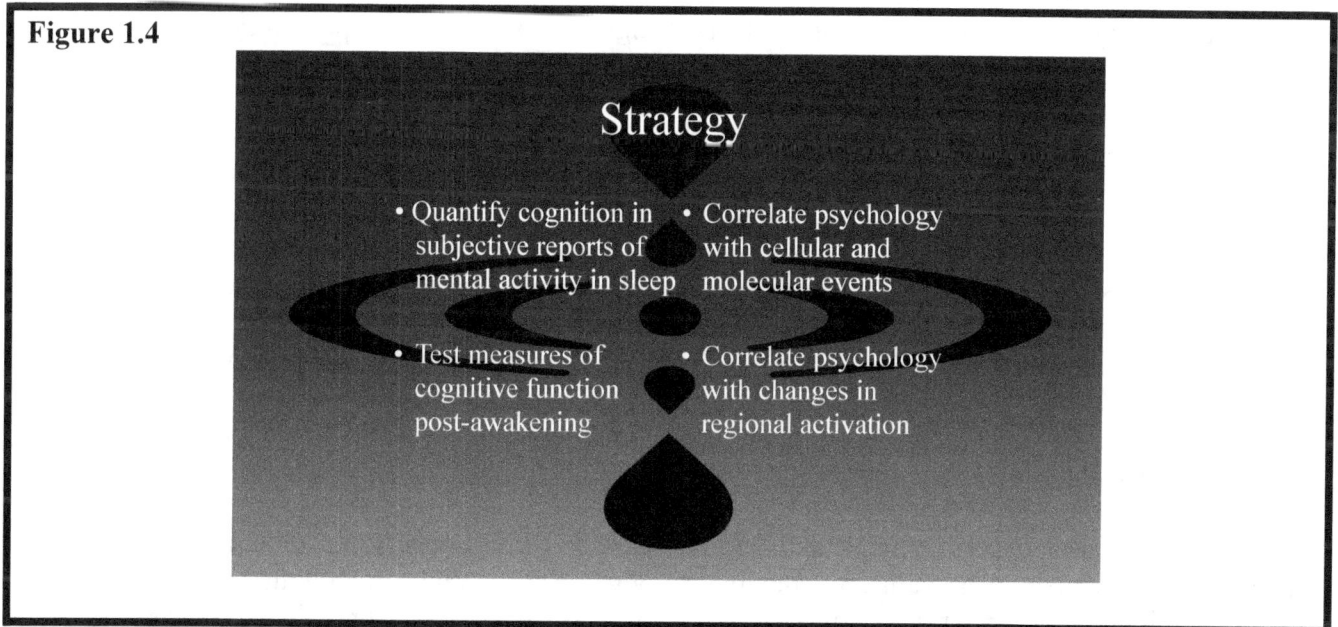

States of Mind and Brain

Despite continuously changing, consciousness tends to be relatively constant for sustained periods of time. These temporally sustained segments are called "states" by scientists who seek to establish working definitions for their psychological features and discern their physiological underpinnings. The best way to appreciate this technical approach is to think of the commonplace reference to states of mind, which refers to the differentiation of waking consciousness. Here, we take a step backward and

downward as we explore the similarities and differences between waking and dreaming.
For example, both waking and dreaming states are vivid which is not surprising since both are associated with brain activation. But while waking is related to the outside world, dreaming is almost entirely internally oriented. This difference is remarkable and demands both theoretical and empirical consideration.

States of mind and brain can be defined rigorously and mathematically. Thus qualitative phenomena, such as the philosophers' famous qualia, are reducible to numbers that are worthy of a physicist's respect. States are the ranges of sets of variables such as subjective visual image intensity (in the case of cognitive perception) and visual area activation (in the case of PET or fMRI brain imagery).
Since the changes in state are marked and robust, the study of consciousness is, in fact, far from impossible, as C-word critics have asserted. These measures are considerably easier to obtain and more easily differentiable than in most cognitive science experiments. The reason for this is a conceptual shift from the traditional stimulus response (which compares small differences within a state — usually waking) to the large state vs. state comparison of consciousness science (for example, REM vs. NREM sleep).

The stimulus-response paradigm has been deeply imbedded in both neuroscience and psychology since the landmark work on spinal reflexes of Sir Charles Sherrington in the late 19th century. So exclusive is the doctrinaire reverence for this paradigm that experimentalists on both sides of the brain-mind hyphen have ignored spontaneous activity and regarded it as noise when it is signal. Even spinal reflexes are state-dependent and can shift from excitation to inhibition. Ask yourself the question, "Why do I not actually run when I am pursued in a nightmarish dream?" This shift in reflex excitability is crucial to understanding the persistence of sleep immobility when the brain is activated in REM.

Before reading further, be sure you understand the difference between reflexes and states. When your doctor taps your patella tendon, he is testing spinal reflex excitability. When he asks you to close your eyes or make a tight fist before he taps again, he is changing the state of your spinal reflex excitability. The brain makes just this sort of alteration when it changes state spontaneously. Consciousness science takes advantage of spontaneous state change when it asks: how and why is this done? This book will suggest some answers to these questions.

Consciousness scientists can identify and measure many other variables to test hypotheses about the similarities and differences between states. With respect to the example given at the beginning of this segment, it is possible to compare as seemingly trivial and simple a measure as report word count length and then correlate this value with the degree of brain activation as physiologically quantified. Not surprisingly, word count is seven times as long in reports following brain activated sleep as from brain inactivated sleep.

First and Third Persons

Our subjective consciousness is unique and private. Thus some philosophers may object to our assertion, in the previous segment, that qualia are reducible. They will assert, quite correctly, that no one can know what I am thinking or whether the red that I see or imagine is the same as the red that you see or imagine. This "first person" consciousness, they say, is irreducible. To emphasize this point, Thomas Nagle has questioned, "What is it like to be a bat?" and replied that the question is unanswerable.

The critical scientist, speaking in the third person, wonders if this difficulty is insurmountable. This is another way of asking if the hard problem of David Chalmers is really that hard? Can it not be softened in the same way that first person qualia are reducible to numbers? Consciousness, some scientists say, is not that unassailable. They will try by several specific means to mount an attack. Some of these means, depicted in Figure 1.4, are listed below:

- Collect reports of conscious states from many first persons
- Quantify the formal features of the reports
- Select the formal features which show ubiquity and robustness
- Correlate formal features with other psychological variables
- Correlate psychological with physiological variables

Sir Charles Sherrington (1857 1950) typifies Anglo-Saxon reductionistic neuroscience. His speculations about consciousness (*Man on his Nature*, 1940) reflect the dualism of his rigorous narrowness and his deeply mystical religious spirit. As such, Sherrington offers a sober counterpoint to the revolution in thinking that occurred within his lifetime and the breadth of interest evinced by his Latin peers. With respect to his science, Sherrington is rightly revered for his careful work on the spinal cord and his conviction that all parts of the brain, including the cerebral cortex, evince functionally significant differentiation. His focus on the spinal cord prevented him from seeing the functional differentiation of the brain stem and more rostral brain structures that mediate conscious states. His "shuttle loom" metaphor for the brain basis of awareness is poetically appealing but has no cogency as a scientific theory.

Two caveats are in order:

As stated in the foreword, formal features are assessed as coarse-grained measures (e.g. vision vs. audition) as against fine-grained measures of content (e.g. what is actually seen or heard). Correlations are not eliminative (e.g. physiology does not replace psychology). Rather, the two domains are separate but equal modes of a monistic reality. By this sort of scientific maneuver, first person subjective experience is rendered tractable (i.e. softened) and made understandable via third person psychology and physiology. States can then be analyzed across and within individuals. One can never be certain that the subjective experience is identical within and across individuals but science can proceed, as we will show below, short of the perfection espoused by philosophy. Anticipating details to be discussed later, recognize that the vision of waking can be compared with the vision of dreaming as a way of better understanding the mechanisms and functions of each.

This pragmatic approach may also reveal the reasons for the separation of the two states of consciousness and the relegation of dreaming to the unconscious with its Freudian assumptions of escape, subterfuge, and hidden meaning. In the following segment, we raise questions about these assumptions and suggest that dreaming may be a protoconscious state upon which waking is built and upon which waking depends.

Psychoanalysis, Behaviorism and Cognitive Science

After Charles Sherrington had enunciated the reflex doctrine and Santiago Ramón y Cajal promulgated the neuron as the essential structural unit of the brain by about 1890, it was then natural for psychologists like William James and psychiatrists like Sigmund Freud to attempt an integration of these two elements into a unified theory of the mind. Consciousness was acknowledged by both James and Freud but not clearly formulated; it was not yet clear how it could be studied scientifically.

In spite of James' caution, Freud moved ahead of science and formulated a theory of dreams, which he regarded as the Royal Road to the Unconscious. This led to the development of psychoanalysis as a theory and as a movement. Psychoanalysis split psychology into a clinical branch, which followed Freud's psychodynamic formulations, and an experimental branch, which was highly skeptical (if not dismissive) of Freud. The most successful experimental work was behaviorism, which denied that consciousness could be approached scientifically.

The most trenchant and vociferous behaviorist was B. F. Skinner (1904–1990) who discovered operant conditioning while a doctoral student at Indiana University. Skinner promulgated behaviorism throughout his long and influential career at Harvard (1948–1990). So scientifically successful was behaviorism and so clinically popular was psychoanalysis that together they forestalled consciousness science for almost a century. Skinner was so vehement in inveighing against a psychology of the mind that consciousness became known as the C-word, never to be pronounced by any serious scholar. Behaviorism was easily adapted to neuroscience as these two movements marched hand in hand on parallel tracks with psychoanalysis, never making a meaningful contact between basic and clinical science. Even cognitive science, wedded to the stimulus response paradigm, failed to incorporate subjective experience into its highly successful agenda. Consciousness, spontaneous activity, and state dependence were seen as uncontrolled variables to be controlled, or even denied, as enemies of scientific progress.

The Origins of Conscious State Science

An important avenue to consciousness science was opened by Hans Berger's 1927 discovery of the EEG with its implication of the brain as electrical generator. The EEG gave rise to the experimental line that included Frédéric Bremer's isolated and deactivated forebrain as well as Loomis and Harvey's periodic brain activation in human sleep, culminating in the clearly endogenous RAS (reticular activating system) and REM. With respect to consciousness itself, the capacity to analyze power spectra in the EEG led to the proposition by Wolf Singer and others that diffuse synchronization by 40 Hz electrical activity could underlie the binding of disparate cognitive elements in consciousness originally restricted to waking but now known to be prominent in REM sleep dreaming, too. These developments are treated in more detail in the subsequent chapter.

Sleep and dream research was clearly incompatible with behaviorism. For obvious reasons, it was initially but unsuccessfully press fit into psychoanalysis. Owing to the discovery of dysfunctional states of sleep and dreaming, they were welcomed into medical science. Cognitive science can and must now expand its scope to include the most important of all mental faculties. This book tries to make clear how and why this should happen.

It should have been obvious that the discovery of the reticular activating system (by Moruzzi and Magoun in 1949) and REM sleep/dreaming (by Aserinsky and Kleitman in 1953) ushered in a new era of consciousness science, but that signal was swamped in the noise of psychoanalysis, behaviorism, and cognitive science and blindered by adherence to the stimulus-response paradigm. The shift to an emphasis on spontaneous state alteration occurred slowly and has only recently approached codification. These empirical advances and their conceptual significance will be revisited in subsequent chapters.

B.F. Skinner's (1904–1990) mantra was that "behavior is determined by its consequences." By this, he meant that the probability of recurrence of any behavioral act was increased by positive, and decreased by negative, reinforcement. He argued that psychology could only become a science by ignoring the mind and consciousness and by restricting scientific attention to the outward signs of brain activity. Skinner's operant conditioning complemented Pavlov's classical conditioning paradigm and together they dominated psychological research from 1935 until about 1950 when the tide began to shift back to an interest in internal states. Psychology was then, and remains today, fixated on memory. Psychologists study learning with experiments structured in the reflex paradigm from which spontaneous brain activity is excluded. At his most polemical, Skinner went so far as to negate conscious awareness from respectable study; his denunciation of the "C-word" (i.e. consciousness) has gradually been countered by mounting evidence that the conscious mind is real and can really be studied scientifically.

Important steps along this way include:
- **Sleep lab quantification of EEG and correlation with psychological measures**
- **Single cell recording of physiological states in experimental animals**
- **Pharmacological manipulation of such states in humans and animals**

- **Brain imaging of physiological states in humans**
- **Neuropsychological studies of brain lesion effects on conscious states in humans**

These and other landmark events will be detailed in subsequent chapters. Here, we only stress the multidisciplinary nature of the work, which helps to explain why conceptual progress was so slow. An additional problem is that each subfield developed in relative isolation from the others. Furthermore, theoretical modeling (especially across the brain-mind interface) was often discouraged by funding agencies. Reductionism was favored over integration. Monism was favored over any dualistic model. Integration was intrinsically difficult as well as risky.

By 1975, modeling had begun but this was still in one domain only, the physiological (in keeping with the reductionism mentioned above). The C-word was still a profanity. It was only after the turn of the 21st century, 50 years after the RAS and REM were discovered, that consciousness clearly entered the scientific picture. The Freudian unconscious mind concept was tenacious until at least 2008. With a challenge to this obdurate theory, it was possible to contemplate virtual reality and dual-aspect monism, ideas which have only assumed theoretical weight in the last five years. Consciousness science is like a newborn infant, full of hope for the future, but still fragile and at risk.

The dualism of René Descartes became the D-word of twentieth century science (the D stands for dualism). Mind was viewed as an epiphenomenal illusion and even those scientists who thought that mind was real had no way of understanding its physical basis in solely structural terms. Exceptions to this conceptual limitation were suggested by sophisticated analysts of the EEG (such as Walter Freeman) but always within the narrow confines of the stimulus-response paradigm of evoked potential experiments.

One of the reasons for the persistence of Cartesian dualism is that religionists believe in the separation of mind and brain. "Spirit" or "soul" is thought by them to emanate from God, only temporarily to inhabit bodies, and to be dissociable from the body at the time of anatomical and physiological death. While I realize that such beliefs arise from brain functions, I reject them as truth. I am thus as allied with Freud's atheism as I am opposed to his dream theory.

The psychoanalytic theory of free association was an attempt to set the mind adrift in the interest of discovering hidden motivation but, of course, no clinical setting could possibly be devoid of suggestive stimuli as Freud had hoped. Seated behind his couch borne subjects, he elicited reports of dreams from which deep meanings might be derived via his interpretations.

Protoconsciousness and Free Will

The next crucial step in conscious state science was to realize that dreaming, and other forms of consciousness were in part spontaneous and in part stimulus-bound. From this realization emerged my theories of protoconsciousness and virtual reality.

Protoconsciousness theory posits that a preverbal way of knowing is universal in all mammals, arises early in intrauterine life, and gradually becomes adult human consciousness as language develops. This preverbal way of knowing consists of a virtual model of the world that is as essential to cognition as the binding together of conscious state components. Protoconsciousness is both genetic and epigenetic. It arises together with the construction of the brain and is further shaped by experience.

The virtual reality concept subserves an understanding of waking consciousness. It is most easily discerned in dreaming, when information about the outside world is restricted to memory. The waking mind, which genetic and experiential protoconsciousness and experience combine to produce, depends upon the brain but, as Karl Friston has insisted, it also has causal effects upon the brain.

This formulation conforms to the definition of dual-aspect monism. It is admittedly dualistic but it replaces the Cartesian spirit (or soul) with a physical but non-anatomical force, the mind, that exerts a power over the anatomical/physiological brain from which it arises.

According to Friston, the waking mind is not only causal but it is partially and significantly free to decide upon what action to take, or not to take, upon the world. It could be that decision-making is largely unconscious, but that does not prove that free will is an illusion. If only a very small part of conscious will is causal, some freedom of choice is guaranteed. It is scientifically impossible to accept the reality of waking consciousness without assuming an adaptive benefit of conscious deliberation.

Sigmund Freud (1856–1939) hoped (in 1895) to create a scientific psychology that was "perspicacious and free from doubt." Recognizing that the neurophysiology necessary to realize his "Project" was not available, he turned his attention to the *Interpretation of Dreams* (1900). By his own account, he was a "conquistador" rather than a scientist and, as such, successfully founded the psychoanalytic movement, which thrived until about 1960. Via his championing of the unconscious mind, and despite major intellectual errors, Freud must be credited for focusing attention on the dual aspect of human consciousness. There is an awake state of consciousness which is normally subjectively isolated from dream consciousness and vice versa. The two states of consciousness may be incompatibly conflicted, as Freud suggested, but are now seen as more often mutually enhancing. The dynamic relationship between waking and dreaming posited by Freud may thus be more positive and essential to health than it is deleterious.

I will say more about these revolutionary and controversial ideas as this text is elaborated and I will treat them in detail in my conclusions in Part IV. I wish to assert here that:

a) Consciousness is real,
b) Consciousness is state dependent, and
c) Consciousness can be studied scientifically.

The scientific study of consciousness that I propose takes advantage of major and significant similarities and differences between waking and dreaming. I will show that these similarities and dissimilarities are robust and accessible to hypothesis testing probes. It may well be that protoconsciousness, the virtual reality model, and free will are difficult to operationalize but they give our scientific efforts the philosophical shape they need and deserve.

Dialogue 1. Definitions

TH: You make light of definitions. How can you explain something that you cannot define?

AH: I do not think its possible to achieve consensus about a definition of consciousness, and I regard disputes about definitions as counterproductive. I have a simple definition of consciousness that I have illustrated as Figure 1.1.

TH: What is it?

AH: Consciousness is our awareness of the outside world, our bodies and ourselves.

TH That seems simple enough.

AH: But it is also indisputable. Critics may raise objections to the simplicity of my definition but I regard it as a good place to start.

TH: What is the next step?

AH: The next step is to assume that consciousness is a mind function which depends strictly and closely upon brain function.

TH Most people would not dispute that assumption even though they cannot imagine how this simple truth might be further elaborated — clearly the mind depends on the brain, but the question is how?

AH: For a long time, I didn't see this simple truth either but now it seems self-evident: the mind must be a brain function because they both change so dramatically together over the course of a 24 hour day.

TH: Is this where sleep and dream science enters the picture?

AH: For me, sleep and dream science *is* the picture. The brain and the mind change consciousness radically over the sleep-wake cycle. If we could begin to understand how and why these functions change together, we could begin to understand consciousness scientifically.

TH: Many people think that dreaming is an unconscious mental process. Are you challenging that assumption?

AH: Yes. I regard waking, sleeping, and dreaming as brain states each of which is associated with a distinctive state of consciousness.

TH: How are waking and dreaming related to each other?

AH: This question will come up again and again throughout the book. For now, just know that they are related in ways we never before imagined.

TH: Oh, come on, don't be such a tease. Give us a hint.

AH: We now consider dreaming to be a necessary precursor of waking instead of the pale and distorted copy of waking that dreaming has previously been taken to be.

TH: Tell me more about your strategy. In particular, what is the *conscious state paradigm* that you make so much of?

AH: The conscious state paradigm is a simple statement of fact: the brain and mind change state together in lawful and informative ways over the diurnal sleep-wake cycle.

TH: To track these changes, don't you need to be both a psychologist and a physiologist?

AH: Yes, and the separation of the two fields explains why their integration has never before been achieved or even attempted.

TH: What makes you think it will happen now?

AH: I am not at all sure that it will ever happen. I only know that it could happen because the concepts and tools are in place for younger hands and heads than ours to do the work.

TH: It seems wrong though.

AH: Science has shown that many of our intuitions are wrong.

TH: It also isn't just the hard question, it's the central question: how do we explain qualia? Everything else in consciousness research is dancing around the edges.

AH: So you're just going to give up? Let's see what science has told us about consciousness and then we can return to this question.

Note: These discussions were written before Allan Hobson and Trevor Harley went their separate ways and therefore bear our initials. The statements clearly communicate the differences in our points of view.

Chapter 2. Historical Background

Man is by nature spiritual. That is to say, human consciousness is intrinsically numinous. It is impossible, even for the committed atheist, to live without some form of belief. We do not know enough to function without some form of religion. It can be argued that even science is a belief system. We feel sure that the sun will rise tomorrow but this is only a probabilistic assumption. The Big Bang was no more unpredictable an event than the end of the world in current evangelistic prophecy. Scientists may know more than non-scientists, and they are right to warn against belief, but there is no escaping the fact that our minds are easily fooled.

The science of consciousness makes it clear why this is so. When awake, we make probabilistic decisions, based as much as possible upon knowledge that is as scientific as possible. But even rationalistic decisions are prone to error. There is too much to know and much that we suppose is known is false.

An easily understood example is that every person seeks libertarian democracy whereas it is clear that many, if not most, people in the world prefer authoritarian rulers who dictate rules that are compliantly followed. Freedom from anxiety about uncertainty is as ardently sought by the religious masses as by the rational scientists and philosopher kings of this world.

Furthermore, when we dream, our consciousness becomes embarrassingly credulous. In dreams, we believe things that could not possibly be true in the light of day. "Renounce your dreams," the rationalist may well advise. But let us suppose, as we now have good scientific reason to do, that REM sleep dreaming is essential to waking. Solid evidence now suggests that dreaming is essential to life itself. This implies that the rational is intrinsically irrational. This irony could be vital, no matter how uncomfortable the proposition makes us. Science, including consciousness science, is at best an antidote to irrational belief.

If it is true that spiritualism and belief are intrinsic to consciousness, it must also be true (as many scholars now maintain) that spiritualistic traits have survival value. The religious opposition to Darwinism can thus be seen as irrational. Evolutionary adaptation depends upon such irrationality. Dreaming and irrationality have been around for a very long time and will not go away in the foreseeable future. While I do not advocate a turning away from science and certainly do not advocate a replacement of science by religion, I do suggest a moderate and harmonious recognition of these two sides of human consciousness.

The difference between religion and science is the difference between faith and skepticism. In the balance of this chapter, we will examine the tug of war that has characterized these two poles of consciousness. As scientists, we will naturally favor skepticism but in doing so we express the faith that science can help us understand consciousness, including its spiritualistic aspects. This is an unproven and unpopular view, however, and we freely acknowledge that the faithful will not heed our call. There is too much comfort in religion to expect that and the current balance of power seems to us to be shifting away from skepticism in the direction of faith.

Anti-science and the resurgence of what we consider to be cultish superstition is an understandable disappointment that the problems of man have not been eased by rationalism. The advances of technology have solved many of the challenges of subsistence without providing spiritual satisfaction. Man yearns for explanation and moral guidance. The best that science can now offer is humanistic guidance. Scientific humanism is what we call a new philosophy that recognizes both brain and mind and appreciates their constant, spontaneous, dynamic, and causal interaction.

The Atomistic Greeks and Romans

The mind and brain are so different that the assumption, which now seems so obvious, that although the mind depends on the brain and the two are in some way inextricably related, hasn't always been made. For the Greek philosopher Aristotle (384–322 BC), sensation was carried out by the heart. It seems incredible that anyone could locate the mind in anything other than the head. The position of our sense organs, the way in which our eyes and ears seem to channel the outside world down into our head, alone suggest that "we" are located in the head. "I" seem to be right behind my eyes.

Pagan philosophy held that dreams were caused by the flying god, Hypnos, who flitted each night from sleeper to sleeper. We have to accept that the ancients had a different way of thinking about the world. In Greek physics and philosophy, everything was made up from a combination of the four classical elements, earth, air, fire, and water. Heat was thought to be the most important element of sentience, and it is told how Aristotle touched an exposed heart and found it to be warm; he then touched an exposed brain and found it to be cold. He therefore concluded that the brain could not be the primary organ involved in sensation but rather placed it in the heart.

Later Greeks, such as Galen (130–200 BC), were confident that the senses were connected to the brain, and took Aristotle to task for his poor understanding of anatomy. Aristotle aside, it has been long recognized that the mind happens inside the head.

Figure 2.1. Head of Hypnos. Greek, 3rd Century BC. British Museum, London. The winged god Hypnos, the son of Morpheus, the god of sleep, was the principal bearer of dreams in Greek mythology. Upon reaching his destination after traversing the night, he removed his wings and transformed himself into a human shadow.

That there is nothing new under the sun and that science and religion have oscillated over long periods of time is readily appreciated by the rediscovery of Titus Lucretius' 44 BC poem *De Rerum Natura* (on the Nature of Things). It has recently been argued that Lucretius, a Roman philosopher who lived two thousand years ago, was both a humanist and a protoscientist whose work was suppressed by religionists throughout the middle ages. His manuscript was found in the library of Benedictine monks who were spiritualistic opponents of Lucretius' materialistic speculation.

The Greeks who inhabited the Ionic coast of present-day Turkey practiced a form of philosophical speculation that was not only materialistic but also dynamic. Many of them were also secular pleasure seekers who followed Epicurus; hedonists celebrate his name to this day. The atoms and ions that we now recognize to be fundamental elements of nervous conduction are named for them and it is not farfetched to suggest that Lucretius imagined them millennia before their physical instantiation was established.

Titus Lucretius (99–55 BC) was a Roman scholar whose classic work, *De Rerum Natura*, used dreaming to advance an argument for the physical basis of mental life. For Lucretius. who followed the Ionic Greeks (especially Epicurus), all natural phenomena, including dreams, had their origin in the invisible atoms that are the building blocks of life. After centuries of neglect, his seminal work has recently been celebrated in new translations and scholarly reviews, which suggest that he may have been the hinge-point of modernism. That a protoscientist should also be seen as a pioneer humanist is significant for advocates of an integrative culture. Among other things, Lucretius described the muscle twitches of REM in sleeping dogs and inferred that the animals were practicing and/or reliving hunting behavior. Not even virtual reality theory is really new; it is 2000 years old.

Whether Lucretius was the first modern thinker or not, he was an avid dream theorist and a keen behavioral observer. Observing the muscle twitches of his hunting dog asleep on the hearth, Lucretius hypothesized that the canine was dreaming of chasing rabbits. Lucretius is thus reasonably thought to be the discoverer of REM sleep and the first to associate it with dreaming. His discovery also anticipates the virtual reality concept and the idea of behavioral rehearsal. No wonder the church was against him. According to medieval Christianity, dreams were messages sent by God to man via angels descending Jacob's ladder. Lucretius was a pagan but a very skeptical thinker. Sleep, dreams, and waking consciousness were not the work of Gods, as many pagans thought, but the manifestations of the internal flux of invisible particles of which everything in nature was composed.

We now see that ionic fluxes are at the root of consciousness, awake and asleep.

Figure 2.2. The Dream of Jacob, Lambeth Bible (Lambeth Palace Library, London). English, 12th century. According to the 28th chapter of Genesis, the sleeping Jacob dreamed that a ladder reached up to heaven and that angels were ascending and descending it. Above the figure of Jacob asleep he is shown standing, anointing with oil the stones that have been his pillow. On the upper right is the sacrifice of Isaac, and at the top of the ladder, holding a scroll, is God the Father.

Descartes and Dualism

René Descartes (1596–1650) was a brilliant French polymath who could not dig himself out of the deep hole of scholastic mysticism. Born and raised an orthodox Catholic, he thought that the brain and mind were two perfectly synchronized watches set in motion by God until separated, at death, so that the body could be buried while the soul had an afterlife. When we say that Descartes thought, we recognize the power of his philosophy: "*Je pense, donc je suis,*" (I think, therefore I am). My identity is my consciousness. My mind is my subjectivity. We cannot disagree with these existential claims as much as we deplore Cartesian dualism.

Descartes was also admirable in two ways: he studied his own dreams and he wondered where in the brain body and soul might meet on their perfectly parallel tracks. Descartes' interest in his dreams was an expression of his conviction that dreams were a part of his consciousness. Descartes' insistence on the primacy of subjective experience is one that modern dreamers should appreciate. For Descartes, nothing in nature was more certain than mental experience and all science must start, not end, there. Today brain science puts consciousness at the end of its concerns. At the rate that it is going, it will never get there.

The pineal gland was Descartes' choice for the mind-brain meeting place. We now know that the pineal gland contributes to coordination of cosmic and corporeal rhythms via its connections with the circadian clock in the hypothalamus. This is the "third eye" of submammalian species because of its sensitivity to light. Consciousness does fluctuate in keeping with the heavens after all as our biological clocks are synchronized with the daily fluctuations of light and energy. That consciousness should be a function of both information (light) and calories (heat) is a deep biological truth.

Mathematics was one of Descartes' many talents. He invented the system of three-dimensional coordinates now used by scientists to probe brain space. He also contributed to analytical geometry, a forerunner of calculus. He shared the goal of modern model builders like Karl Friston, who describe the mind, including waking and dreaming consciousness, in mathematical terms.

The philosophical theory of a "Cartesian theatre" where all consciousness components come together is thesis to the antithesis of a diffuse distribution of such components. The idea of a little man (or homunculus) who watches the show is the *reductio ad absurdum* of Descartes' theory. Yet we cannot say exactly how consciousness is unified if there is no theatre. If consciousness is a physical but immaterial force, then the force field is the modern theatre of the mind. It is tempting to suppose that Descartes might have been pleased by this idea were he alive today.

The Rise of Skeptical Empiricism

If Descartes was the father of rationalist philosophy, he also set the stage for the enlightenment shift of the 18th century to skepticism and empiricism. This became experimental psychology as we now know it. Descartes was sure that the only certainty was his own thought but how could he be sure that his thought was not misleading. It was, and Descartes' dualism is now dead except to the most ardent religionists who insist that the soul survives the body. Immanuel Kant credits David Hume with alerting him to the dangers of pure reason and his critique was the result. Both philosophers were scientific and insisted that all theory be tested empirically, that is, by experiment. Kant's apriorism is now testable in fetal animals that exhibit primordial REM sleep in the womb. Do fetuses dream? Do subhuman mammals dream? We cannot know, but that they have built-in brain activation long before they are born, there can be no doubt.

John Locke's hypothesis that the brain was a blank slate to be written on only by postnatal waking experience was clearly in error. Locke was probably also wrong in contending with Gottfried Leibnitz that sleep was a null state of the mind. Leibnitz' postulation of *"petites perceptions,"* when the mind was an apparent blank slate in sleep, has proved more than correct. As Hans Popper reminded John Eccles, a falsifiable hypothesis is as much the essence of good science as a verifiable one. Sleep lab studies have revealed that the perceptions of dreams are anything but *"petites,"* so Leibnitz was wrong, too. Why, we might ask, were such experiments so long in coming? A scientist doesn't really need an EEG to perform awakenings to test for the possibility that dreaming is intense and occurs in finite real time.

The answer echoes the resort to the empiricism of Hume and Kant. Because dream consciousness is so securely isolated by amnesia from waking consciousness, it was not appreciated that the brain and its mind were periodically activated in sleep. Sleep was thus wrongly assumed to be unconscious until the middle of the twentieth century. Even Sigmund Freud missed the experimental boat and, like the pre-Humean Kant, wandered into the never-never land of pure reason. What similar errors are we perpetuating today? What facile observations are we not making because we are wedded to rational assumptions? How can we pass tests, get a degree, and practice a profession without becoming hopelessly blindered to the obvious uncertainty of our theories? These are rhetorical questions that only you can answer. Our advice is to play smart but be dumb.

Immanuel Kant (1724–1804) attacked the idea that knowledge could advance by reason and experience alone. His *a priori* concept posited a built-in propensity to think, feel, and be. As such, Immanuel Kant is a very significant forerunner of protoconsciousness theory, the idea that in prenatal life the foundations of consciousness, including sensation, emotion, and self are laid down before we are born. Kant was scrupulously scientific and absolutely committed to the experimental test of all inductive reasoning. Greatly influenced by David Hume (whom he praised for arousing him from "dogmatic slumber"), Kant's metaphysics were rooted in the physical sciences. Although he called his philosophy "transcendental idealism," he may more properly be considered a transcendental materialist.

Associationism and Experimentalism

The nineteenth century realized the enlightenment philosophers' encouragement of empiricism. An important bridge figure was David Hartley who, like his predecessors, wrote wide-ranging essays on the nature of man. But Hartley set the stage for an experimental tradition that continues to enrich consciousness science to this day. Associationism begat learning theory and it begat semantic priming (the associative facilitation of memory for words) which is enhanced by REM sleep. Dreaming has long been recognized as a hyperassociative state of consciousness.

The easiest way to understand associationism is to recognize your own use of it in a memory search. The deliberate focus on a related memory item will bring to mind related items until the sought item "pops out" of your non-conscious memory. Hartley surmised that the mind consisted of a web of related words, concepts, and emotions. When Sigmund Freud encouraged free association and interpreted dreams, he was applying a clinical variant of Hartley's law. When a contemporary psychologist determines the time taken to recognize the association between words, he can endeavor to define the brain networks underlying associations.

Taken a step further, the effect of the changes of state of consciousness on both brain and mind can be quantified. REM sleep dreaming is hyperassociative because of a shift in the excitability of specific brain networks. It remains to explain the increases in learning capacity known to be accorded by sleep. It was the growth of physiological knowledge that prompted Hermann von Helmholtz and his colleagues to sign their famous pact against vitalism. They wanted to clear the air of the vestiges of spiritualism where non-physical entities, like the soul, were invoked as causal agents rather than figments of overheated imagination. Speaking of overheated, Helmholtz's physiology was directed not only at information management but at the necessity of all living things to regulate energy. To assure yourself of what this means, think of your body temperature as a "vital sign" and recognize the pains you must go to keep warm (in winter) or stay cool (in summer). A simple self-observation experiment is to feel the heat generated by your body under the covers of your bed.

The experimental psychology that William James instituted in America was imported from the German laboratory of Wilhelm Wundt, a student of Helmholtz. Wundt was a tireless experimenter and a passionate integrator. Among many other genial notions, Wundt supposed that the difference between waking and dreaming consciousness was determined by the weakening of some neurocognitive forces and the strengthening of others. We will later see how prescient a scientific theory this was.

Darwin and Evolution

There is no direct connection between the science of consciousness and the theory of evolution but the indirect impact is momentous. According to Darwin, mankind arises from a state of nature having no relation to extra-material forces. It is within the context of the Darwinian paradigm that the modern theory of consciousness is situated. Consciousness is a product of the evolutionary development of the brain and has its functional significance in its adaptation to the environment.

A conscious organism has survival success because it enables an animal to master a great number of challenges including, in the case of humans, self-understanding. In this respect, man is indeed the king of the beasts, especially if he learns to accept and control his instincts. We have already stressed the contributions that Darwin made to a truly biological theory of emotion. Animals warn each other of their feelings to achieve intraspecific equilibrium. But now we explore the possibility of rising above that important insight to consider other ways in which evolution and consciousness science may interact to the mutual benefit of both paradigms.

As we will detail further below, the evolutionary evidence indicates that REM sleep evolved separately in birds and mammals. Temperature control (the energetic component) and sensorimotor integration (the neurocognitive component) are the very significant products of this evolutionary advance. This insight builds on the foundation of Helmholtz's free energy concept, providing a strong connection between such evanescent states of consciousness as dreaming and brain activation in sleep, but also anchors sleep research in the solid soil of the associative learning paradigm. These connections give a strong theoretical foundation to consciousness science. This foundation can be modeled mathematically and yields testable experimental hypotheses.

For example, we can now explain (and further explore) the relationship between sleep quality and quantity to dietary calorie management. Steps in this direction have already been taken: sleep curtailment renders animals (including humans) diabetic. This surprising finding of Eve Van Cauter ties sleep science to medicine and health. The recovery from cold induced sleep curtailment should be REM and dream rich. Experiments to test this hypothesis have not yet been performed. They are technically challenging but feasible. If Charles Darwin slept well, it was in part because of his heat loss resistant bed covers.

Charles Darwin (1809–1882) is best known for his theory of evolution but he made a major contribution to the science of consciousness by his insightful observations of animal behavior. He recognized that mammals communicate their emotional state behaviorally as a way of achieving territoriality, mate selection, and intraspecific rapport. The pleasure that bind pet and master is thus a sign of non-verbal but precise communicative ties that are very likely to be subjectively appreciated as protoconsciousness by all mammals. Darwin's seminal observations and inferences imply that consciousness is not an all-or-none phenomenon restricted to humans but is rather a progressive trait already present in infrahuman animals. This theory has profound moral implications even as it encourages the humane use of animal subjects in the study of human consciousness.

Pre-modern Neuroscience: Cajal and Sherrington

After Darwin's paradigm shift, brain science flourished. In about 1890, Santiago Ramón y Cajal enunciated his neuron doctrine and Charles Sherrington advanced his spinal reflex model. The neuron was thereafter regarded as the structural unit and the reflex as the functional unit of the brain. These two principles are the twin pillars of modern neuroscience. Cajal, who dabbled in hypnosis, feared that consciousness would never be understood, while Sherrington speculated that the dynamic activity of the brain was akin to a shuttle loom, implying but not explicitly proposing spontaneous activity.

It is ironic to note that Cajal's neurone doctrine successfully replaced the idea that the nervous system was a syncytium or reticulum of uninterrupted cells. The dethroned champions of the syncytial reticulum model were able to explain synchronization because their brain was intrinsically unified; when one neurone fired, all the rest fired, too. Not until more than a half century later did the discovery of the reticular formation of the brain stem provide a model for the activation of the neuronal brain and its mind in conscious states. Cajal considered the brain to be a hopelessly impossible "wearisome labyrinth" at which he threw up his hands.

Cajal, however, was interested in hypnosis and kept a dream journal. His intense opposition to Freudian dream theory has recently come to light.

Santiago Ramón y Cajal (1852–1934). The father of modern neuroscience was a rebellious, anti-authoritarian student who challenged scientific assumptions and retained throughout his career his early interest in art and literature. These traits made him a daring and multitalented scholar who claimed to have produced his classic *Histologie du Système Nerveux* in his attic with a $25 microscope and two boxes of rat brain slides. He forbade photography and, holding that active involvement with data was essential, required that his students illustrate their histological observations with camel hair paintbrushes. Among his artistically inclined and imaginative pupils was Rafael Lorente de Nó, a pioneer in the integration of anatomy, physiology and behavior. Although Cajal despaired of ever solving the hard problem, he actively pursued hypnosis and, while fulminating against the pseudoscience of Sigmund Freud, kept a dream journal after his retirement from science. His autobiography extols a naturalistic appreciation of life as well as a dedication to rigorous science.

Timeline of Modern Consciousness Science (1780–1945)

	Neurobiology	Biology and Behavior	Psychology
1780			Mesmer and Hypnosis
1800			Hartley and Associationism
1850	Helmholtz Pact Against Vitalism	Claude Bernard Experimentalism	Wilhelm Wundt and Experimental Psychology
1859		Darwin's Origin of Species	
1861			Dream Science Alfred Maury Hervey de Saint-Denys
1873		Expression of Emotion in Man and the Animals	
1875	Alexander Caton The Electrical Brain		
1890	Cajal Neurone Doctrine Sherrington Reflex Doctrine		William James' Principles of Psychology
1900			Sigmund Freud Interpretation of Dreams
1912	Graham Brown's Paired Half Center Model		
1919			J.B. Watson BEHAVIORISM Classical Condition
1927	Ivan Pavlov	Learning Theory	
1928	Hans Berger's EEG		Brain-Mind States
1929		Walter Cannon Homeostasis	Psychosomatic Medicine
1933	Adrian and Matthews Neuronal activity & EEG		
1935		Tinbergen, Lorenz and von Frisch ETHOLOGY (Biology of Behavior)	
1936	Loomis and Harvey EEG Sleep Cycle		
1938			B.F. Skinner Operant Conditioning
1945	Hodgkin and Huxley Ionic Basis of Action Potential		

Also understandably, Sherrington never really looked higher than the spinal cord for his physiological understanding of consciousness. The shuttle loom metaphor for mind that he introduced in *Man on his Nature* is charming but unscientific. The state dependence of spinal reflexes (which we now know to be brain stem mediated) was suppressed by the decerebration to which Sherrington subjected his experimental animals. Decerebration rendered the animals spastic but their reflexes were unmodulated by descending influences. The state dependence of spinal reflex activity was not suspected for half a century even though Thomas Graham-Brown, a student of Sherrington, warned, as early as 1912, that the oscillatory or clock-like nature of interacting neurons was being underestimated.

It is difficult to quarrel with the scientific success of the Cajal-Sherrington juggernaut. They deserved the Nobel Prize that they shared in 1909 and the thousands of sound papers and books that followed continue to command respectful attention. The upper brain is forbiddingly complex and the experimental analysis of its anatomy and physiology is difficult enough without complicating it by such subtle and scientifically unappetizing realities as spontaneous activity and rhythmic oscillation. Consciousness became the forbidden C-word in part because neuroscience did so well without it. At the same time, psychology and philosophy of mind were stagnant and unfulfilled.

Freud's Failed Project

To understand the appeal of Cajalian neurons and Sherringtonian reflexes, it is useful to consider the *Project for a Scientific Psychology* written by Sigmund Freud in 1895. Freud was imbued with dream consciousness and hoped that he could model it and its difference from waking in anatomical and physiological terms. He drew neuronal circuits and endowed them with reflexive connectivity before, to his credit, he recognized the impossibility of the task he had set himself.

Meanwhile, the neurologist Freud was becoming the psychiatrist who would insist that his 1900 psychoanalytic dream theory owed nothing to his failed 1895 *Project for a Scientific Psychology*. Many scholars, on both sides of the brain-mind hyphen, see the influence of Freud's neurobiology upon his psychology. This is not surprising since both arose at the same time from the same neurologist's cerebrum.

I will now outline four of the most egregious examples of the neurobiological errors which tainted the psychoanalytic model of the mind:

- FALSE: ~~Neuronal conduction can flow in both directions.~~
  TRUE: **Neuronal conduction always flows from cell body to axon.**
- FALSE: ~~Neuronal activity is exclusively reflexive.~~
  TRUE: **Neuronal activity is predominantly spontaneous.**
- FALSE: ~~Neuronal activity functions to isolate the brain from unwanted invasion by high voltage inputs.~~
  TRUE: **Neurons transduce external information to low voltage signals.**
- FALSE: ~~Once information, from whatever source, enters the brain, it can never escape and it erupts in dreams.~~
  TRUE: **Memory gradually deteriorates and is lost to recall in dreaming or in waking.**

Inspired by these and other misconceptions, Freud's dream theory wrongly ascribed dream content to the dynamic repression of unacceptable infantile wishes. Dreaming may represent unpleasant feelings and memories but there is no reason to regard them as otherwise unacceptable impulses. These ideas were designed to establish a therapeutic technique by which the mind could be relieved of internal distortion by means of free association and dream interpretation.

Figure 2.3. Freud's neurons. In his *Project for a Scientific Psychology*, Freud depicted a neuronal circuit that conducted imposed stimulus-energy along one of two paths. Path alpha (α) to gamma (γ) led to discharge of the imposed energy as motion, while the "side-path" (a to b) stored the energy for later discharge – for example, in dreaming or in a psychological symptom.

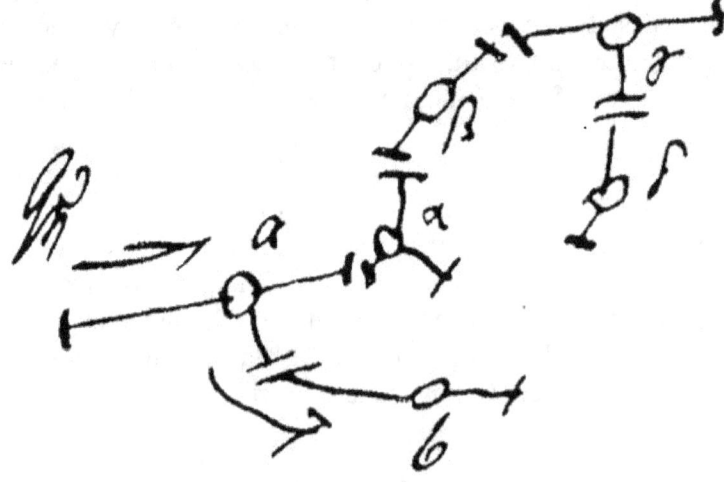

In spite of these flaws, Freud must be credited for calling attention to what he called the unconscious and for thus opening the door to what many now regard as an alternative state of consciousness. Abandoned now is the idea that dreaming protects the waking conscious mind from invasion by repressed infantile wishes. Instead, dream consciousness may be viewed as a positive and essential provider of waking consciousness with a virtual reality model of the world.

The virtual reality model is in part genetic and in part a reflection of individual post-natal experience. No doubt, this new theory will be refined or jettisoned as consciousness science further evolves.

Dialogue 2. History

TH: How long has a quest for integration of brain and mind been going on?

AH: At least 2500 years.

TH: That's a long time. What makes you optimistic about solving this age-old problem now?

AH: Twentieth century discoveries have created revolutionary opportunities. That's what this book is all about.

TH: Who best anticipated the modern view?

AH: The Ionic Greeks who lived in colonies on the coast of what is now western Turkey were materialists even though they were not experimentalists.

TH: What did they think? Who were their intellectual leaders?

AH: We will focus here on Epicurus, who gave his name to the pursuit of pleasure as part of his atomistic philosophy.

TH: If I call myself an epicurean to justify my self-indulgence, am I celebrating this early view of my brain-mind?

AH: Yes. In 500 BC, Epicurus theorized that we were made of tiny particles that we now call ions in his memory. He was an Ionic Greek.

TH: How amazing. Why were these prescient insights lost?

AH: Christianity was anti-materialist. It promised life after death and it vaunted suffering over pleasure. Christianity did its best to suppress pagan thought.

TH: Are you saying that the dream and consciousness theories that we now entertain were suppressed by the church?

AH: Yes. For over one thousand years, the work of Lucretius, a Roman apologist of Epicurus [99–55 BC], was secreted in the library of a monastery.

TH: Is there nothing new under the sun?

AH: Genuine novelty did not occur overnight in the Renaissance but the modern intellectual and experimental tradition did begin then.

TH: Why has it taken another 500 years to study sleep and dreaming experimentally?

AH: The growth of Renaissance humanism and of enlightened rationalism was not accompanied by scientific hypothesis generation and by formal experimentalism until the end of the nineteenth century.

TH: Science, as we know it today, is still very young.

AH: There was another obstacle to progress: the subjective experience of sleep oblivion was taken to imply that brain activity ceased when we closed our eyes at night.

TH: Didn't Gottfried von Leibnitz argue with John Locke about this?

AH: Yes, but even such modern scientific giants as Charles Sherrington and Ivan Pavlov were, like Locke, fooled by their own sleep-related conscious state alterations. No wonder that Sigmund Freud also missed the boat. He should have read Leibnitz.

TH: Freud wrongly supposed that dreaming occurred only in the minute before awakening. I am beginning to see why you regard 1953 as a watershed year.

AH: Not only was the molecular structure of DNA described by Watson and Crick in that year but brain activation in sleep was also reported to be associated with dreaming by Aserinsky and Kleitman.

TH: After two millennia of slumber, the world woke up that year to the resounding echoes of Ionic Greece.

Chapter 3. Philosophical Considerations*

Think about — or even better — look at, a red rose. You could probably describe the rose to another person. But now look at the color red. What does it look like? How would you describe its properties to mention somebody else? Describing the redness is extremely difficult; you might try analogies with other colors ("it's like orange") or mentioning other things that share that color ("it's the same color as a sunset or fire"), but these don't get at the essence of redness.

Pinch yourself hard (but not too hard). How would you describe the pain to somebody else? How would you describe it to someone who mysteriously never experienced pain before? You can't.

Qualia and the "Hard Problem"

So there appear to be some aspects of consciousness that are private and unique. The term *qualia* is used to refer to the part of conscious experience that is wholly personal and subjective and that cannot be communicated to others, such as the quality of red and the nature of pain.

Psychologists can study the details of the perceptual process, but qualia, the essence of things, seem forever out of reach. For many philosophers, qualia are the central issue of consciousness research. The philosopher, David Chalmers, has famously talked of the "hard problem" of consciousness: What are qualia? Why do we have them? And how do they arise? (Really that's three problems, but they're closely related.)

So intransigent is the hard problem of qualia that many philosophers (and scientists) have thrown up their hands in despair and given up. The philosophical position that the hard problem of consciousness cannot be resolved by humans is called *mysterianism* (named after a slightly obscure 1960s rock band). For example, the philosopher and mysterian Colin McGinn has said that consciousness is "a mystery that human intelligence will never unravel." A weaker position is that consciousness is not within the grasp of present human understanding, but may be comprehensible to future advances in science and technology. The list of mysterians is long and contains some notable names: Noam Chomsky, Martin Gardner, Colin McGinn, Thomas Nagel, Steven Pinker, and John Searle.

The most notable opponent of mysterianism is Daniel Dennett; we will return to his argument against it later. We finish this section with an observation that if mysterianism is correct, we are left with perhaps the sweetest and most profound paradox of all time: the human mind is ultimately unable to comprehend the human mind.

Dualism

Bricks, chairs, birds, stars, viruses, and all physical objects are made out of molecules and atoms and their constituents, down to quarks and photons. We call this type of physical stuff *matter*. What is the mind made out of? We could argue that ultimately it, too, is based on matter — and this position is the one that almost every scientist now adopts. We then have to explain how consciousness, thinking, awareness, and qualia all arise from matter — or if we're mysterians — say that they do but we don't know how. This approach, that there is just one type of matter, is called *monism*.

There is an alternative view, called *substance dualism*, that mind is a fundamentally different type of physical entity to matter. This idea was first formulated in detail by the French philosopher René Descartes (1596–1650), and is sometimes called *Cartesian dualism* in his honor. Descartes believed that while the body was a machine, the mind, or soul, was non-material. The mind does not occupy space in the way that matter does, and the laws of physics do not apply to mind stuff.

One big problem for dualism is how a non-material substance (mind) influences a material one (matter). Descartes suggested that mind and matter came together in a small organ deep in the brain called the pineal gland. He came to this conclusion on the basis of a mistaken understanding of the anatomy of the brain; he recognized the pineal gland as a discrete organ, but thought that it was at the center of all things. He recognized the importance of nerve cells in sensation, but thought they were tubes full of "animal spirits." In any case, identifying a location doesn't solve the problem of how one influences the other. And what does the rest of the brain do?

Dualism has never been popular among psychologists, and currently very few would own up to being a substance dualist; it's just too defeatist. Psychology and neuroscience can provide a long and still-growing list of successes where the cognitive and biological bases of mind have been revealed, so that it is clear that brain and mind are very closely related. And it's not even clear how dualism solves the hard problem anyway, so we've introduced a new type of material but still are left with the same old problem.

Substance dualism has been persistently influential in religion, however, in the concept of the soul. The *soul* is a religious rather than scientific concept. The soul is the mind material constituting an individual; it's what's left after we die and it goes up to heaven (or down to hell), or gets reborn again. It's what experiences and what makes us *us*. Psychologists don't talk about the soul any more; we argue that all we need is a mind, and the mind arises from the brain. There isn't any evidence for mind substance — a few might mention survival of the soul after death, or reincarnation, or parapsychological phenomena such as telepathy or clairvoyance, but there is no good scientific evidence for any of these. And the more we've learnt about psychology, the clearer it's become that we just don't need any special mind stuff; the mind belongs to the brain. There's nothing else. We believe that when we die, that's it. Without a brain, my mind will fade away (very quickly and painlessly, we hope); there will be no smoke-like soul that drifts up to the sky. Sad, but that's the way it is.

Reductionism

How then are brain and mind related if they're not different sorts of stuff?

We are surprised by how many people are surprised by the idea that all there is to mind is the brain. One of us gave a talk recently to a group of family doctors; he was amazed that they expressed amazement at this simplicity. When pressed on what else could be the case, they then looked a bit dazed; no one likes to admit they don't have a soul anymore, but what's the alternative? If we don't have a mind made of mind stuff, we've got to accept that all we need is a brain. Somehow the mind and consciousness must emerge from the brain.

We have heard people say of someone who has had a stroke: she couldn't walk or move her left arm, but, she said, "Her mind's all right!" People think of the mind as being the central bit of them, their "soul," whereas their brain isn't.

There are many different levels at which we can study nature. At the bottom, we have quarks and the quantum realm. Physicists are also concerned with protons, neutrons, and electrons, and how these combine to form atoms and then molecules, and the laws that govern their interaction. Chemists are concerned with the behavior of matter at a higher level: what are the macroscopic properties of matter and how are they derived from the atomic structure? Life is based upon organic molecules, mostly large molecules containing carbon, such as amino acids and proteins. Neuropharmacologists examine the chemical basis of behavior, such as the group of chemicals called neurotransmitters that mediate communication between nerve cells, or neurons. The brain is made up of billions (approximately one hundred billion) richly interconnected nerve cells (called neurons), and contains many distinct structures, such as the cerebellum, two cortical hemispheres, and the limbic system, which comprise substructures such as the hippocampus and amygdala. We're now at the level of the neurologist, neuropsychologist, and neuroscientist. But then we jump to a level that's not easy to describe in terms of a physical basis: belief, thought, language, memory, attention, and consciousness. We're now at the level of cognition. Then we interact with other people, we form social groups, our behavior is affected by others, and social structures and societies seem to have a life of their own. We're now at the level studied by social psychologists and sociologists. The level at which I exist at is the cognitive level and beyond. The jump from brains, where we talk about neurons and neurotransmitters, to cognition, where we talk about beliefs and perception, seems to be a huge one.

René Descartes (1596–1650) was a French philosopher who was a brilliant mathematician and practiced the self-observational analysis of his own dreams. He also made an attempt to incorporate in his philosophy the scholastic religion of the Middle Ages in his philosophy. He asserted that the mind and brain were two perfectly synchronized watches set in harmonious motion by God. Any causal connection between them was thus an illusion. Even those who celebrate "spirituality" would no longer accept this theory, which António Damásio has rightly called *Descartes' Error.* It is nonetheless difficult to accept the tenets of dual-aspect monism, which hold that, yes, there are two physical modes — one, the immaterial mind, and the other, the material brain. These two modes depend causally upon each other and both cease to function after death. This idea is so unwelcome that many modern thinkers are Cartesian dualists in spite of themselves.

The idea that we can explain higher-level processes of any sort completely in terms of lower level processes is called *reductionism*. Reductionism works very well in the physical and biological sciences; the interaction of complex molecules reduces to an explanation in terms of their constituent atoms and the laws that govern them, and these eventually reduce to protons, neutrons, and electrons, and then to quarks and other elemental particles. A reductionist would argue that consciousness is eventually explicable in terms of the laws of physics.

Eliminative materialism takes this material position to the extreme; saying that certain mental states are really illusions, and all that really matters is the biological level. We have already eliminated the soul from psychological discourse; perhaps there are other things we can eliminate as well, such as, perhaps, qualia.

What is the appropriate level for studying how the mind works? The mind, the brain, or both? The brain has nerve cells and structures, while the mind has mental states. We don't talk about our amygdala having feelings or the occipital cortex seeing a chair; *we* see chairs and have feelings. The "I" or the "We" are the agents of seeing and feeling. I will now argue that the most acceptable philosophical theory is dual-aspect monism, the idea that mind and brain are two aspects and levels of analysis of a single unified physical system.

We noted above that dualists believe mind and matter are fundamentally different sorts of material. There aren't many overtly dualistic scientists around these days. We have also just seen that eliminative materialists believe that all that matters is the biological level. Most psychologists are somewhere in between. The position we argue for is *dual-aspect monism*. In our view, both the brain and the mind are physical; the brain is a structure and the mind is a state of the brain. A brain state is not to be found in gross brain anatomy but it is no less physical. It is to matter (the brain) as energy is to matter. This position claims that mental and physical are just two aspects of the same substance. An example would be a computer running a program; whether it's electrons in a printer circuit, or a program to merge photographs into a panorama, just depends on how you look at it.

First and Third Persons

You can't know what another person is thinking, and they can't know what you are thinking: our subjective consciousness is *unique* and *private*. Thus philosophers will assert, quite correctly, that no one can know whether the red that I see or imagine is the same as the red that you see or imagine: my qualia are private. This "first person" consciousness they say is *irreducible*. To emphasize this point, Thomas Nagel has asked, "What is it like to be a bat?" and replied that the question is unanswerable. Their little lives, with even a different means of perception to us (sonar), are completely unimaginable to us.

As already pointed out, the critical scientist, speaking in the third person, wonders if this difficulty is insurmountable. Consciousness, some scientists say, is not an unassailable scientific goal and they will try to by several specific means to mount an attack on its apparent irreducibility. Some of these means are listed below:

- **Collect reports of conscious states from many first persons.**
- **Quantify the formal features of the reports.**
- **Select the formal features, which show ubiquity and robustness.**
- **Correlate formal features with other psychological variables.**
- **Correlate psychological with physiological variables.**

Two caveats are in order. First, formal features are assessed as coarse-grained measures (e.g. vision vs. audition) as against fine-grained measures of content (e.g. the details of what is seen or heard). Second, correlations are not eliminative — that is, physiology does not replace psychology.

This sort of scientific maneuver can make first-person subjective experience understandable via third person psychology and physiology. States can then be analyzed across and within individuals. However, one can never be certain that the subjective experience is identical within and across individuals but science can proceed, as we will show, short of the perfection espoused by philosophy. We can get a long way without worrying about qualia too much. The alternative is to throw our hands up in despair and stop work.

This pragmatic approach derives from our objection to the separation of the two main states of consciousness and the relegation of dreaming to the unconscious with its assumptions of escape, subterfuge, and hidden meaning.

Some Thought Experiments

What makes consciousness so difficult for psychologists (and neuroscientists, philosophers, and anyone else for that matter) to study? We will try to answer that question with a few *thought experiments*. A thought experiment is philosophical — we carry out an experiment in our minds and see what our intuitions tell us about the results, and what those intuitions then tell us about our theories. Thought experiments are particularly useful for exploring the ideas of consciousness.

Here's a first thought experiment, and it's about my neurons and me. Imagine you have one neuron replaced electronically ... and then another ... and then another ... When would you start to worry? Suppose science progressed a bit, and you had your whole hippocampus replaced with a silicon implant, would this bother you? (OK, science has to progress quite a bit for this to occur, but it's not inconceivable) This idea is a version of Parfit's Question (named after the Oxford philosopher of mind, Derek Parfit): if you had to keep your physical or personal identity, but not both — which would you choose? Parfit makes use of reasoning about tele-transportation (now you see why these have to be thought questions). Suppose science has progressed to the point where we can put you to sleep, and put you into a tele-transporter. It then breaks you down, atom by atom, and relays the information at the speed of light to a receiving station, say on Mars, which reassembles you, from locally available materials, atom by atom, so that you are an exact atomic copy of the original. Are you still the same person? You will certainly think you are.

Imagine the receiving device goes wrong and produces two identical copies. When they wake up, they will both think they are the same person — you.

The value of this thought experiment is to illuminate what we think about personal identity and its relationship to the body. Most people would rather have the same identity than the same body. Here's another thought experiment. Mary, the clever scientist, knows everything there is to know about physics, psychology, and neuroscience (Jackson, 1982, 1986); in particular, she has complete knowledge of the physiology of the color system (for which reason this thought experiment is sometimes known as the Knowledge Argument). Now being surrounded by inquisitive but evil scientists, she grows up in a black and white house. When she is 35, she is given a red rose. Does she learn anything new, and what does she say? Does she say, "Given my knowledge and expertise, that's exactly how I expected a rose to look"? Or does she have no idea before she sees it, and when she does she just says "wow"? Most people think that she learns something new by seeing the red rose, and maintain that this example shows that qualia exist independently of the laws of physics.

This example gets to the heart of the *hard problem of consciousness*: qualia are those irreducible qualities of experience, such as what makes red red or pain painful. Any complete theory of consciousness must give an account of why we have qualia. It is not sufficient to say that the purpose of pain is to make us avoid aversive stimuli — we also have to explain why pain feels the way it does. Scientists differ on the extent to which they think we can have a complete theory, how long it might take, and even how big a worry it is if we can never give such an account. Note also that if Mary learns something new on seeing the red rose, physicalism must be false — that not everything can be reduced to the laws of physics.

Another particularly revealing thought experiment involves the concept of zombies. A zombie (or, if you prefer, a robot) is like us, but happens not to be conscious. Zombies are behaviorally indistinguishable from us. Do you think such a zombie could exist? If you answer Yes, you must think consciousness is an inessential extra; it is *epiphenomenal*. If you answer No, then you think that we cannot do without consciousness, and so it follows that consciousness plays some vital role. But what could that role be? Virtually everyone asked this question answers that zombies defined in this way can't exist, but no one can say what vital role consciousness plays.

David Chalmers (1966–) is an Australian thinker who has become famous for his formulation of the "hard problem," namely explaining how qualia (or subjective mind) can arise from the material brain. Some skeptical colleagues hold that the problem is not only hard but philosophically insoluble. For others, the problem is not only not hard but non-existent. In their view, mind is physical but immaterial and the "hard problem" consists only of understanding this paradox. Chalmers' arguments carry weight because he is seductive, eloquent, and because many hard-nosed neuroscientists believe that mind is an epiphenomenon and therefore neither real nor physically causal. When mind is likened to a physical force (like gravity), the hard problem softens considerably. Some might even say that it disappears altogether but the fact remains that the "force of gravity" analogy is not yet quite up to the solution of the mind-brain problem.

A related issue is how do you know if your best friend is really a zombie or not? How do you know if someone is conscious? How would you decide if a pheasant is conscious? One obvious way is to ask to what extent do these things behave like us. If your best friend acts like you and speaks like you, has beliefs and shouts ouch when you stick in a pin in her, then as long you don't believe zombies can exist, she's conscious. If you do believe zombies can exist, you're in trouble. Whether you think a blackbird is conscious is more difficult: you make a judgment based on what you know about blackbird physiology, particularly their brains, and how they behave. The more like us something is, the more likely we are inclined to think that it is also conscious.

What do these thought experiments tell us? First note that there is no "right answer" to any of these questions — it all depends on your intuitions. Second, that there's a "hard problem" in consciousness research which asserts there is something ultimately private about our experience so that no amount of research and experimentation will cast light on it. Science hits the wall at qualia. "In that case," asks Ma, "why is this chapter so long?" Because although we might ultimately be able to say nothing about qualia, cognitive psychology can say a great deal about ideas that are very close to it, and for most people they would fall under the rubric of consciousness and awareness.

Emergence

One idea that has become influential over the last few years is the idea of complex systems, and particularly, for our purposes, that consciousness is an *emergent property*. Complex behavior can emerge from a large number of very simple interactions, where order comes from the bottom up. Examples of such emergence of large-scale structure coming from aggregated small-scale behavior are the hills of termites and the flocking behavior of birds and fish. The complex behavior of flocks emerges from three simple rules: a clumping force keeping them together; a separation force so they don't get too close; and an alignment of velocity and direction. These rules are entirely local in that each bird need only look at its immediate neighbor rather than think about the behavior of the whole flock. The rules were implemented in a computer program called "Boids" by Reynolds.

Many other complex systems show emergence: for example, the World Wide Web and the social behavior of crowds both demonstrate emergence.

The definition of "emergence" isn't completely clear-cut. It's useful to distinguish two types. We could have a very strong emergence, sometimes called "magical emergence," where properties emerge that are in principle unpredictable from knowledge of the lower levels. An example would be a property that we could not predict of a complex system in principle even though we had complete understanding of all the laws of physics. Most people seem to use the term emergence in a different way, to talk about computational tractability. That is, in practice we can't predict high-level behavior from knowledge of the lower level because the computations are too difficult and too time-consuming. Perhaps by the time we get around even to starting them the initial conditions have changed. We can call this "practical" emergence.

The boundary between magical and practical emergence isn't that clear-cut, either. A commonly cited physical example of emergence is the properties of water: we would not expect the physical properties of water, how it behaves as a fluid, and its freezing and boiling points, from knowledge of hydrogen and oxygen alone. On the other hand, no one would say that there's anything magical about water, or that it defies the laws of physics. We might say that the feel of water is special, but while that feel depends on the physical properties of water, the sensation of its feeling is a psychological issue.

There are in fact four aspects to emergence: collective self-organization, non-programmed functionality, interactive complexity, and predictions that require computer simulation or actual computation. One particularly interesting question is the extent to which consciousness is an emergent property; can we explain consciousness in terms of the interaction of many, many (one hundred billion) nerve cells? But this solution might not help us very much. Suppose we succeeded in building a very complex simulation of the mind. We have an enormous computer program that interacts with the world in real time, receiving inputs and acting upon the world. Suppose it shows every sign of being conscious. We say that its consciousness has emerged as a consequence of a complex system interacting with the world in real time. What then have we really achieved? How would we be any better off? In what sense have we solved any problem? Simply saying something is emergent is merely a label. We haven't shown how you get from one level to another. In what sense does saying consciousness emerges from complex interactions in a very complex system explain, say, qualia or self-awareness? It's inadequate as a complete explanation.

The idea of consciousness emerging from complex computer systems is common in science fiction: we have Skynet in the *Terminator* films, and the network of the ship's computers in the *Star Trek: The Next Generation* episode "Emergence." The brain is complex, operating in real time in a complex world: how could it be anything other than conscious?

Daniel Dennett (1942–) is an American philosopher with broad scientific interests including the evolutionary theory of Charles Darwin. Dennett's (1991) book "*Consciousness Explained*" presents his "multiple drafts" (widely distributed as against sharply localized) hypothesis. His opposition to the dualism and central control ideas of René Descartes marks a major and welcome shift in philosophical thinking. With respect to sleep and dreams, however, Dennett has, until recently, fostered the claim by the Australian philosopher, Norman Malcolm, that dreams are responses to waking brain activation (1952). According to Malcolm, dreams are back-projected in time and have only illusory, independent, real-time existence in sleep. Laboratory studies reveal this error to be a good example of why philosophy needs both physiology and psychology as well as logic and introspection. Although no one would now want to resuscitate Cartesian dualism, the scientific study of lucid dreaming does indicate that consciousness is localized as well as diffuse. When I am lucid, one frontal lobe part of me watches while another occipital lobe part of me dreams. The idea that any self has multiple instantiation (or drafts) is voiced in Dennett's theory of multiple personality.

The Cartesian Theater

The philosopher Daniel Dennett argues that most theories of consciousness suffer from the illusion that there is a space in the mind in which something happens; it's as though we're watching a television screen or theater arena with consciousness displayed on it. Dennett calls this space the *Cartesian theater*. The problem with the Cartesian theater view is that there is no one place in the brain where things are brought together (which is one reason why the search for a neural correlate of consciousness has proved so fruitless). Furthermore, who or what is watching the screen? We have here a problem of circular reasoning: if consciousness is a screen in our mind on which everything is brought together, it must be observed by something or some one who is called a *homunculus*, or little man. But what then is happening in the mind of the homunculus? We later discuss an alternate theory when confronted with recent evidence of selective frontal lobe activation in what is called "lucid dreaming."

"Cartesian theater" is not meant merely to be a cute phrase. Dennett uses it to point out that modern psychology (and philosophy) still suffer from the remnants of Cartesian dualism; although we say that the brain is all there is, we still think of the mind as a separate thing, a separate arena. Part of the illusion of the theater is fostered by the dominance of our conviction that we are watching a screen in our minds on which the percepts are played. What, though, is the alternative?

The Multiple Drafts Model

We know that the brain is a massively parallel information processor: we do many cognitive things at once. Consciousness, though, is strikingly serial. How do we resolve this apparent contradiction? We note that any experience is really made up of many threads, running in different parts of the brain, corresponding to different aspects of the experience, and not necessarily occurring in synchrony. (We know from Libet's studies how difficult it is to take our intuitions about the timing of conscious phenomena at face value.) Yet we have the experience — or think we do — of having a coherent experience, so where does it all come together?

Consider watching a person ringing a bell at some distance; we see the visual scene (it is difficult to avoid lapsing into Cartesian theater speech here), and hear the bell ringing and assemble everything into one percept. Dennett argues that we are constantly constructing and reconstructing our experience as multiple draft copies. In effect, we are conscious of our memories. This view might seem radical, but remember Libet's half-second lag: we are always lagging behind reality, and interpreting the past. Not all percepts make it to a sufficient strength of processing to cause an updating of our current model of the world — a new "now." Whenever information comes to the brain, it enables us to update our model of the world, which it does immediately; our past is over-written. The present, the now that "we" are watching, may be nothing more than an illusion. As we will argue at length anon, consciousness is virtual reality.

The Global Workspace

Another psychological model that treats consciousness as a psychological compendium is the *Global Workspace Theory* (GWT) of Bernard Baars. GWT is based on the idea of multiple draft copies of working memory, with the contents of consciousness being whatever element that is most active at any one time. I discuss this selective aspect of GWT as attention in Chapter 15. As the principal components of working memory are the phonological loop and visuo-spatial scratchpad, the contents of consciousness are predominately verbal and visual, depending on what is most in focus at the time. Baars uses a theater stage analogy, with a bright beam moving across the stage, picking out elements of the display of which we are then aware. This description makes it sound just like an explicit version of the Cartesian theater which, as we have just seen, Dennett ridicules so much, but in spite of the theater analogy, Baars denies that his is a *Cartesian* theater. Crucially for Baars, there is no little person, no homunculus observing the scene. Hence, Baars argues, there is no implicit dualism in his model. Baars also makes use of a "behind the scenes" analogy in a theater, too; whereas in real life we have stagehands and production assistants, in the mind, we have unconscious processes shaping what becomes conscious, while staying themselves out of the spotlight.

Models such as GWT do not address the "hard problem" of consciousness. Nevertheless, Baars and other workers in the tradition of GWT models argue that GWT shows the relation between cognition and consciousness, particularly demonstrating the cognitive functions of consciousness, and provides constraints upon future theories that might address the hard problem.

Neither Dennett nor Baars considers the physiological basis of their multiple drafts and global workspace concepts. We have already mentioned the 40 Hz EEG synchronization findings of Wolf Singer as a potential unifier and we will later discuss the neuronal evidence for recursion by which multiple drafts could become dominant percepts and concepts in waking and dreaming consciousness.

Death

We all come to the same end: certain death and the eventual cessation of our consciousness faces us all (at least for the foreseeable future). We tend to think of death as being a switch, when in a moment we pass from a state of life to one of death. In death, everything is switched off; the heart and organs no longer function, and the brain is no longer active. In practice, the transition is usually not that clear cut. Medical science, with life support systems, enables us to keep organs going where in the past, death would certainly have followed. The key moment is when oxygen stops being circulated through the body: cells start dying very soon afterwards. Neurons are particularly susceptible to oxygen starvation and die very quickly, leading within a few minutes to irreparable brain damage.

Given the conflict between being able to keep the body alive through life support (sometimes for years) and the need to provide closure to relatives and make organs available for transplant, there has been much debate about providing a precise legal and medical definition of death. The definition is not static, having changed in response to changes in technology. For some time, the definition was the cessation of heart function, but we can now often restart the heart and keep it pumping, so now the definition centers on the idea of *permanent and irreversible brain damage,* a point called *brain death.*

A person with brain death, but with heart and lungs still working supported by machinery, is said to be in a *coma.* But even taking brain death as a criterion is problematic because, as we know, the brain is not a simple thing: should we take the whole-brain as being dead (showing no electrical activity and no signs of activity, such as even basic reflexes or eye responses)?, or just the brain stem, without which a person is incapable of maintaining basic vegetative functions? Hence, in practice, the way death is declared may vary from country to country.

We believe that death is the end of us, although our genes and legacy may live on. Sadly, there is no robust scientific evidence for survival or reincarnation. At the moment of death, our consciousness fades forever.

Dialogue 3. Qualia

TH: You don't seem to be very worried about qualia and the hard problem.

AH: I think they're real concerns but if we worry about them too much we'll waste valuable research time. We should just get on with doing the science.

TH: But aren't qualia the central issue for consciousness research? When we stop worrying about them, aren't we left with major issues to tidy up?

AH: I think that qualia need to be operationally defined and measured as we have done with dream content. If you insist that qualia are irreducible, then I agree, there can be no science of consciousness and you are free to remain in your study clucking or go to church and praise God.

TH: But the hard problem is that qualia can't be measured or operationally defined. However, I'm reluctant to dismiss all of psychology as a consequence: we have clearly learned a great deal around the margins (about the brain, different types of coma, the timing, just to a mention a few examples). Possibly I am too pessimistic and too optimistic. I notice that you haven't given an answer to the hard problem: you've just said that IF it's true THEN everything is pointless. And I am still struck by the fact that I cannot describe the pain of my gout in my toe to you.

AH: To the extent that qualia are unmeasurable, they remain beyond the reach of science. I am quite happy to admit that. But who would have guessed that dream form might be explained by physiology?

TH: I fear you might be exaggerating the extent of the success of physiology, even in the study of dreams. As ever we seem to come back to the use of words; what do you mean by dream form? Psychology and physiology have been much less successful at explaining the content of dreams, and as much a failure with dream experience as they have with waking experience. The only coherent attempt to explain dream content on a large scale was by Freud, and we both reject Freud as unscientific — you perhaps more than me. It seems that I am prepared to allow much of what we'll call for now "mysticism" in consciousness research, with the cost that some crucial aspects that are core issues can never be adequately addressed. To that extent, I am a pessimist — a mystical pessimist. You seem to be much more optimistic about how far psychology and physiology can go.

AH: Why worry about these thought experiments? That's all they are. You don't even know the answer. Perhaps Mary will say, "Oh that's just as I expected" of the rose. And perhaps your zombies can exist.

Chapter 4. Brain Anatomy and Physiology

Neurons and glia are the structural units of the brain. At last estimated count, there were thought to be about 100,000,000,000 neurons and many more glia. *Neurons* do the electrical information processing of the nervous system. *Glial cells*, as the name implies, are the "glue" of the nervous system, providing physical protection, support, and nutrition for the neurons. In this and the following section, we will make a few salient points about neurons and how they communicate. These points are focused on the mediation of consciousness, and are no substitute for a more complete knowledge of neurobiology, a forbiddingly complex but foundational scientific subject well beyond the scope of this text.

As already pointed out, Santiago Ramón y Cajal won the scientific debate with the German "reticularists" who had solved the connectivity problem by denying its existence. Instead of a syncytial, unified net, as the reticularists proposed, the brain was shown to be composed of countless unitary and discrete elements, each surrounded by a semipermeable membrane. The gaps between the neurons, called synapses, could be crossed by molecular ferryboats of great chemical diversity — called *neurotransmitters*. This discovery allowed a rich diversity of functions of relevance to consciousness to be made clear in the century of investigation that followed the discovery of neurons. Paramount among those clarifications was the synaptic efficacy alteration fundamental to understanding how the brain-mind could alter its state of consciousness and thus lead to a scientific theory of consciousness itself.

Each of Cajal's membrane-surrounded neurons could be electrically activated, either spontaneously or in response to inputs from other neurons. This means that the brain activation necessary for consciousness is intrinsic but also subject to modification by extrinsic forces, some of which come from the outside world. Depolarization of the neuronal membrane led to the all-or-none action potential via the summation of smaller potentials of quantifiable magnitude. Depending on the balance of excitation and inhibition and the recently discovered process of neuromodulation, the whole brain proved capable of a wide range of states. The net result is a physical system with the diversity worthy of consciousness and its state vicissitudes.

The relationship of neuronal electrochemical diversity to the electroencephalogram (EEG) is still limited to correlation, but it is clearly the substrate of EEG activation in waking and REM sleep. Both states are correlated with high net rates of neuronal discharge allowing us to move from neurons to brain waves to conscious experience in waking and REM. As far as activation is concerned, waking and dreaming consciousness are undifferentiated.

Differentiation between waking and dreaming consciousness is realized by two neurophysiological mechanisms that derived from Cajal's neurone doctrine. One was the distinction between excitation and inhibition discovered by Sir Charles Sherrington and the other was chemical modulation at the synapses discovered by Sir Henry Dale (both were Nobel laureates).

We wish to underline the philosophical point that neuronal anatomy and physiology do not eliminate or diminish psychology. Knowing how we wake and dream only explains the brain substrate of these processes. It also teaches us to beware of speculative mystical notions that postulate worlds other than the physical. The physical world has already revealed itself to be surprisingly rich and this is just the beginning of the story. Most important is that psychology itself may now be construed in physical terms such that mind and brain can be viewed as interdependent entities each of which is causally linked to the other. Out of this philosophy emerges the precious faculty of free will. We are free to be determined and determined to be free.

Figure 4.1. Neurons.

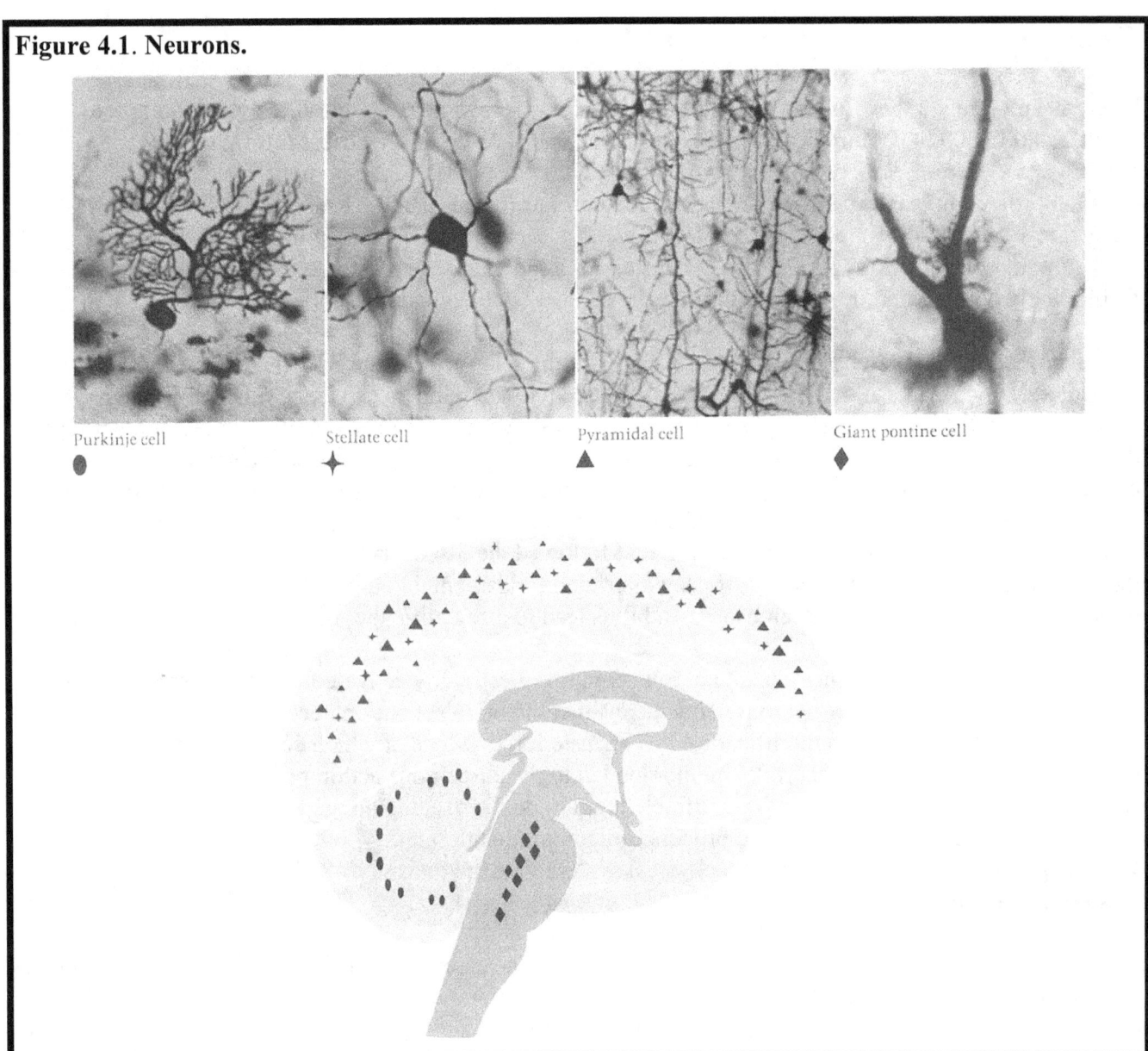

Purkinje cell   Stellate cell   Pyramidal cell   Giant pontine cell

Connectivity: The Axons

Each of the 100 billion neurons in the human brain sends out electrical signals via fibers called *axons*. The electrical signals from each neuron make synaptic contact with at least 10,000 other neurons. Since neurons discharge at a rate of 2–50 times every second, that means that the brain generates information at a rate of at least 10 to the 27th power bits per second, an astronomical figure that gives one a sense of the brain's generative power. Rather than wondering how the brain could give rise to consciousness, we might well ask how such a powerful dynamic structure, receiving input from the world and acting upon the world in real time, could not be aware of that world and itself. We don't yet know exactly how this happens. This is the still obdurate "hard problem."

Neuronal axons are well-insulated wires, assuring the spatial precision of delivery of the electrical signals. Each axon is separated from others by a sheath of protective material called *myelin* under which the electrical signals pass. When a neuron discharges, the electrical signal is rapidly conducted to its myriad synaptic endings. There, one of at least one hundred chemical neurotransmitters is released; the neurotransmitter molecule is ferried across the synaptic cleft and reconverted into another electrical signal in the post synaptic neuron. William James wondered why consciousness was not a "great buzzing confusion?" We must wonder, too, and our job is to try to answer this question.

Figure 4.2. Neuronal Conduction.

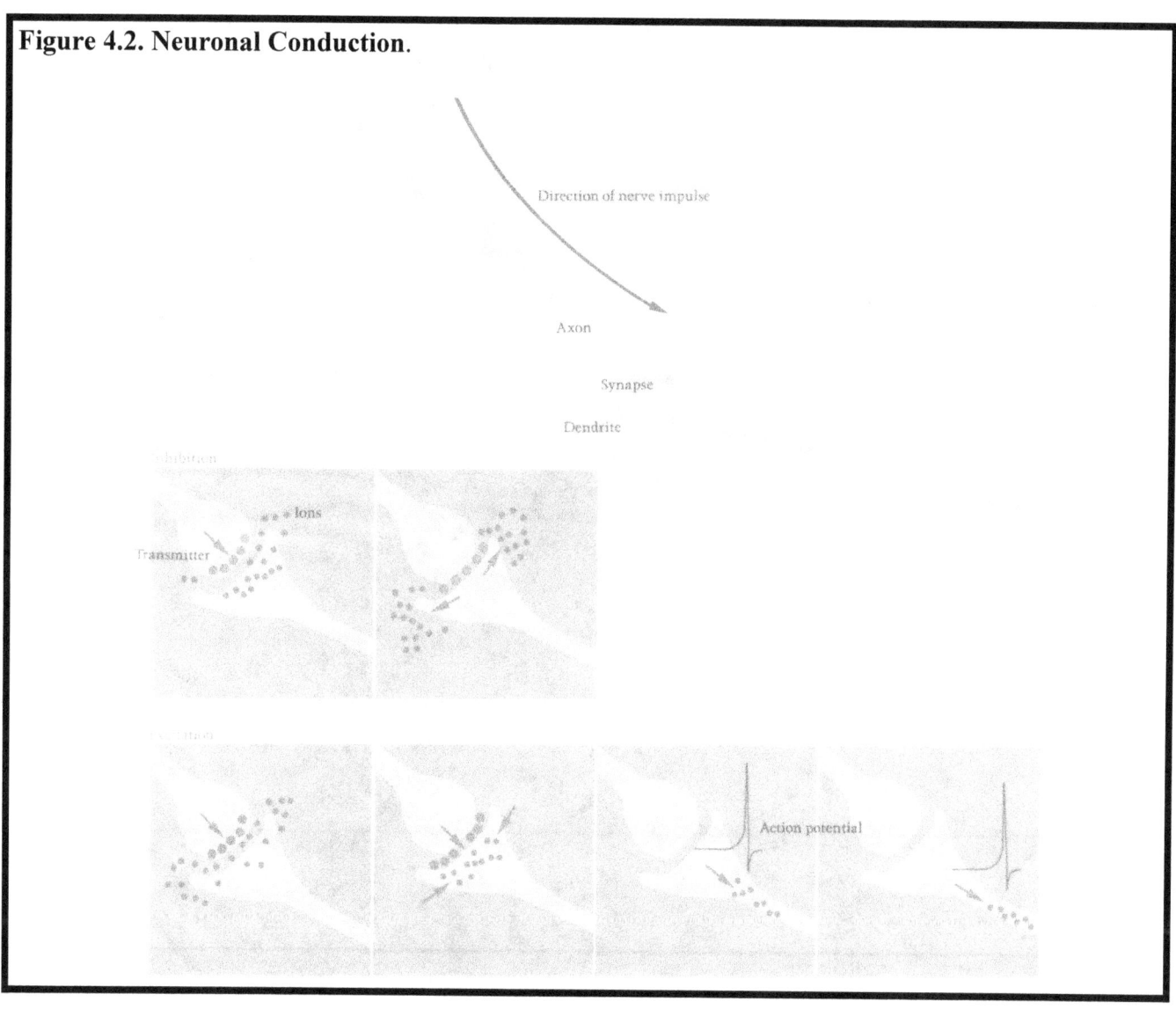

We can now begin to add in facts that help us understand the differences between waking and dreaming consciousness, both of which share high levels of brain activation. One fact is that the activation patterns differ; we will see exactly how when we consider brain imaging. A second fact is that activation may be excitatory and/or inhibitory; we will see exactly how when we consider the motor paralysis of REM sleep which prevents us from actually running when we dream of doing so. A third fact is that the chemical consequences of activation differ in ways that transcend excitation and inhibition; neurons that turn on or off may thus determine entirely distinct activation consequences.

To decipher the complexity of axonal anatomy, the new kid on the neuroscience block is the *Human Connectome Project*, an attempt to reconstruct the precise wiring diagram of the brain. Via a painstaking application of electron microscopy, it is possible to track individual axons as they snake their way through the brain. Every synapse can be detected and characterized to quantify more precisely the number and type of connection made. The tissue blocks investigated are infinitesimally small and the number of synaptic contacts infinitely large. Unemployment will only impede progress in this field if the government runs out of money.

Synapses and Receptors

Synapses guarantee versatility via their sign (excitatory, plus or inhibitory, minus), and the class of their chemical message (neurotransmitters and neuromodulators). Neuromodulators act by diffusing through relatively large areas of the brain and by their unique receptor dynamics. Each postsynaptic neuronal membrane is studded with proteins that bind with presynaptic chemical messengers: the more a given messenger is seen, the less sensitive is the receptor response. This neuronal habituation may underlie the rise and fall of activation in the brain with consequences for the intensity and sort of conscious experience.

One theory holds that the transition from wake to sleep is mediated by receptor *desensitization* and the transition from sleep to wake is associated with receptor *resensitization*. The alternation, within sleep, of NREM and REM phases, is similarly mediated. Recognition of receptor dynamics is thus helpful in understanding the vicissitudes of consciousness that occur spontaneously and indispensable to an explanation of clinical and street drug effects.

Sebastian Seung is a Korean-American scientist educated in the United States who worked in the Department of Cognitive Science at MIT in Boston before becoming a professor at Princeton. He has set himself the daunting task of identifying each and every synapse in the human brain. If there are 100,000,000,000 neurons and each one contacts 10,000 others that means that there are at least ten trillion synapses (10,000,000,000,000) to be identified. We will probably die before Seung completes his project but he is certainly right when he claims that, "I am my connectome." (No wonder so many people give up on the brain's complexity, read a good book, and go to church!) To demonstrate and quantify his "connectome," Seung cuts thin slices of brain and looks at them in the electron microscope. In this way, he can see all of the synapses. There are lots of them and they are of many different types. What these connections may have to do with consciousness will be addressed in the balance of this book.

The neurons of the brain are not only made more versatile by the variety of their receptors but they are made more efficient as well. Each neuron can recover its chemical messengers by a re-uptake mechanism that removes them from the synaptic cleft before they are enzymatically digested or diffuse away. In this way, the molecules can be used again, obviating the need to manufacture new chemicals in the cell body and pipe them down the axon via protein transport to the synapses. Important savings in time and metabolic energy are thus made.

Second messengers relay the effects on synaptic physiology and receptor dynamics from the membrane of the neuron to its nucleus. This mechanism effects a link between the outside world, via the conscious states of the brain-mind, to the genetic command center of each neuron. As yet poorly understood, it is already evident that this chain of events is capable of translating instantaneous phenomena, like the membrane depolarization of the neurons (on a time scale of milliseconds) to the much longer scale of days, weeks, and months on which we experience our moods as critical background elements of our consciousness. This amplification is over a million fold in the time domain.

Neural Network Theory

One way to respond to the overwhelming complexity of the brain is to simulate its functioning. *Neural network theory* assumes that there are inputs and outputs which are specifiable but that most of the brain consists of what are called "hidden layers" which cannot be observed directly. A mathematical model can then be constructed to test hypotheses about what the hidden layer must do to explain the link between inputs and outputs. Some models are capable of learning how outputs could be related to inputs. The inputs and outputs are not limited to external reflexes like the Sherringtonian "knee jerk" reaction, but may also consider entirely internal transactions of the kind that pertain to consciousness.

A good example is the Helmholtzian efferent copy model that posits a quantitative relationship between, say, an eye movement command (taken as an input and measurable as eye movement direction, speed, and positional change) and a cortical potential change (taken as an output and measurable as an EEG or scanning image alteration). Waking and REM sleep differences can be studied to deduce hidden layer alterations. These abstract deductions can then be compared with animal model data on the reasonable assumption that (whether or not non-human animals dream) the inferences pertain to models of human consciousness.

Many scientists are made uncomfortable by modeling of any kind and are suspicious of neural network modeling in particular. They believe there is no substitute for real physiology, no matter how laborious and time-consuming it may be. These scientists tend to be skeptical that any psychological reality, including consciousness itself, can be studied in a scientifically respectable manner. Advocates of neural network modeling defend their approach by insisting on rigor in the mathematics of their work.

A cogent example of a scientific approach to this issue is the network model of the bizarreness of dream consciousness compared with the relatively greater continuity and congruity of waking consciousness. The real physiological basis of this robust psychological difference can be modeled using neural network theory.

Specialized Circuits in the Upper Brain

The major anatomical difference between humans (who are thought to be fully conscious) and other mammals and birds (whose consciousness is probably more limited) is the enormous enlargement of man's cerebral cortex. It stands to reason that this anatomical fact is responsible for the psychological traits of man. This section presents several specific features of human cortical anatomy of relevance to this general hypothesis.

The first feature of interest is the massive extent of the human cortex. There are over 50 clearly demarcated cortical regions, each with its own name and number. By means of infolding (or *convolution*), the extent is far greater than the already massive volume and large superficial surface area of the human forebrain would suggest.

The visual cortex is particularly well understood. The many millions of cortical neurons are layered such that extrinsic inputs make synaptic contact with layer II while outputs to distant targets arise from layer IV. Other layers are "hidden," allowing interaction of the neurons in a vertical column, and horizontal, via connections, between columns. Signal processing within these columns has led visual scientists to propose that these columns are functional units encoding orderly sequences of the "receptive fields" of an animal's visual sensory surround.

Figure 4.3. Neural Network Diagram.

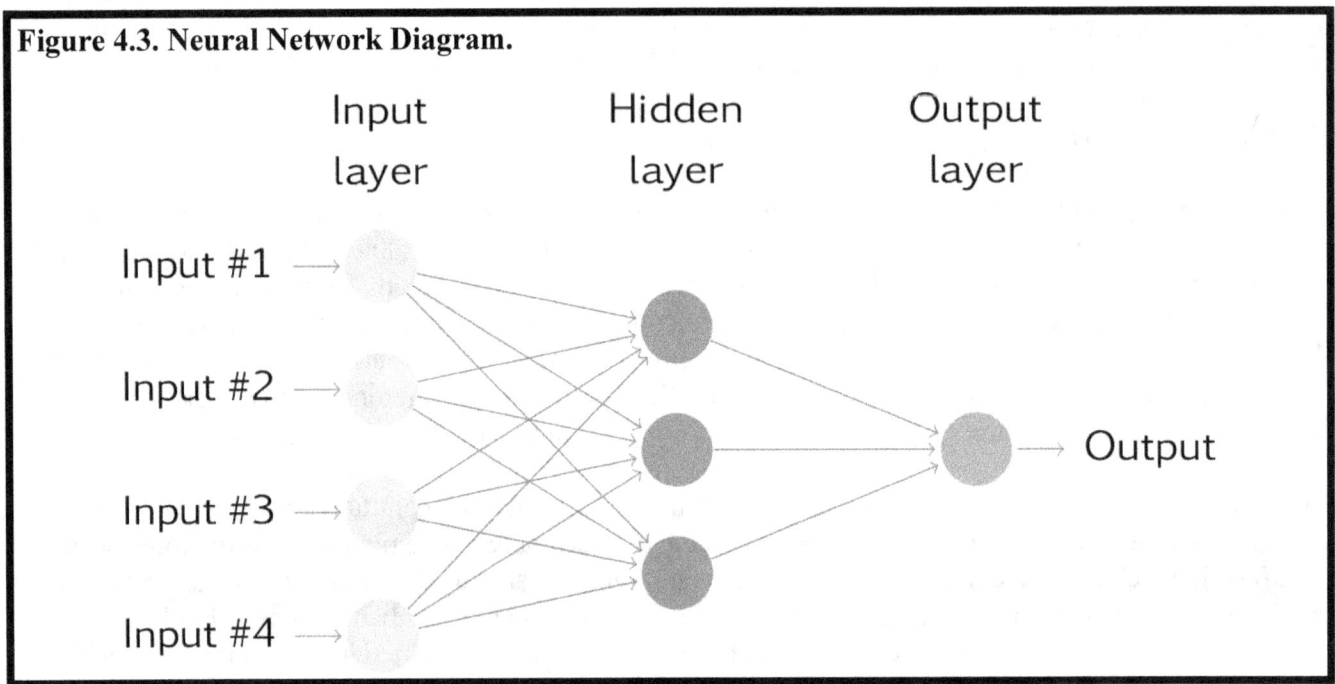

Cortical neurons broadcast their information to other cortical regions for integration (such as those, for example, which are sensitive to movement) and broadcast for further processing in the elaboration of imagery. This feature is germane to our understanding of endogenous vision, such as that of eyes-closed waking visual imagination and dream vision, both of which are internally generated in the absence of external inputs. Naturally these connections require a massive number of axonal fibers that are beginning to be specified and quantified via the Connectome Project. This anatomy is consistent with Dennett's multiple drafts theory of consciousness.

One well-studied fiber bundle is the massive corpus callosum, which connects the neurons on one side of the upper brain to the other (as is necessary for depth perception in vision, for example). Accidental, surgical, and experimental interruption of the corpus callosum results in the disintegration of consciousness, and this has suggested localization of speech, thinking, and even mood to one hemisphere or the other. This defect is called the *split-brain syndrome*. A related interconnection of relevance is that tying emotion centers in the limbic lobe to the frontal cortex associated with alterations of consciousness in meditative states.

The thalamus is the sensory gateway allowing the cortex access to selected input from the outside world and internal signals from other parts of the body and brain. The thalamocortical system governs selective attention in waking and exerts an all-or-none censorship of both external and internal data in NREM sleep when consciousness is obtunded if not eliminated altogether. The experimental work of Hans Berger and Mircea Steriade has already been mentioned in this connection.

The anatomy of the cerebral cortex and its myriad connections currently occupies thousands of neuroscientists and could continue to do so for centuries. With respect to consciousness, the experiment of nature constituting waking, sleeping, and dreaming affords an informative and strategic alternative to the pathway mapping studies described in acres of storyboard reports at annual neuroscience meetings.

David Hubel (1926–2013). While still a medical student at McGill University in Montréal, David Hubel worked with Herbert Jasper on a single cell recording system for the cat brain. Their goal was to test the hypothesis that the brain turned off in sleep. When they found, instead, that the neurons changed their pattern of discharge, Hubel changed his focus to the classical reflex paradigm in which he discovered the physiological mechanisms of external vision. Internal vision was studied only much later.

Specialized Circuits in the Lower Brain

The excitability of spinal cord and upper brain systems is controlled by circuits originating in the brain stem. Some of these circuits project upward, others downward, and a few go in both directions at once. Integrative aspects of this anatomy relevant to conscious state control will be addressed in the subsequent section. We have already discussed the reticular activating system and emphasized its role in turning up the engine of consciousness. Now we turn our attention to reflex excitability and introduce the function of chemical modulation.

A major advance in understanding the role of brain stem neurons in the control of conscious state was the discovery of aminergic neurons there which projected widely to other parts of the brain. Swedish scientists, including Kjell Fuxe, used fluorescent dyes to light up these cells and their widespread axonal domains. Among the neurons so illuminated were: the norepinephrine and serotonin systems of the pontine brain stem; the histamine system of the hypothalamus; and the dopamine system of the midbrain. A similar approach was used by Marsel Mesulam in identifying the cholinergic neurons. What all of these systems have in common is relative spatial restriction of their cell bodies of origin but diffuse spatial amplification of their axonal domains. They are thus ideal conscious state modulators, as we will discuss in Chapter 8.

The activated brain of waking and dreaming consciousness must modify its reflex excitability such that the external world has access to the brain (in waking) and is excluded from it (in REM sleep). Likewise, in waking, movement is allowed, while it must be prevented in REM. These changes are affected by the turning on (in REM) of circuits mediating inhibition of both sensory input and motor output. The neurons that are selectively enhanced in REM include those of the *medullary reticular formation* that sends fibers downward in the spinal cord. There they interact with the spinal motor neurons and render them unlikely to respond to motor commands from the cerebral cortex. We may dream of running but are, in fact, actively paralyzed. Otherwise, sleep would be interrupted and the dreamer might be injured. This unwanted possibility is in fact a danger for persons with REM sleep behavior disorder as described in Chapter 20.

Other brain stem neurons mediate sensory blockade at the level of both the spinal cord itself and more central relay synapses by a mechanism called *presynaptic inhibition* (which has an opposite effect to synaptic excitation). These sensory and motor changes make it possible to alter information access to and from the activated brain such that consciousness changes from waking to dreaming.

Figure 4.4. Schematic Drawing of spinal cord transection showing mechanism of pre-synaptic inhibition from Pompeiano experiments. Cross-section of the lumbar spinal cord showing the muscle afferent pathways entering the dorsal horn via the dorsal root. The cell bodies are in the dorsal root ganglion and the endings in the cord are monosynaptic excitatory (Ia), disynaptic inhibitory (Ib), and polysynaptic excitatory (I1) with respect to the motoneuronal efferents in the ventral horn. Cutaneous afferents and contralateral projections are not shown.

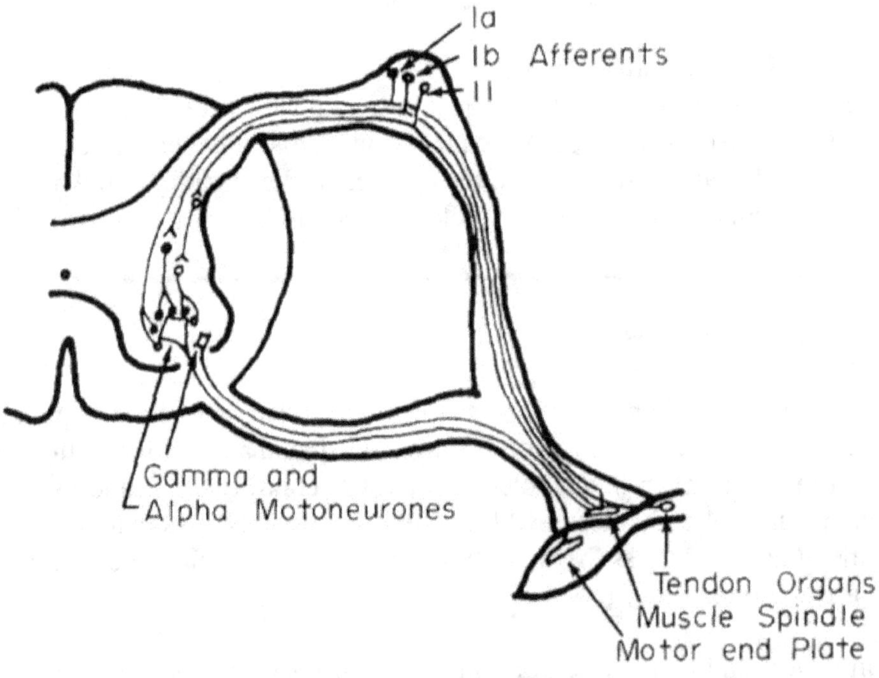

In addition to the synaptic gating that reverses the information source of consciousness, there is radical alteration of the mode of operation on the information. This chemical modulation of the entire brain is effected by neurons of the brainstem and basal forebrain that send fibers both up and down to the cortex, cerebellum, and spinal cord. When they all are active, the brain awakens. When some of them cease firing, the brain sleeps and dreams. We will specify the details of this modulatory mechanism anon but here mention only one consequence: the separation of one state from another and the poor memory for dreaming when we are awake.

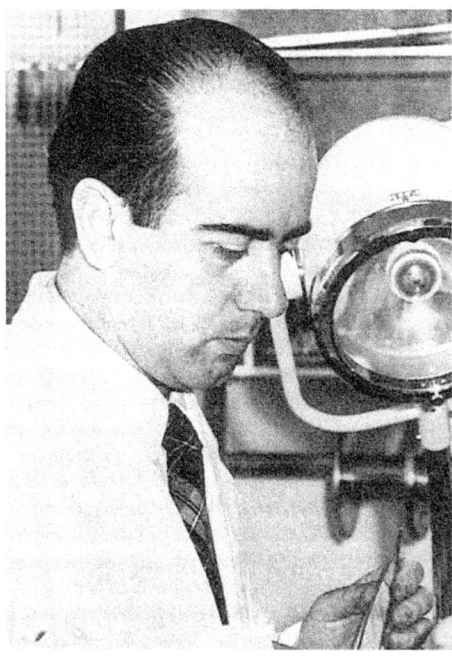

Ottavio Pompeiano (1927–2008) was an Italian neurophysiologist who used the techniques of classical Sherringtonian reflexology to specify the mechanisms of the spinal cord mediation of several Factor I elements. On the assumption that these changes were orchestrated by the brain stem, together with the veterinarian physiologist Adrian Morrison, he conducted studies of changes in the vestibular system during REM and pioneered the Factor M mediation of these changes by acetylcholine. A prolific writer, Pompeiano edited and published his scientific findings and theories in the Italian Archives of Biology and many international books and journals. Pompeiano thus advanced the intrinsic control of conscious states, first promulgated by his mentor Giuseppe Moruzzi, quantified as Factor A of the AIM model.

Integration

The ability to simultaneously activate, gate, and modulate the cortex assures the temporal integration of fundamental functions that are at the root of state dependent alterations in the conscious state. How are such widespread functions so flawlessly orchestrated?

A first principle is the massive interconnectivity mentioned above, now specifically augmented by the ubiquity of modulatory neuronal projections. Chemical neuromodulation underlies spatial unification in the time domain. The neuromodulatory systems of the brain stem fail utterly as point-to-point information reporters but their central location and widespread distribution makes them ideal synchronizers. All parts of the brain enjoy an identical microclimate at every minute of the night and day and the 100 billion neurons beat to the march of the same drummer.

A second principle is the flip-flop switch properties of conscious state control neurons. We met this concept when we considered the circadian rhythm of rest and activity earlier. Now we find that the same rule applies to the NREM-REM sleep cycle. Whether they are REM-on or REM-off neurons, the collateral connections of these brain cells to each other render their state changes exponentially rapid and synchronous. Put another way, when one population of neurons begins to augment or diminish its activity, all the neurons in that network follow suit. This is particularly true of the sleep-wake transition, whose virtual instantaneity is of obvious survival benefit. When it is time to wake up, we have an alarm clock built into the brain.

A third principle is the continuous range of time constants, which is related to the gamut of neuronal sizes in brain stem populations. Large cells with long axons can conduct impulses afar while smaller neurons tend to conduct impulses locally. As far as consciousness is concerned, this size principle subserves the sensorimotor integration of which we are not aware but without which we could not possibly be aware. Large cells mediate the phasic (fast) coordination of eye movement with both postural tone and cortico-thalamic adjustment while smaller cells mediate the tonic (slow) activation and deactivation that underlie both dreaming and waking.

These three principles help us to understand the brain basis of our conscious states and to begin to discern the similarities and differences between waking and dreaming.

The Neural Correlates of Consciousness

Where then in the brain is consciousness located? The Neural Correlates of Consciousness (NCC) concept refers to the part or parts of the brain essential for consciousness.

We will see that the reticular activating formation is involved in mediating arousal, but that does not mean that consciousness is located in it, or in the thalamus, any more than we can conclude a television picture is contained in the plug or in the on/off switch. We can however gain some idea of how consciousness is related to the brain by looking at different types of conscious states.

We have seen that sleep is a complex state involving many brain structures, but particularly the frontal cortex, the thalamus, the reticular formation, and the aminergic nuclei of the brainstem.

Francis Crick argued that particular circuits are important for visual awareness, which he supposed arose from reverberation between the thalamus and certain layers of the visual cortex. Vision is important, but there is more to consciousness than visual awareness.

Different types of general anaesthetic agents involve different brain structures and neurotransmitter systems. They all involve the thalamus and frontal cortex. Surprisingly little is known about how general anaesthetics work. Brown and Schiff have suggested that entering a state of general anaesthesia is more like entering a coma than falling into a deep sleep. However, studies by Ralph Lydic and Helen Baghdoyan on the effects of anaesthetic chemicals on aminergic and cholinergic neurons suggest that a sleep model of anaesthesia may also be informative. Sleep and coma are on a continuum after all and both are altered states of consciousness.

As pointed out in Chapter 23, recent research has focused on the differences between coma and persistent vegetative states (PVSs), on the one hand, and minimally conscious states (MCSs) on the other. Coma is the deeper state; in coma, the person is totally unresponsive to external stimuli, including light and pain; comatose persons do not show sleep-wake cycles, and do not initiate action. In a persistent vegetative state, the person is coma-like but does display sleep-wake cycles. In a minimally conscious state, on the other hand, the person, although largely unresponsive, does occasionally show some limited awareness of his surroundings.

Table 4.1. Physiological Basis of Cognitive Features of REM-Dreaming.

Function	Nature of difference	Causal Hypothesis
Sensory input	Blocked	Pre-synaptic inhibition
Perception (external)	Diminished	Blockade of sensory input
Perception (internal)	Enhanced	Disinhibition of networks storing sensory representations
Attention	Lost	Decreased aminergic modulation causes a (decrease in) signal to noise ratio
Memory (recent)	Diminished	Because of aminergic demodulation, activated representations are not stored in memory
Memory (remote)	Enhanced	Disinhibition of networks storing mnemonic representations increases access to consciousness
Orientation	Unstable	Internally inconsistent orienting signals are generated by cholinergic system
Thought	Reasoning ad hoc Logical rigor weak Processing hyperassociative	Loss of attention, memory, and volition leads to failure of sequencing and rule inconstancy Analogy replaces analysis
Insight	Self-reflection lost (failure to recognize state as dreaming)	Failure of attention, logic, and memory weaken second- and third-order representations
Language (internal)	Confabulatory	Aminergic demodulation frees narrative synthesis from logical restraints
Emotion	Episodically strong	Cholinergic hyperstimulation of amygdala and related temporal lobe structures triggers emotional storms, which are unmodulated by aminergic restraint
Instinct	Episodically strong	Cholinergic hyperstimulation of hypothalamus and limbic forebrain triggers fixed-action motor programs, which are experienced fictively but not enacted
Volition	Weak	Top-down motor control and frontal executive power cannot compete with disinhibited subcortical network activation
Output	Post-synaptic inhibition	

We conclude that consciousness is not located in any one part of the brain, but depends on the sustained activation of broad portions of it. This conclusion is consonant with the multimodular outline detailed in Table 4.1. Consciousness is as complex as the brain that generates it.

Dialogue 4. Brain Structure and Conscious States

TH: When I was a student, I saw the brain as the fist of a boxer with an arm attached to it.

AH: The boxer's glove was the upper brain seen from the side and the arm-tail was the spinal cord. That was a good image to start with. I will try to help you see the brain as an AIM cube.

TH: Before you do that, what else should I know about brain anatomy?

AH: That the spinal cord receives from and sends messages to the skin and muscles of the body below the neck.

TH: These are the reflexes that Sherrington and Freud wrote about between 1890 and 1925.

AH: The anatomical plot was thickening all the while, but no one knew how to integrate the data.

TH: What should be added to my simple picture?

AH: The boxer's glove-like upper brain has been subdivided into the cortex at the top, a subcortical thalamus which relays information to the cortex and other subcortical structures thought to mediate emotion (the limbic lobe), hormone release (the hypothalamus), and modulation of movement (the basal ganglia).

TH: What holds all these parts together?

AH: Between the upper brain glove and its spinal cord tail is the brain stem. This intermediate position makes it a kind of command central for the upper brain above and the spinal cord below. It reaches up and down at the same time to be sure that the boxer's glove brain knows what its arm and lower limbs are doing and vice versa.

TH: The brain stem sounds like an internal book-keeping agent. How does it work?

AH: It has its own reflex functions which must be tied to those of the spinal cord below and the upper brain above. For example, it houses twelve cranial nerves, three of which command eye movement; each and every eye movement must be taken account of by the upper brain and the spinal cord.

TH: That's a tall order. It must be a challenging task.

AH: The brain is a many-splendored thing. In addition to the usual electrical signals that the brain employs for its own point-to-point communication, it produces chemical messages that set the states of the brain.

TH: Can you provide a simple example to help me understand the chemical setting of brain state?

AH: Memory is enhanced (in waking) and impeded (in sleep) according to whether or not certain brain stem chemical messengers are released.

TH: Your AIM model is designed to explain how the brain stem works to guarantee integration and control state.

AH: Factor A (discussed in Chapter 6) represents the strength of activation; it is the brain's power supply. Factor I is for input-output gating; it raises or lowers the excitability bar of reflexes (as discussed in Chapter 7); Factor M is for chemical modulation; it dictates how information is processed (we assert in Chapter 8 that this is the most significant new insight in Conscious State Science).

TH: What a relief. Now I have to remember only three things about the brain in order to be a consciousness scientist. I think I will junk my image of a boxing glove and its tail. To replace it, I will turn my brain-mind to a consideration of how the three AIM factors may work together. I never really liked the boxing glove metaphor anyway. I've often thought that pictures of the brain and spinal cord look like aliens floating in space so I welcome the soft earth-landing you provide.

Chapter 5. Deafferentation

There is a John Hughlings Jackson neurologist and a Sigmund Freud psychiatrist in all of us. We love to think the brain is nothing but a collection of centers and that the psyche, too, can be split into parcels and its functions localized in various parts of the brain. More realistic distributed systems models are less easily understood. They seem vague and thus appeal to us less than simple-minded theories. The concept of a state, implying the whole brain-mind, is the most unappealing example of a distributed model. We might say, "Please don't bother me with complexity. I would much rather tackle one system and examine one and only one state at a time."

Don't Believe Everything That You Read

The history of the quest for a consciousness widget seated in the front row of a Cartesian theater is a graveyard of logical errors, simplistic and well-meaning, to be sure, but nonetheless misleading. A good example is the "radio tube fallacy" which dates from the pre-transistor era. Remove a vacuum tube (or a transistor), find that the radio makes no sound, and conclude that the missing part is the source of the missing function (whereas it is the speaker that makes the sound). Brain lesion studies often fall afoul of this fallacy. In the previous chapter, we warned of the same danger when you pull the plug and the radio falls silent. The noise is not in the plug.

The easiest way to understand this problem is to consider stroke, the accidental destruction of brain cells by a limitation of blood-borne oxygen owing to circulatory failure. When a patient can't articulate speech after damage to Broca's area (or understand speech following damage to Wernicke's region), we jump to the conclusion that we have localized two language centers. But have we really made any progress in understanding language and are we really sure that Wernicke's and Broca's areas are centers as against crossroads of a far more widely distributed system? We are correct in taking the center hypothesis seriously but we need to be critical. It is difficult to be optimistic and critical at the same time. Is the cup half full or half empty? It is both.

The distinction between the two explanations (centers vs. crossroads) is as valid for intentional experimental lesions as it is for accidental brain injury. Since all areas of the brain are an admixture of neurons and axons, the disconnection of parts in a diffuse, distributed network is always a consequence of any form of brain damage, be it natural or experimental. This problem can be mitigated in experimental studies by damaging neurons with kainic acid (which kills only neurons vs. electrical heat which kills axons, too). Animal models are, of course, not available for studies of speech because non-human species are relatively speechless. But because all mammals sleep and all mammals have REM in their sleep, it is possible to learn a great deal about the contributions of the brain to what we call protoconscious experience. Remember that the term protoconsciousness is introduced to signal the antecedence (in evolution and development) of consciousness. We theorize that consciousness arises gradually, not in a sudden jump (see again Figure 1.2).

Electrical stimulation incurs similar limitations in the experimental effort to localize function. It is impossible to activate brain cells without, at the same time, activating fibers of passage (and vice versa, it is impossible to activate axons without also exciting neurons). In 1949, Moruzzi and Magoun were wise enough to realize that the brain stem was the locus of neurons that activated the forebrain when they attempted to excite the axons passing downward from the motor cortex to the spinal cord. Epilepsy is an experiment of nature in which an excitatory focus can be localized but it is a disease that can also help us to understand natural sleep as the product of spike and wave EEG signatures (such as PGO waves). In these cases, the brain stimulates itself, allowing clinicians and scientists to observe rather than cause or treat the process under study.

The concept of how an experiment should be defined comes into focus here. Since we are interested in spontaneous activity, we must beware of eliminating phenomena of interest by intervening too vigorously. Stimulation and lesion experiments, by definition, disrupt the brain because they are performed intentionally to induce damage. Only when we have an idea of how the brain-mind self-organizes should we thus perturb the central nervous system. The premature introduction of drugs with potent and unnatural effects on the brain-mind is a perturbation that is as undesirable in clinical medicine as it is in basic science research. Consciousness science is as much watchful waiting (for nature to reveal her secrets) as it is the forced confession of fabricated alibis.

Experiments of Nature

It is, of course, immoral to inflict experimental brain damage on human subjects but nature herself is sometimes cruel rather than moral. Stroke and seizure disorders are all too common human mishaps but they are often instructive. Mental illness may also teach us about the defects of consciousness that we wish we could fix. The scientist must learn to take advantage of these unfortunate accidents until we know enough to remedy them definitively (or at least help patients compensate for irremediable defects). This section provides general guidelines for the study of experiments of nature.

Giuseppe Moruzzi (1910–1986) trained as a motor physiologist with expertise in cerebellar and vestibular science. Moruzzi was recruited by the U.S. Air Force after World War II to investigate the costly loss of consciousness by fighter pilots who blacked out when subjected to sudden acceleration or deceleration. His 1949 discovery (with Horace Magoun) of the reticular activating system was followed by an immersion in sleep science through which he became a world leader. Moruzzi combined his experimental skill and expertise in the motor physiology of the brain stem in a meticulous series of lesion and stimulation studies which demonstrated the importance and complexity of the lower brain in determining the state of the presumed forebrain site of conscious experience. His theory of a medullary mechanism for the active control of slow wave sleep remains untested to this day. His love of books is reflected in his University of Pisa collection of historical classics, which deserve careful study as the modern origins of consciousness science.

Natural structural damage to the brain by stroke, like an experimental lesion, is usually irreversible because dead neurons are not replaced. The damaging effects are dynamic and difficult to interpret. When should the damage be evaluated? At ten days, ten weeks, ten months or ten years after the insult? Is the defect to be regarded as evidence that the damaged brain region is a "center" normally responsible for the lost function? Or is it rather a locus of interconnection between distant points in a network? The late Norman Geschwind (1926–1984), a Professor of Neurology at Harvard Medical School, is responsible for the concept of "disconnection" to advance and specify the latter interpretation. Geschwind's caveat is part of the general trend in scientific philosophy regarding distributed networks with multiple nodes of synchronous interaction.

Norman Geschwind (1926–1984) was a neurologist at Harvard Medical School in Boston. Among many other important intellectual achievements was the articulation of disconnection theory. By means of disconnection following cerebral damage, the brain and its mind are less integrated than they once were. According to Geschwind, the syndromes of neuropsychology discussed in Chapter 18 are as much the result of the disconnection of neuronal centers as they are of the destruction of the centers that talk to each other. We have stressed that the thalamocortical systems function because of reciprocal connections. When these connections are interrupted (or blocked as in natural sleep) consciousness is altered or lost. Geschwind also contributed to our understanding of aphasia, left handedness, and personality. Owing to his wide-ranging interests he founded behavioral neurology, the beginning of the reunion of psychiatry and neurology that this book seeks to foster. Norman Geschwind was a clinician who understood that nature was performing experiments without informed consent and he took advantage of that unfortunate truth.

Of course there really *are* centers: the oculomotor nuclei of the brain stem are genuine centers but they are also parts of networks that transmit information to other parts of the brain about the eye movements that they command. A famous example is Hermann von Helmholtz's patient, whose visual field shifted when his paralyzed eye could not move to the side on command. Such patients suffer from both center destruction and disconnection. In normal REM sleep, eye movement commands are executed and information about them is processed as part of dream consciousness.

Neural excitability is dynamically controlled so as to allow just enough activation of the brain networks responsible for behavior and consciousness. In a major seizure, excitation spreads like wildfire throughout the brain resulting in convulsions and loss of consciousness. More localized hyperexcitability, such as that typical of temporal lobe epilepsy, causes a brief lapse in consciousness and alterations in perception, emotion, and memory that are very similar to those of dreaming.

Temporal lobe seizures have in fact been called "dreamy states." A hypothesis to be considered and further evaluated is that dreaming is, among other things, the subjective awareness of a modified seizure of the brain stem and temporal lobe. In this case, we move beyond mere localization to a global concept of neural network excitability variation in the mediation of conscious states. In this connection, a "seizure" is as normal (and abnormal) as dreaming itself. Conscious states cannot be sharply delimited as normal or abnormal. Do we all not have a healthy seizure when we dream in REM?

The notion of dreaming as a healthy seizure can be extended to mental illness, which has long been thought of as a "waking dream." This is because, from a formal viewpoint, dreams are psychotic by definition: they comprise both hallucinations and delusions. Many scientists object to this comparison because dreaming is normal (read universal) while psychosis is abnormal (read relatively rare). We do not blur this distinction to be provocative. Rather, we are opportunists who want to better understand consciousness by the strategy of formal comparison of all natural conscious states. Sleep and epilepsy have common formal features. This implies shared physiological and psychological mechanisms.

Antiepileptic medication is increasingly used in the treatment of major mental illness. It is surprisingly effective, implying that neural hyperexcitability may be part of the psychiatric problem. This is a good example of how patients can be helped whether or not heightened neural excitability is a deep cause of the propensity for hallucinations and delusions. In any case, it is easy to see how debilitating dreaming during waking might be. Hybrid states are those in which the normal firewall between dreaming and waking breaks down.

Bremer's Brain Stem Transections

Frédéric Bremer (1892–1982) was a Belgian pioneer of consciousness science. Like most of his peers, he supposed that sleep was the result of decreasing external stimulation although his experimental work strongly suggested the internal mechanisms that were later discovered by his students and followers. In articulating his theory of deafferentation he was as significantly wrong as Sigmund Freud was in proposing that dreams were caused by repressed infantile wishes. As the philosopher Carl Popper said to the physiologist John Eccles, a hypothesis is useful to science if it can be proved to be incorrect.

Total transection of the neuraxis is radical but it avoids many of the disadvantages of experimental lesion and stimulation experiments. The results can be interpreted with confidence because brain regions can be completely separated from each other both neuronally and axonally. The only possible communication across a transection is humoral, via the cerebrospinal fluid and/or the blood. We have already criticized the deafferentation hypothesis that Frederic Bremer, the Belgian neuroscientist, set out to confirm in the early 1930s long before the reticular activating system or REM sleep had been discovered.

It is important to realize that when Bremer worked in the 1930s, all brain function was thought to be reflexive, that is, to emanate from sensory stimuli arising in the outside world. The stage was also set for Bremer's work by the 1918 epidemic of influenza that left many survivors both somnolent and Parkinsonian by destroying what are now known to be dopamine-containing neurons of the substantia nigra in the midbrain. We now recognize that the spontaneous activity of these and other aminergic neurons create waking and dreaming consciousness from within.

Hoping to understand waking and sleep, Bremer created the *cerveau isolé* (or isolated forebrain) by severing the brain stem just rostral (anterior) to the midbrain reticular formation. This is the best known and most easily understood of his two classical preparations. Following surgery, the cats were comatose with persistent EEG slow waves. Bremer misinterpreted this result by assuming that it was due to the elimination of activation from external stimuli rather than the lessened intrinsic activation of the reticular formation.

Figure 5.1

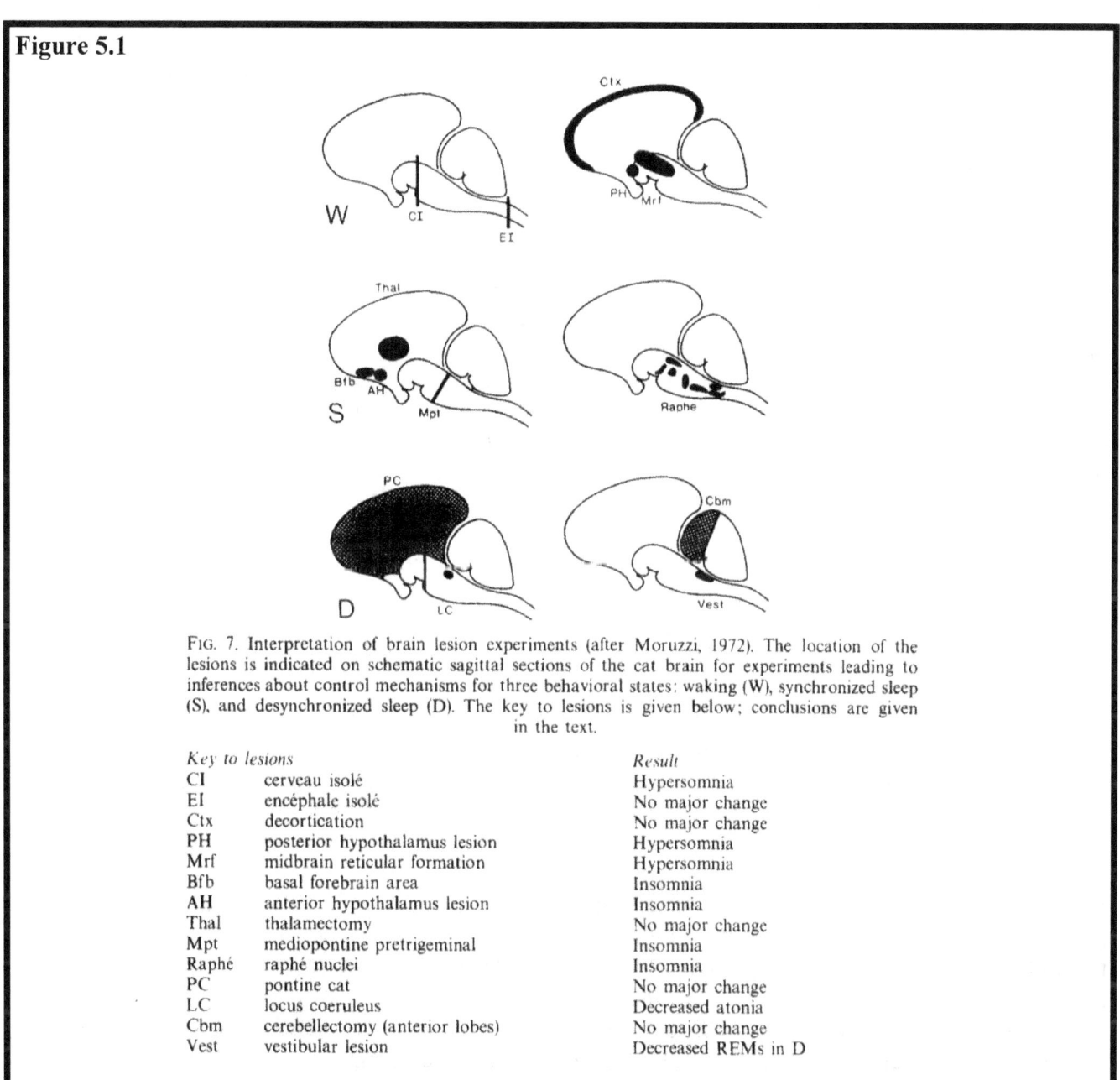

FIG. 7. Interpretation of brain lesion experiments (after Moruzzi, 1972). The location of the lesions is indicated on schematic sagittal sections of the cat brain for experiments leading to inferences about control mechanisms for three behavioral states: waking (W), synchronized sleep (S), and desynchronized sleep (D). The key to lesions is given below; conclusions are given in the text.

Key to lesions		Result
CI	cerveau isolé	Hypersomnia
EI	encéphale isolé	No major change
Ctx	decortication	No major change
PH	posterior hypothalamus lesion	Hypersomnia
Mrf	midbrain reticular formation	Hypersomnia
Bfb	basal forebrain area	Insomnia
AH	anterior hypothalamus lesion	Insomnia
Thal	thalamectomy	No major change
Mpt	mediopontine pretrigeminal	Insomnia
Raphé	raphé nuclei	Insomnia
PC	pontine cat	No major change
LC	locus coeruleus	Decreased atonia
Cbm	cerebellectomy (anterior lobes)	No major change
Vest	vestibular lesion	Decreased REMs in D

To test his own erroneous deafferentation theory, Bremer then transected the CNS at a much lower level, the junction of the medulla and spinal cord. This *encéphalé isolé* (isolated brain) was surprisingly hyperalert, a result mysterious to scientists and explained by Moruzzi, who postulated a still undemonstrated EEG synchronizing influence in the medulla. To save his doomed theory of deafferentation, Bremer claimed that the brain of the *encéphalé isolé* cat was hyperalert because it was activated by external stimuli entering the brainstem from the face and head via the trigeminal nerve. Moruzzi put this idea to rest by observing that cats were hyperalert even when the transection was anterior to the trigeminal nerve. These and other experimental preparations are depicted in Figure 5.1.

The difference between the *cerveau isolé* and the *encéphalé isolé* preparations remains unexplained but one conclusion survives. The brain self-activates and is thereby potentially supportive of both wake and dream consciousness. In the case of waking consciousness, external sensory signals are processed and acted upon. In REM sleep, external stimuli are excluded and motor output is prevented. Dream consciousness is not only virtual, it is entirely synthetic. Confidence in this theory derives from the most modern and definitive transection of them all, Jouvet's pontine cat, discussed in Chapter 6. The brain stem is surgically cut caudal to (below) the reticular formation, which eliminates the forebrain and its activation. But it leaves intact the eye movement generators (a part of AIM, factor A) and the spinal cord inhibition (AIM, factor I) as well as the aminergic and cholinergic modulators (comprising factor M of AIM).

The Loss of Dreaming after Stroke

If dreaming depends upon activation and modulation of the cortex, it might be possible to prevent dreaming by stroke damage to the forebrain. To check on this possibility, the South African neuropsychologist, Mark Solms, working in London, asked 300 successive patients whether their strokes had caused any change in their dreaming. Two brain areas, when damaged by stroke, caused a cessation of dreaming. One was the parietal operculum, a convergence zone of fibers from primary (e.g. visual) and associative (e.g. non-specific temporal) areas. The other was the deep frontal white matter, which connected the emotional to the volitional brain. These findings need to be confirmed with sleep lab recordings and experimental awakenings but are of great interest because of complementary findings from other sources.

The loss of dreaming following parietal operculum damage had earlier been reported in two independent studies, one by Martha Farah (at Carnegie Mellon University in Pittsburgh) and the other by Cristiano Violani and Fabrizio Doricchi (at the University of Rome). Both of these studies collated the results of many case reports in the literature. Disconnection (or interruption of cross roads) was the interpretation that they shared with Solms. In order for humans to dream, it would appear to be necessary to recruit a widely distributed cortical network.

Even more intriguing was the importance to dreaming of the deep frontal white matter. Dreaming is strikingly emotional (with anxiety, aggression, and elation predominating) and avolitional (dreamers say that the plot is not under their voluntary control). These psychological features of dreaming are frontal brain functions, which apparently depend on the integrity of the deep frontal white matter.

Both areas found critical to dreaming in the Solms study are uniquely and selectively activated in the normal REM sleep of subjects monitored by PET and fMRI imaging methods. In other words, the areas that light up in normal REM are the same as those whose destruction leads to the loss of dreaming. This result is a compelling confirmation of the hypothesis that dreaming depends upon these brain structures. Of course we would like to know if parietal or frontal white matter stroke victims lose REM or if they have REM without dreaming. We would also like to be sure that dreaming is really lost (as against only lost to recall).

Mark Solms (1961–) is a South African neuropsychologist who has become a leading defender of institutional Freudianism. He is a Trustee of the Neuropsychoanalysis Fund in London and the Director of the Arnold Pfeffer Center at the New York Psychoanalytic Institute. According to Solms, the negative scientific verdict against Freud is invalid. Instead of being driven by random (or chaotic) brain noise, dreams are both motivated and over-determined sources of therapeutically crucial information. Dreaming is a cortically mediated phenomenon and the cortex is essentially independent of the brain stem. Solms has not rebutted the arguments against Freud's theory of dream symbolism or the untenability of Freud's disguise-censorship mechanism of dream bizarreness. Rather, he has retreated to defend the guardian of sleep tenet of psychoanalytic dream theory. He promises to soon present evidence that good sleep is associated with abundant dreaming.

Figure 5.2. Human brain with Solms' stroke lesions.

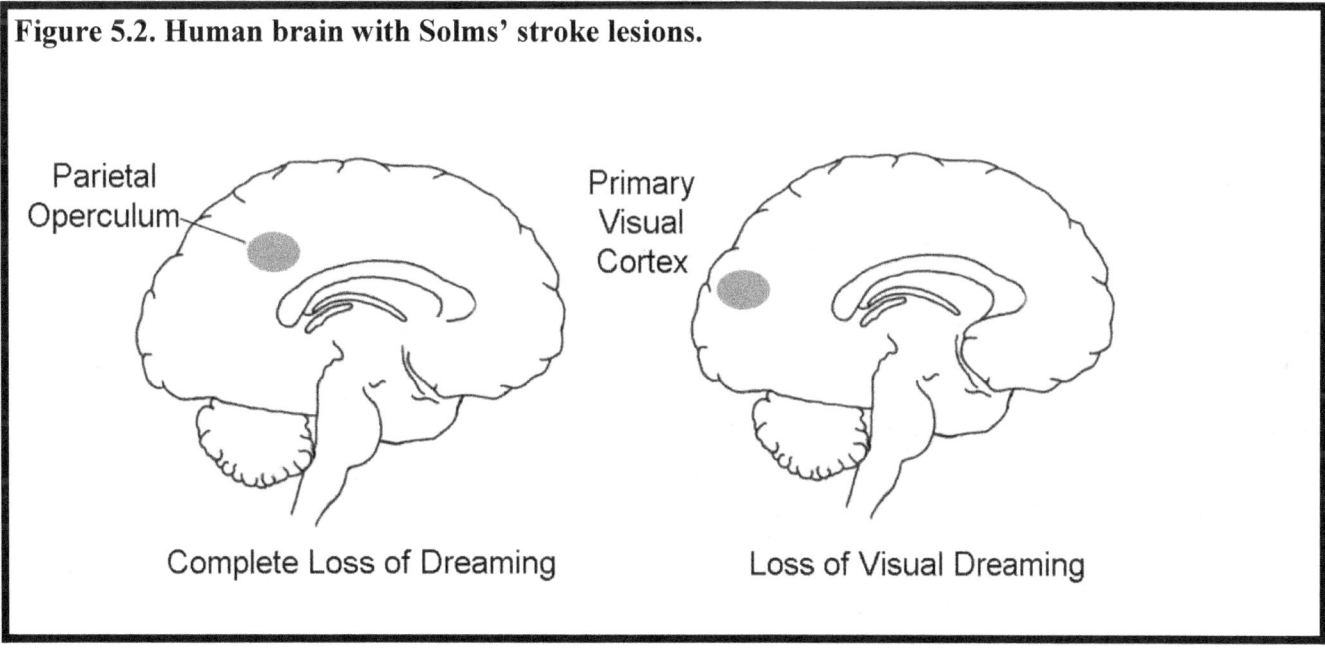

From 1930 until 1950, a common human brain-mind intervention was prefrontal lobotomy. In the pre-psychopharmacological era, this operation reduced psychosis, allowing patients to live at home. When Mark Solms investigated their case histories, he found reports of cessation of dreaming as well as cessation of psychosis. This fact is relevant to our subsequent discussion of dreaming and psychosis but we focus here on its significance to the neuropsychology of dreaming per se. To dream normally, the frontal brain must be connected to the parietal and occipital brain.

Functional Deactivation

The hypothesis that the PGO waves of REM sleep were held under inhibitory restraint during waking by the raphé nucleus (serotonin containing neurons of the pontine brainstem) was tested by Raimond Cespuglio in Michel Jouvet's Lyon laboratory. Cespuglio designed and implanted a thermode (a thin rod whose temperature can be controlled) which, when activated, cooled the raphé nucleus and turned off its cells (as normally happens only in REM sleep). The result was, as predicted, the immediate release of PGO activity which persisted as long as the raphé was inactivated by cold and subsided when the raphé was rewarmed. By means of this reversibility, Cespuglio was thereby able to avoid the reservations leveled against Dana Brooks, a neuroscientist from Cornell Medical School, that his surgical isolation of the raphé had nonspecifically damaged the brains of his cat subjects.

Brooks, together with Emilio Bizzi, then a post-doctoral fellow in the laboratory of Giuseppe Moruzzi in Pisa, had pioneered the experimental study of PGO wave neurobiology. Brooks and Bizzi showed that it was possible to induce waves if the electrical stimulation of the pontine reticular formation was delivered in the transition period between NREM and REM when high voltage PGO waves are spontaneously present. This finding is analogous to the observations of José Calvo, who stimulated the amygdala and indicated that the pontine tegmentum becomes a "center," functionally speaking, during REM. From a Darwinian point of view, we theorize that this change of state, of which dreaming is the human correlate, benefits the mammalian brain. The idea is that REM is normally incompatible with waking and that waking depends upon serotonin secreted throughout the brain by the raphé neurons.

The Brooks experiment, which Cespuglio's thermode cooling technique replicated and improved upon, was a transection of a sort, but instead of being total and transverse (such as with Bremer's isolé cuts), it was partial and longitudinal. Brooks cut the fibers connecting the raphé nuclei to the PGO generator and observed PGO wave release in waking. What the cats experienced is anybody's guess, but we cannot resist speculating that this may have been the feline form of lucid dreaming (or dreaming waking). The point here is that the experimental investigation of the brain basis of consciousness has a rich history in sleep science. These experiments are depicted in Figure 5.3.

Pharmacological Studies

A potent way of influencing consciousness without making a hole in the brain or buzzing it with electrical or magnetic energy is to introduce chemicals expected to influence its brain-dependent conscious states. By far the most frequent intervention of this sort is surgical anaesthesia. Literally thousands of states of unconsciousness are created daily around the world by means of gases, which render neurons unresponsive to stimuli that normally mediate sensations, including pain. Little is known about the effects of anaesthetic agents on specific physiological systems but the notable fact is that the brain-mind can be taken off line at will. This section will focus on the more subtle and informative studies of the specific brain systems that regulate conscious states.

Ever since the description of the pontine aminergic neurons by Kjell Fuxe and Anica Dahlstrom in 1962, attempts have been made to model Factor M of AIM (the model described in Chapter 9). This epoch in consciousness science is a parable of instructive insight as to how and how not to proceed. We will use it to adduce general principles, hoping to imbue our student readers with a guide to the perplexed and to suggest a sense of patient comfort with this difficult subject.

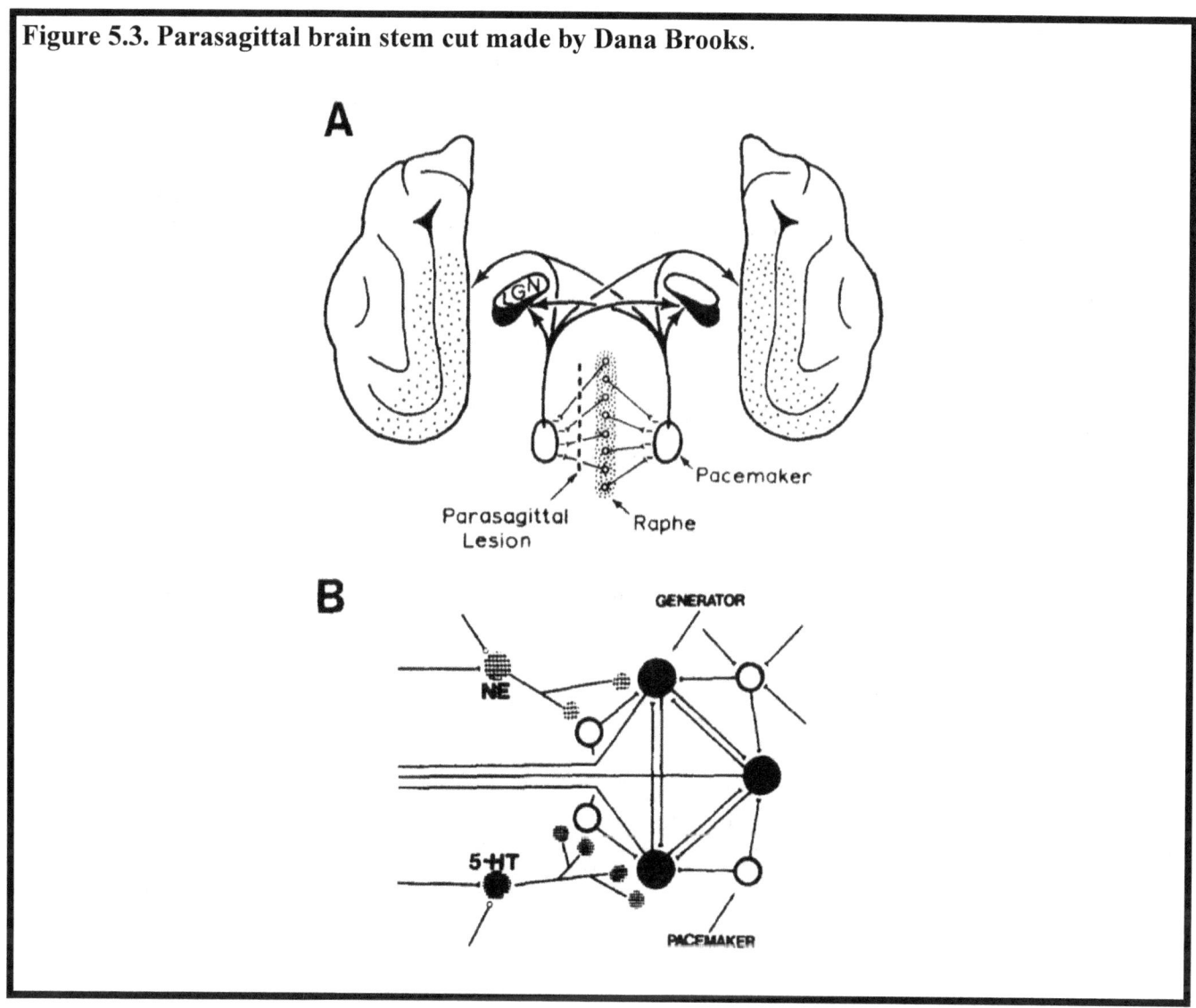

Figure 5.3. Parasagittal brain stem cut made by Dana Brooks.

The first principle is that any chemical that is introduced via a peripheral route (and that includes, per force, all psychoactive medication and street drugs) is distributed widely throughout the body and the brain, making the interpretation of effects extremely difficult.

The *second principle* is the converse of the first: the chemicals used by the brain-mind to create and regulate its own states rely on precise cellular and subcellular mechanisms that cannot be simulated by any but the most sophisticated interventions.

From these two general principles emerge two cautions:

1) Taking any street drug can be dangerous.
2) Giving a drug to an animal in the hopes of learning something may be a fool's errand.

To illustrate the first point, we cite briefly the cases of two colleagues' sons who took LSD, thought they could fly, and fell to their death from the roofs of high buildings.

To emphasize the second point, we will review experience with drugs used to investigate Factor M of AIM to test the hypotheses that:

1) *Norepinephrine*, manufactured by the locus coeruleus, mediates REM sleep while
2) *Serotonin*, manufactured by the raphé nuclei, mediates NREM sleep.

Massive doses of drugs were administered to raise or lower the levels of these two chemicals in the brain. Measurements showed that the drugs were sometimes effective in mediating the desired effects on the brain but the side effects on recipient animal's behavior made confident interpretation of the effects impossible.

Later, we will review the cellular evidence that both hypotheses were wrong. To test the alternative theory that both brainstem norepinephrine and serotonin mediate waking and suppress sleep while brain stem acetylcholine mediates REM, local microinjection studies have corroborated the REM enhancement by acetylcholine.

Inadvertent Surgical Effects on Consciousness

In the same heyday of neurosurgery that produced the cessation of dreaming following prefrontal lobotomy and caused the hemispheric mental dichotomy of split-brain patients, the devastating effects on memory following temporal lobectomy for epilepsy were reported. Some patients, like the famous H. M. studied by Brenda Milner at the Montréal Neurological Institute, had been subjected to bilateral excision of the temporal poles to prevent seizure spread from one side of the brain to the other. The surgical ablations of the medial temporal lobe extended into the anterior hippocampus, a structure now known to be essential to short-term memory.

Following surgery and for the rest of his life, H.M. could not form a new memory. He failed to recognize or name his clinical helpers. Worse yet, H.M.'s amnesia was enduring. Instead of having debilitating temporal lobe seizures, he was rendered socially incompetent by this well-meaning but ill-advised treatment. Thanks to Brenda Milner and others, H.M. has taught us more about ourselves than most intentional experiments. By means of careful, exhaustive investigation, Milner was able to demonstrate a distinction between H.M.'s procedural memory (which was intact) and his semantic memory (which was deficient). This distinction commands attention both because it is fundamental and because it differentiates learning enhancement by NREM (semantic) and REM (procedural) sleep.

H.M.'s performance of a motor task improved with training. He could learn a new skill, but he had no recollection of having learned that skill even on the days when he was trained by Milner herself. The obvious conclusion was that while the temporal lobe and hippocampus were crucial to the formation of semantic narrative memory, motor skill learning must take place elsewhere in the brain. Another key finding was that semantic memory was impaired even for events experienced just prior to surgery. Long term memory was only intact for events that had occurred several years before H.M.'s operation.

The data that emerged led to a dynamic formulation of semantic memory. Recent experience is encoded and stored in the hippocampus/temporal cortex. With time, the memory was transferred out of that region to some other part of the brain.

It is now clear that this dynamic process can occur in sleep whether or not the sleeping subject has any conscious awareness of dreaming. To our knowledge, H.M.'s sleep was never monitored, but there is no reason to suppose that it was abnormal. Since he could not encode narrative aspects of his experience, he had no information to transfer from one part of his brain to another but his skill learning could be consolidated because it was not stored in the hippocampus/temporal lobe anyway.

Consciousness cannot be reduced to memory. Dreaming, which is often as amnesic as was the post-operative H.M., proves this point. H.M. was conscious, but amnesic. Dreaming is often considered unconscious whereas it is reasonably considered to be a conscious state separated from waking by amnesia and from the outside world by active inhibition of sensory input and motor output.

Awakenings

Comatose or highly unresponsive patients can be aroused in two ways relating to the AIM model. One is to increase the aminergic drive via the administration of drugs. The other is to electrically stimulate the thalamus. In both cases, humans long thought to be hopelessly unconscious can be aroused. These morally problematical results force us to take consciousness science more seriously. The results are morally problematical because there are patients that have been regarded as living "vegetables" and for whom "pulling the plug" has been considered ("Pulling the plug" is a casual way of referring to mercy killing or euthanasia). Science may have more to offer than was once supposed.

"Awakenings" will certainly rouse the hopes of caretakers as well as relatives.

Oliver Sacks (1933–2015) was a neurologist with a strong penchant for psychiatry. By virtue of his captivating writing style, he did more through his books and New Yorker magazine articles than any of his contemporaries to make the connection between neurology and psychiatry implicitly clear to educated laypersons. He was enviably free of speculation. As a clinician, he is best known for Awakenings, the account of emergence from coma of patients with Parkinson's disease when treated with the aminergic drug, L-dopa, a precursor of the brain activator, dopamine. For Sacks, who worked as a psychiatrist/neurologist, the brain and its mind were seamlessly connected, however. His life was committed to informal experiments with himself, his companions, and his patients. He was even reconciled to his own recent death from melanoma, which caused the blindness he described dispassionately in his autobiographical book, *On the Move: A Life.*

The first dramatic awakening occurred when comatose Parkinsonian patients were given dopamine for their motor symptoms and woke up. Dopamine had been recognized as a neurotransmitter in 1957 by Arvid Carlsson of the University of Lund in Sweden whose academy rewarded its favorite son the Nobel

Prize in the millennial year 2000. Dopamine, which we would now consider a neuromodulator, has been of interest to the science of motivation, learning, and memory, as well as to psychosis models and antipsychotic medication.

In the early 1960s, Anica Dahlstrom and Kjell Fuxe described the projections of dopamine neurons to more restricted areas of the forebrain than the other biogenic amine systems delivering histamine, norepinephrine, and serotonin, but their function in brain activation is similar. Dopamine is not yet taken into account by AIM because it appears to track factor A in being activated in both waking and REM sleep rather than differentiating those states.

A second dramatic awakening was more recently evoked by electrical stimulation of the central nucleus of the thalamus of minimally conscious patients by the Cornell Medical School neurosurgeon, Nicholas Schiff. The Schiff team has followed up on their discovery by demonstrating activation of the c-fos gene throughout the cerebral cortex following clinically effective thalamic stimulation. Patients who are minimally conscious have been classified by the neuroscientist Steven Laureys, of Liège University in Belgium, as lying on a continuum between NREM sleep and irreversible coma. This places them in the left lower corner of AIM state space (see illustrations in Chapter 9). When stimulated electrically, these patients can be moved into the waking domain via the increase in factors A, I, and M.

A third well known example is the dramatic "awakening" produced in patients with encephalitis by Oliver Sacks. Sacks, in a classic book (later made into a film), tells the tale of victims of an outbreak of the virus *encephalitis lethargica*, in 1918. The cause is still unknown, although the leading theory is that the symptoms resulted from a strong autoimmune system response to an infection by streptococcus. This strep throat-like infection lead to destruction of regions such as the basal ganglia and substantia nigra. In the most severe cases, patients were left in a catatonic state, unable to move or talk, for decades. Many others were left with *post-encephalitic Parkinsonism*, a state resembling Parkinson's disease, due to damage to the dopamine neurons of the substantia nigra.

Sacks describes his attempts to treat patients in the Beth Abraham Hospital in the Bronx, New York with the drug L-DOPA; in some of the catatonic cases, it produced an "awakening" after forty years of inactivity. The drug had serious side effects, including psychosis, tics, excessive libido, and freezing of movement. Furthermore, the patients eventually developed a tolerance for the drug, and in spite of increasing the dose (with yet more side effects), all patients eventually lapsed back into their catatonic state.

Dialogue 5. Experiments of Nature

TH: Brain-mind scientists do not inflict damage on their fellow man but they do learn from natural disasters.

AH: Coma and lesser impairments of consciousness have always been of interest to neurology. Now that interest is shared by psychiatrists and psychologists who can test their theoretical concepts clinically.

TH: Since the classical studies of von Economo of encephalitis lethargica, we have come a long way. No longer are we confined to narrow regional assumptions. Many aminergic neurons project to the entire brain much in the way that reticular formation cells were supposed to do.

AH: The dopamine neurons of the substantia nigra are damaged in Parkinsonism and encephalitis lethargica giving an M axis value in the AIM state space. Dopamine is clearly important, but what role it plays in normal state dynamics is not yet clear.

TH: Not to mention its hypothetical role in mediating the hallucinations of psychosis, which we consider later in Chapter 21.

AH: If I were younger, I would certainly focus on dopamine. As yet, believe it or not, there is no good study of the discharge activity of dopamine-containing neurons over the sleep cycle.

TH: Dopamine may link bodily movement with the virtual action of thought. That's an inviting prospect.

AH: We have drug data to add to the mix because we can influence dopamine systems chemically as is done in clinical settings.

TH: This topic is the central theme of the book and film called "Awakenings."

AH: Electrical stimulation of the midbrain has also been shown to bring patients back to conscious waking after years of comatose absence.

TH: Values of M may set the level of responsiveness in a chemically specific way. I am grateful not to be responsible for my own M values.

AH: Transparency is a blessing in disguise. But you can still make a contribution via your dream journal. Do you ever dream of using your word-processor? I don't, despite the long hours that I spend at the keyboard each day.

TH: No, but I do dream of making presentations. I think that's because they're like school exam: stressful and social.

AH: I encourage people to self-observe after natural accidents as well as in their normal lives because we need good subjective data to compare with the biological fruits of the powerful new brain imaging technologies.

TH: More and more scanners are being made available for research and clinical studies.

AH: We live in a Brave New World. Subjective experience is the essence of consciousness. For neurobiology to realize its full potential, subjective experience must be taken seriously by science.

Chapter 6. A = Activation

We can distinguish two basic states of consciousness: waking and sleeping, with sleeping being frequently pervaded by dreams. Are these phenomenological states distinguishable physiologically, too? It turns out that they are similar but differ in many significant ways. In this and the subsequent three chapters, we build a three dimensional model of conscious states called AIM. The first factor, A, stands for activation, which, as we will show, is both robust and quantifiable. It becomes the x axis of the AIM cube.

Figure 6.1. AIM Factor A, Activation. Factor A can be computed as either the mean frequency or the reciprocal of EEG amplitude.

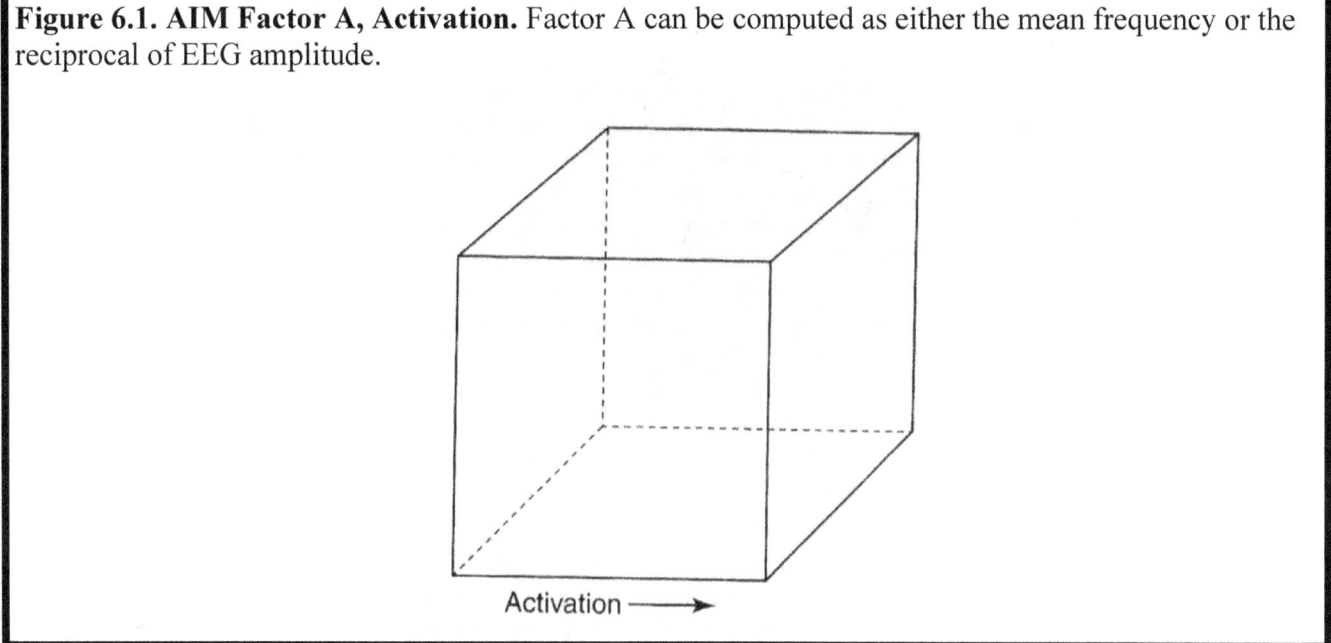

Activation ⟶

The Discovery of the EEG by Hans Berger

The power of the brain is electrical, and the electrical activity of the brain can be recorded and measured as voltage. The key discovery was made in 1927 by Hans Berger, a neuropsychiatrist working in Jena, Germany. Berger attached electrodes to the scalp of his subjects and recorded the microvolt signals on moving paper. Berger's discovery of the EEG stemmed from Einthoven's earlier description of heart muscle electrical rhythmicity, the well-known EKG. The muscle artefacts that Berger's critics feared might be responsible for his "brain waves" can be recoded as the electromyogram (EMG) and eye movements can be recorded as the electro-oculogram (EOG). Together, these and still other discrete measures constitute the polysomnogram (PSG) that is used in sleep science and sleep medicine today. Discriminating within and across PSG channels provides a quantitative, objective tool for consciousness science.

Berger had a very difficult time convincing his critics that the *electroencephalogram* (EEG) was of brain, not artefactual muscle origin. His claims were aided by the fact that the brain waves that he recorded changed character from low-voltage fast to high-voltage slow when his subjects fell asleep. This pattern was reversed by stimulated or spontaneous arousal.

From the outset, it was thus clear that the state of consciousness changed when the electrical power of the brain rose and fell. Recently, it has been suggested that consciousness *is* the power of the brain. According to this theory, mind is physical but immaterial — like the flow of electrons in any current, or like the production of a magnetic field by moving electrical charges. But this idea is relatively recent. Berger's discovery was affirmed in the early 1930s by Edgar Adrian and Brian Matthews, English physiologists who were recording brain electron flow with the cathode-ray oscilloscope, an instrument used to pick up the extra and intracellular currents caused by the action potential discharge of individual neurons. The exact relationship of neuronal action potentials to the EEG is still a focus of active scientific research.

Hans Berger (1873–1941) was a German psychiatrist working at Jena at the height of the organicist era, which originated in Munich at the hands of Emil Kraepelin. Berger was inspired by the Dutch scientist Willem Einthoven, who discovered the EKG. Berger wondered if electrical signals of brain origin were recordable from scalp electrodes. They were. Berger's observation that the brain waves that he recorded changed from low-voltage fast (when his subjects were alert) to low-voltage slow (when they became drowsy) helped to convince him and the world that the EEG was a valid and useful scientific tool. The activation of the brain, now known to be a substrate of waking and dreaming, is a cardinal feature of conscious states and was present at the beginning of the modern scientific era.

The EEG reveals several basic types of electrical activity, and these come to the fore in different mental states. Electrical activity is classified by the underlying rhythm, in terms of cycles per second (Hertz, or Hz); it is as if the brain were "pulsating" with a regular beat. The thalamus, a symmetrical structure near the center of the brain connecting the cortex to the lower regions of the brain, plays an important role in regulating the sleep-wake cycle and the brain's electrical activity.

Berger's EEG is now analyzed according to the frequency of the waves. They help us understand the conscious states with which they are associated.

- *Gamma waves* have a frequency of about 40 Hz. They are prominent in both waking and REM sleep dreaming consciousness.
- *Beta waves* have a frequency of about 13–30 Hz. Beta waves are associated with normal, alert, eyes-open consciousness in waking.
- *Alpha waves* (sometimes called Berger's waves in honor of their discoverer) have a lower

frequency than beta waves: 8–13 Hz. They appear to originate in the posterior cortex and are associated with a calm, relaxed state with the eyes shut in waking but they may be observed in REM sleep when subjects dream.

- *Theta waves* have a frequency of about 4–8 Hz. They originate in the cortex and hippocampus and are associated with very deep relaxation, meditation, drowsy states, and light sleep. They are as prominent in REM as they are in waking. Children show proportionately more theta, and abnormally high levels of theta are associated with certain types of brain damage.
- *Delta waves* have a frequency of less than 4 Hz. They are very slow, and associated with deep sleep; hence this type of sleep is usually called *slow-wave sleep* (SWS). SWS is marked by a relative paucity of rapid eye movements, and so is also called NREM (non-REM) sleep.

At first glance, we appear to have five basic types of wave ranging from gamma and beta waves (very alert), through alpha (relaxed) and theta (drowsy) to delta (deep sleep). However, the picture is more complicated and interesting than that. As Berger observed, the low-voltage fast gamma, beta, and alpha patterns were a function of alert waking.

Furthermore, the function of gamma waves is much debated. In the face of deep skepticism, Francis Crick (who with James Watson won the Nobel Prize for unraveling the molecular structure of DNA) worked later on consciousness with Christof Koch. Echoing Wolf Singer, Crick and Koch have argued that gamma waves play a role in unifying consciousness — particularly visual awareness — by binding together different regions associated with the same visual stimulus. That is, different regions of the brain process the many modular components of conscious states. Although widely separated from each other, they are linked together by their common 40 Hz activation. The focus of the gamma wave originates in the thalamus, and the waves sweep across the forebrain to and fro several times a second, unifying stimuli at the center of attention. Further support for this idea comes from the observation that if the thalamus is damaged, the 40 Hz gamma waves do not appear, and the person lapses into a deep coma.

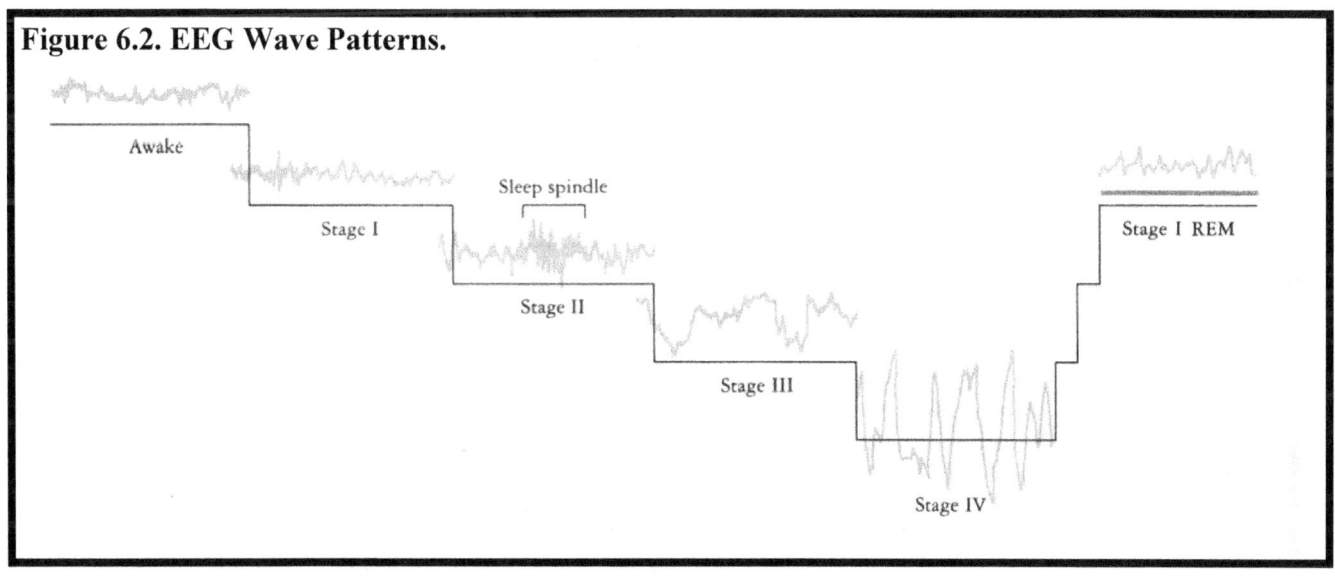

Figure 6.2. EEG Wave Patterns.

Awake

Stage I

Sleep spindle

Stage II

Stage III

Stage IV

Stage I REM

It is no longer held that all of sleep is characterized by very-low frequency (delta wave) oscillations. All night EEG recordings show an approximately 90-minute rhythm of brain activation and deactivation identified since 1953 as the REM-NREM cycle. We make much of this fact in our treatment of consciousness in this book.

NREM sleep has been divided into four stages.

- *Stage 1* occurs at the beginning of sleep, as electrical activity moves from alpha to theta. People awakened in this stage think they have been fully conscious, and sometimes hear spoken voices and may experience their limbs moving or jerking suddenly (*hypnic* or *myoclonic* jerks). There are very slow, rolling eye movements. This stage is often called the *hypnagogic phase*.
- In *Stage 2*, the sleeper is still relatively easy to awaken. The emerging slow wave activity of Stage 2 is interspersed with sudden bursts of electrical activity called *K-complexes* and *sleep spindles* (about 13 Hz lasting for half a second). These bursts of sudden activity seem to be involved in the brain blocking peripheral stimuli so as to maintain sleep.
- *Stage 3* (spindles and slow waves) is the transition between Stage 2 (spindles) and
- *Stage 4* (slow waves but no spindles).

Because there are no striking differences between Stages 3 and 4, modern thinking treats them alike. NREM dreaming, which tends to be less bizarre and more idea-based than REM sleep dreaming, occurs in Stages 2, 3, and 4. NREM sleep is characterized by body movement as the brain shifts from NREM to REM sleep.

REM sleep (also called emergent Stage 1) shows frequent rapid eye movements (hence the popular name) combined with muscle paralysis. REM sleep is correlated with a high proportion of bizarre dreaming. It is sometimes called paradoxical sleep because the electrical activity of REM sleep is very similar to that of waking.

The transition from NREM to REM sleep is marked by the presence of ponto-geniculo-occipital (PGO) waves. PGO waves originate in the brainstem, and are linked with the initiation of the rapid eye movements.

Figure 6.3. Three Nights of Polygraphic Sleep (a) and 12 Nights of REM (b).

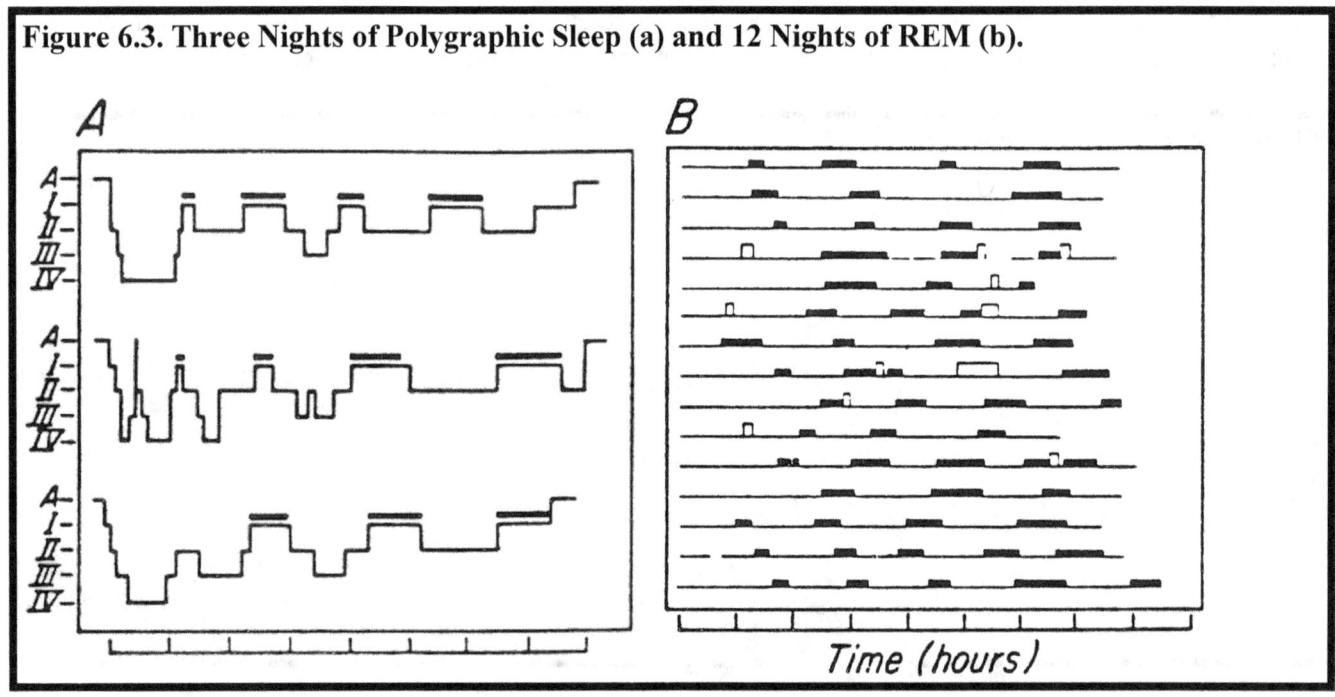

The most important general implication from Berger's research is that the brain and consciousness are periodically activated and inactivated by electrical power. We measure this power as Factor A (activation in the AIM model presented in Chapter 9). The EEG has not only permitted scientists to analyze the brain's spontaneous activity but also its response to stimulation as evoked potentials to stimulation in the reflex paradigm.

In the light of these subsequent studies, the results of Bremer's experiments reviewed in the previous chapter could be due to internal and spontaneous activation rather than the withdrawal of external stimulation (deafferentation) or, at least, a combination of the two mechanisms. The alterations in reflex excitability are measured as factor I in the AIM model. If this were so, initiating sleep could be an active process, with the control of the sleep-wakefulness cycle located somewhere between the two brainstem transections made by Bremer. Later research showed the importance of the structure called the *reticular formation*, the most important of a set of structures forming the *reticular activating system*. For the scientific study of consciousness, the reticular formation and its subdivisions are the most important structure in the brain stem. The reticular formation projects to the thalamus, which we have already seen plays a central role in maintaining consciousness and alertness.

Electrical stimulation of the reticular formation instantly awakens a sleeping animal, while lesions to the formation produce persistent sleep (Moruzzi & Magoun, 1949). Later work by Jouvet showed the particular importance of the raphé nucleus; damaging the raphé nucleus results in several days of insomnia for the cat.

Bremer and Deafferentation Revisited

So strong was the reflex paradigm that even so distinguished a neurobiologist as the Belgian Frédéric Bremer believed that waking consciousness was a response to external light. We have reviewed these findings in the previous chapter. We repeat them here for emphasis. Recall that Freud's dream theory was advanced 30 years before Bremer's time and was based on the same erroneous reflex assumptions. Bremer's theory of sleep was called "deafferentation" to denote the idea that only the stimulus decline at sunset was determinant of sleep and the return of light at sunrise determinant of the next day's waking. Bremer went so far as to ascribe the difference between midbrain transection (the *cerveau isolé* with EEG slowing) and lower medullary transection (the *encéphalé isolé* with EEG speeding) to sensory input via the trigeminal nerve. We have seen that his ideas turned out to be wrong, although the transections were fortuitously informative in another way.

The contrasting hypothesis, of intrinsic and brain deactivation at sleep onset, was first advanced by Giuseppe Moruzzi, who transected the brain stem in front of the trigeminal nerve and observed a state as hyper-alert as that of Bremer's *encéphalé isolé* preparation. In a major paradigm shift, Moruzzi concluded that sleep and waking were caused by internal and active brain mechanisms. Consciousness was internally controlled by structures in our brain, not by external stimuli, such as light.

For two decades, the major experimental tools were surgical manipulation and electrical stimulation of the brain. Both were potent but each had major shortcomings. Surgery caused side effects that made many of the observations quite non-specific, while electrical stimulation was difficult to localize because applied current spread widely throughout the brain. But these effects left little doubt that Cajal's "wearisome labyrinth," the brain stem, contained the secret of conscious state control. Exactly how this was accomplished was still uncertain a half century later.

In the late 1930s, the outbreak of World War II sidelined and truncated the careers of electrophysiologists, as they were recruited for the development of electronic weapons such as radar. The good news was that computer technology was simultaneously advanced, making the cellular revolution in neuroscience possible. By this means and sophisticated pharmacological chemistry, the precise role of neurons in conscious state control was made dramatically clear. This advance led to the recognition that one reason that the brain stem was so important in regulating conscious states was the presence there of chemically coded neurons. Their contribution is formalized and measured as factor M of the AIM state space model.

The Reticular Activating System of the Brain Stem

Giuseppe Moruzzi and Horace Magoun made their breakthrough 1949 discovery by accident. They wanted to identify neurons projecting from the motor cortex to the spinal cord by stimulating the medulla and backfiring the cortical neurons of interest. To quantify the level of arousal in the brains of their animal subjects, they recorded the EEG, and were surprised to observe EEG activation when they stimulated the medulla of the brain stem. Realizing that there might be an intrinsic system of brain activation, they found that the most potent arousal followed stimulation of the midbrain reticular formation.

One reason why Cajal threw up his hands at the bewildering complexity of brain stem neurons was that reticular formation cells received multiple inputs and projected too widely to be obvious components of reflex pathways. Rather, they were ideally suited for "nonspecific" activation — just what was needed to turn the brain on in a general and reliable sense no matter what specific sensory signal was being processed. Here again, the science of consciousness needed this apparent non-specificity to advance. Subsequent research has shown that reticular neurons actually convey specific as well as non-specific activation, particularly in those sensorimotor integration circuits that coordinate posture and eye movement.

Moruzzi and Magoun followed up on their findings with a variety of stimulation and lesion experiments which were thereafter complemented by cellular level studies that tied their relatively crude interventions to the classical tradition of Cajal and Sherrington. Their work can therefore be counted as the beginning of the modern science of consciousness. Many peers thought that despite the methodological shortcomings, they deserved a Nobel Prize for their discovery. They put brain activation on the scientific map and showed clearly that consciousness was a variable function of the brain. The physical nature of consciousness, long suspected, was clearly established.

REM Sleep and Dreaming

For the first half of the twentieth century, psychology and physiology ran on the separate tracks of mind and body. Moruzzi and Magoun's reticular activation system began to unite them but their approximation was firmly established by Aserinsky and Kleitman's 1953 report in the journal *Science* of the correlation of spontaneous, periodic activation of the brain in sleep with dreaming. The early psychophysiological investigation of dreams tended to be psychoanalytically oriented and perpetuated the idea that dreaming was unconscious. In contrast, it has recently been proposed that there are two kinds of consciousness, the traditional state of waking, and via a new way of thinking, of dreaming. Subsequent chapters will further explore this line of thought.

REM sleep was noticed by accident as Aserinsky's child subjects fell asleep in habituation to an attention task in waking. The rapid eye movements, which gave this state of sleep its name, were, like Moruzzi and Magoun's EEG activation, seen in the EOG at sleep onset in the children and at 90-minute intervals throughout sleep in the adult subjects. These adult REM epochs lasted from five to fifty minutes and occupied a total time of 15–25% of sleep, which comes to roughly 1.5 hours per night. When the adults were aroused experimentally, they reported hallucination-like dreaming 90% of the time after awakenings in the middle of REM periods with intense eye movement.

REM sleep and dreaming were thus far from incidental to arousal and certainly not a response to infantile wishes but an essential aspect of the physiology underlying consciousness, broadly conceived. Our dream amnesia, our inability to remember dreams, is in part responsible for our failure to recognize dreaming for what it was, and for the failure of consciousness scientists to take advantage of this misunderstood state. Now we can be sure that no theory of consciousness can afford to overlook dreaming. In fact, dreaming may hold a key to our understanding of conscious awareness.

Besides alerting us to the promise of consciousness science, the discovery of REM sleep ushered in the modern era of sleep medicine. Already in their brief *Science* article, Aserinsky and Kleitman noted that cardiac and respiratory activity were also brain-mind state dependent. Now virtually every bodily function is known to fluctuate, sometimes pathologically or even fatally, in sleep. Along with now obsolete notions of psychology, the idea that sleep was only restful and never harmful must be abandoned — we need think only of insomnia and conditions such as sleep apnea, to which we will return.

Horace Magoun (1907–1991). An important method in consciousness science is the stereotaxic technique by which a three dimensional target in brain space may be established and attained. A pioneer in the refinement and application of stereotaxis was Horace Magoun, working at Northwestern University near Chicago. An epochal discovery of modern consciousness science was the fruit of the collaborative research of Horace Magoun with the Italian sleep expert, Giuseppe Moruzzi, on the reticular activating system. A paradigm shift from reflexive to spontaneous change in brain activity was thus attained. On the strength of this discovery, Magoun was named head of the Brain Research Institute at UCLA, which became a center for studies of the brain basis of conscious states. Among others, the French neurosurgeon Michel Jouvet studied at UCLA and the stereotaxic method was crucial to the success of all that followed.

Non-REM Sleep

The exclusive correlation of REM sleep and dreaming claimed by William Dement was challenged by David Foulkes, who found that reports of "mental activity in sleep" also occurred together with high voltage slow EEG activity in about 50% of laboratory awakenings. Since 75–80% of sleep is of this type, loosely defined dreaming occurs throughout the night. This observation has important implications for theories of consciousness and supports Leibniz's idea that "*petites perceptions*" were continuous overnight.

Most of the sleep in the first half of the night is profoundly deep. It is difficult to awaken subjects and they are often demented as well as unresponsive when they are eventually awakened. The EEG is impressively high-voltage slow in what has been called Non-REM or NREM sleep. Actually, eye movements are about 1/3 as frequent as in REM but are difficult to discern because the EOG signals are lost in the fronto-cortical EEG slow wave activity. For this reason, we prefer the name slow wave sleep (SWS) to the misnomer NREM. Later in the night, the differentiation of REM and SWS decreases both physiologically and psychologically but the phenomenological distinction between the mental states nonetheless persists.

The most sophisticated physiological research on slow wave sleep was performed by Mircea Steriade, a neurologist from Romania who emigrated, via Paris, to Québec in Canada. Steriade was not interested in dreaming and eschewed any interest in consciousness, but his elegant intracellular recordings revealed that the access of brain stem and spinal inputs to the cortex were blocked at the level of the thalamus, a relay nucleus for sensation and attention in the waking state.

Bremer's postulate of deafferentation was thus supported by Steriade's work, but the deafferentation turned out to be central, not peripheral, as Bremer supposed. The brain has its own way of controlling the source of information (and the way that the information is processed). Steriade's mechanistic model of information gate control is based upon work on the brainstem conscious state control system that will be discussed in a subsequent chapter.

Figure 6.4. REM Sleep is characterized by EEG activation, rapid eye movements and diminished muscle tone.

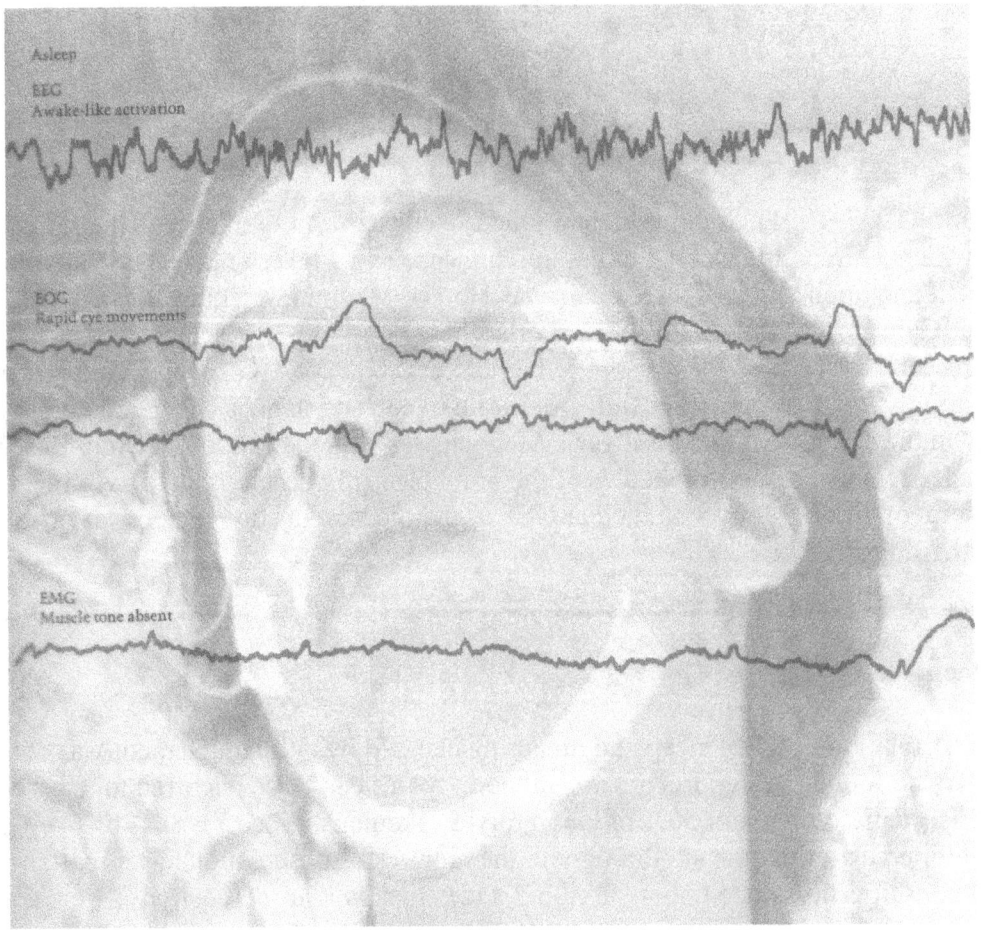

Eugene Aserinsky (1921–1998) was a physiologist with a research interest in attention. His child subjects quickly grew tired of his stimulus arrays and drifted off to sleep. He wired the kids up with EEG electrodes to record brain activation and measured their eye movements as a sign of their alertness. When he reported to his supervisor that when his young subjects fell asleep, they showed EEG activation and eye movements, Nathaniel Kleitman suggested that they make all night recordings in adult subjects. Aserinsky and Kleitman reported in 1953 that adult human sleep was punctuated at 90-minute intervals by REM and that these REM intervals were not only regularly periodic but occupied 15–25% of sleep time and were correlated with intense dreaming. Thus consciousness could occur in sleep as well as waking and in both states it was associated with brain activation and eye movements. As Louis Pasteur remarked, "In the field of observation, chance favors the prepared mind."

The Circadian Rhythm

We now know that the body follows an approximate 24-hour internal cycle called the circadian rhythm. Although the cycle is internal, it is entrained by external cues. These cues are called *zeitgeber*s (German for "time giver"), such as light and temperature. People deprived of natural light tend to fall into a slightly longer sleep-wake cycle, of about 25 hours, but there is now debate as to whether this lengthened cycle merely reflects exposure to artificial light, which might have some cycle-lengthening effect.

Thomas Graham Brown worked in Sherrington's lab at Oxford in 1912. Noticing fluctuation in the excitability of reflexes (even in Sherrington's spinalized animals), Graham-Brown proposed that the oscillator (or neuronal clock) was as much the functional unit of the brain as the reflex. This proposal was the beginning of a new way of looking at things that were directly relevant to the spontaneous change in conscious state. Needless to say, Sherrington resisted this challenge. Time has proved them both right: both reflexes and oscillators are functional units of the brain and they act together in an harmonious manner.

Mircea Steriade (1924–2006). A native of Romania, Steriade trained as a neurologist before emigrating via Paris to Québec, Canada in the early 1960s. He is respected for his technically challenging intracellular recording studies of neurons in the cat thalamus and cortex which revealed the mechanisms of activation, deactivation, and reactivation of those elements in waking, NREM, and REM sleep respectively. Thanks to Steriade, we now have a detailed cellular and molecular understanding of how the brain changes state. Rigorous and conservative, Steriade eschewed psychological interpretation of the sort which we champion here. As we stood on his shoulders peering into the realm of consciousness, he collaborated with Robert McCarley on a book length review of brainstem neurophysiology, which has become a classic reference work.

To model the neuronal oscillator, Graham-Brown theorized the push and pull of two paired components, one of which was excitatory, the other inhibitory. When one component was turned on the other was off and vice versa. There was, at the time, no direct evidence for this model, but now we know the oscillator comprises the REM-on (excitatory) and the REM-off (inhibitory) neurons of the pontine brain stem. Unfortunately for Graham Brown, World War I broke out and he was drafted as a military psychiatrist and thereafter lost to physiology. But his idea of neuronal oscillators came alive again when biological clocks were proposed to be at the root of human behavioral rhythms. The leaders of this new science were Jürgen Aschoff and Rütger Wever of the Max Planck Institute in Munich, Germany. They built an underground bunker with the purpose of isolating human subjects from time cues like light and temperature as well as social signals like the mechanical clocks that synchronize work and mealtimes.

Perfection in eliminating time cues from human subjects proved to be very difficult and their work's imperfection has been criticized. It is nonetheless accepted that people share with all living things an intrinsic tendency to alternate rest and activity every day. Within the rest phase of the cycle are body temperature minima and sleep. Within sleep, there is nested the NREM-REM cycle. All of these wheels within wheels would have comforted Thomas Graham Brown and might even have won over Charles Sherrington, who was forced to speculate poetically in *Man on his Nature* through his magic shuttle loom metaphor for consciousness.

Thomas Graham Brown (1882–1965). Charles Sherrington must have resented the gadfly critique of his student, Thomas Graham Brown, who challenged the master's idea that the reflex was the functional unit of the nervous system. Graham Brown argued that the oscillator or biological clock deserved consideration and went so far as to posit "paired half-centers" for this role. Time has proven both correct: there are reflexes and there are paired half-centers which are both building blocks of the CNS. Neither can be considered as equivalent to Cajal's neurons of which both reflexes and clocks are composed. For consciousness science, it is important to recognize both as units of functional organization. In conscious states, the reflexes are set by oscillators such that the subjective experience of each state is distinctive. It is ironic to note that in World War I, Graham Brown became a military psychiatrist whose prescient insights have shaped consciousness science as we know it today.

Whether it is really possible to split apart body temperature and sleep rhythms, there is no question that, outside the bunker, they rise and fall together with a periodicity of less than 24 hours. The robust human circadian rhythm is reset each day by time cues, both natural and artificial, which act on the circadian oscillator in the suprachiasmatic nucleus in the hypothalamus. That oscillator is temperature insensitive, which suggests that its own rhythmicity is intracellular and is not a function of Graham Brown neuronal half centers. But its interaction with other clocks that regulate consciousness may obey his 100-year-old rule.

The daily timing of physiology and psychology is coordinated with the daily light cycle by an internal clock in the *suprachiasmatic nucleus* (SCN) of the hypothalamus, just above the pituitary gland at the base of the forebrain. The SCN is connected directly to photosensitive ganglion cells in the retina of the eye. The SCN is in turn connected to the pineal gland, which secretes the hormone melatonin. Levels of melatonin are highest at night. The temperature of the body also varies systematically across the day, with the minimum occurring at night. This well-studied set of mechanisms is a special variant of the universal synchronization of organisms with information and energy fluctuations in the outside world.

The circadian control of waking, slow wave sleep, and REM is now studied at the level of cells and molecules such that a connection between consciousness and both external and internal information and energy can be specified. Whatever consciousness really is, it is certainly rhythmic.

Figure 6.5. The NREM-REM sleep cycle is synchronized by the circadian clock.

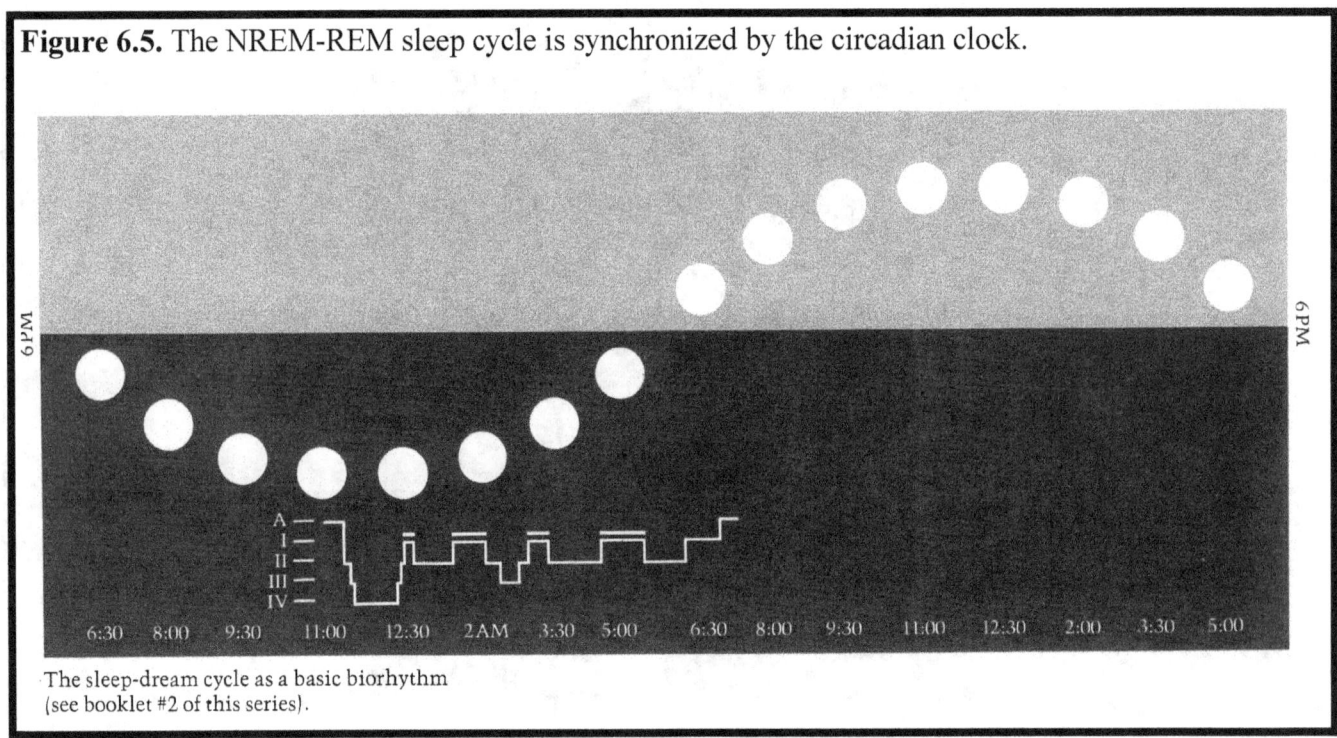

The sleep-dream cycle as a basic biorhythm
(see booklet #2 of this series).

The connections of the SCN with the lower brain stem controllers of conscious state have been investigated by the Harvard neurologist Clifford Saper. Saper models the hypothalamic oscillatory system as a "flip-flop switch" by means of which the brain is precipitately converted from waking to slow-wave sleep, from slow-wave sleep to REM, and from REM back to waking. The flip-flop switch model captures the tendency of the brain-mind to change state dramatically and rapidly. Once a change of state begins, it self-accelerates to some other state without the confusion of gradual state change and without the risk of hybridity.

The human body does not respond well to getting the internal circadian cycle and external day-night cycle out of phase with each other, as shown by shift working and jet lag. Jet lag is subjectively unpleasant, marked by strong physiological symptoms and an inability to sleep at the "correct" times (as signified by external light and dark). One treatment for jet lag involves taking melatonin shortly before the times you wish to sleep. It is also important to get as much natural light as possible during the day in your new time zone, particularly in the morning.

Brain physiology enriches psychology by providing causal information of direct relevance to the understanding of subjective experience.

Tore Nielsen is a dream research psychologist in Montréal, Canada. He explains the paradox of NREM dreaming as the expression of covert REM processes. In other words, REM is active even in so-called non-REM sleep. Nielsen is a tireless sleep laboratory experimenter who correlates mentation reports with physiological data in search of evidential support for his theory. In terms of the AIM story recounted here, Nielsen's work finds resonance with my observation that there are, in fact, many REMs in NREM sleep and that brain is only 50% deactivated at its lowest point in NREM sleep. I conclude that all state differences are quantitative, not qualitative. Day dreaming is a universal case in point.

Dialogue 6. Activation

TH: You make a lot of activation. It becomes factor A in your AIM model. Please unpack this for me.

AH: Activation is the key to unlocking the consciousness door because it is at once both a cognitive and a cerebral process. Psychology and physiology are not competitive or mutually exclusive. They are cooperative and complementary.

TH: Psychology is hard enough to understand without expecting me to master physiology, too.

AH: Take it easy. Relax and restrict your knowledge of physiology to those details which inform the study of consciousness. You don't need to know everything.

TH: That's a relief. Learning the complex patterns and frequencies of the EEG is difficult enough for me. I get lost when you talk about neurons and the AIM model.

AH: Think of the EEG as an overpriced voltmeter. Begin by simply recognizing that consciousness in waking and dreaming is low-voltage fast (meaning activated) while most of sleep is high-voltage slow (meaning deactivated).

TH: I now recognize the importance of Moruzzi and Magoun's 1949 discovery of the reticular activating system (RAS). The RAS acts like a brain power supply. It turns on the upper brain.

AH: In order to be awake, our brain and its mind must be activated.

TH: The same principle applies to REM sleep dreaming.

AH: Both states of consciousness are characterized by high levels of brain activation but we now know that different neuronal/cognitive modules are activated which help us to understand the differences between waking and dreaming.

TH: Let's stick to activation for now. I think of it as a bit like temperature: when things are hot, they are jumping up and down, and very accessible. When they are cold, they are inert.

AH: Like hibernators, they burrow in their nests and stay out of sight.

TH: How do you see activation as both physiological and psychological?

AH: Activation corresponds at the brain level to things like rapidity of neural firing. This signals the amount of energy generated by the nervous system. But that physiological activation corresponds to psychological activation, the level of consciousness, for example.

TH: Activation is a pretty ubiquitous concept in cognitive psychology, from the activation of ideas, to the activation levels of words when they're being selected, to the activation of individual elements in neural network models.

AH: One of my goals is to unify all of these apparently disparate concepts. I am a reductionist in attempting to explain the greatest number of psychological and physiological phenomena with the fewest simple philosophical assumptions.

Chapter 7. I = Input-Output Gating

Freud's (1895) *Project for a Scientific Psychology* and his (1900) *Interpretation of Dreams* were limited by what was then known about neuroanatomy and neurophysiology. Over a century later, neuroscience is still dominated by the neuron and reflex doctrines and it is easy to see why. Sherrington's work was foundational and is still useful, up to a point, in understanding the brain basis of consciousness. Reflex activity is an important part of the global CNS state. It is equally important to realize how limiting and inadequate the reflex paradigm is because it ignores spontaneous activity.

The goal of this chapter is to introduce findings and concepts that led to the development of AIM factor I. In the AIM model, I stands for Input-Output gating. Because the neurophysiology is so clear and convincing, the focus in this chapter is on muscular atonia but readers should be aware that sensory input as well as motor output is actively blocked in REM. No wonder that dream consciousness ignores the outside world and devotes its attention to internal concerns.

Consciousness has access to the outside world in waking but not in sleep, especially in REM sleep, when ambient information is actively excluded. In waking, the brain is privy to the environment as it must be to serve as a guide to behavioral action. In dreaming, the human subject creates an entirely imaginary virtual reality for reasons that are not yet entirely clear. In the waking state, we can act upon the world intentionally; in sleep, we are either effectively paralyzed or we move spontaneously without any intention to act upon the real world. So at some level the brain is deciding on how deeply we process inputs, and what actions we can effect on the world. We focus here on the difference between the quasi-real and the artificial as a way of better understanding both waking and dreaming consciousness. There are several easily understood ways to measure this difference.

Figure 7.1. AIM Factor I, Input-Output Gating. AIM Factor I can be measured as the reciprocal of H-reflex amplitude.

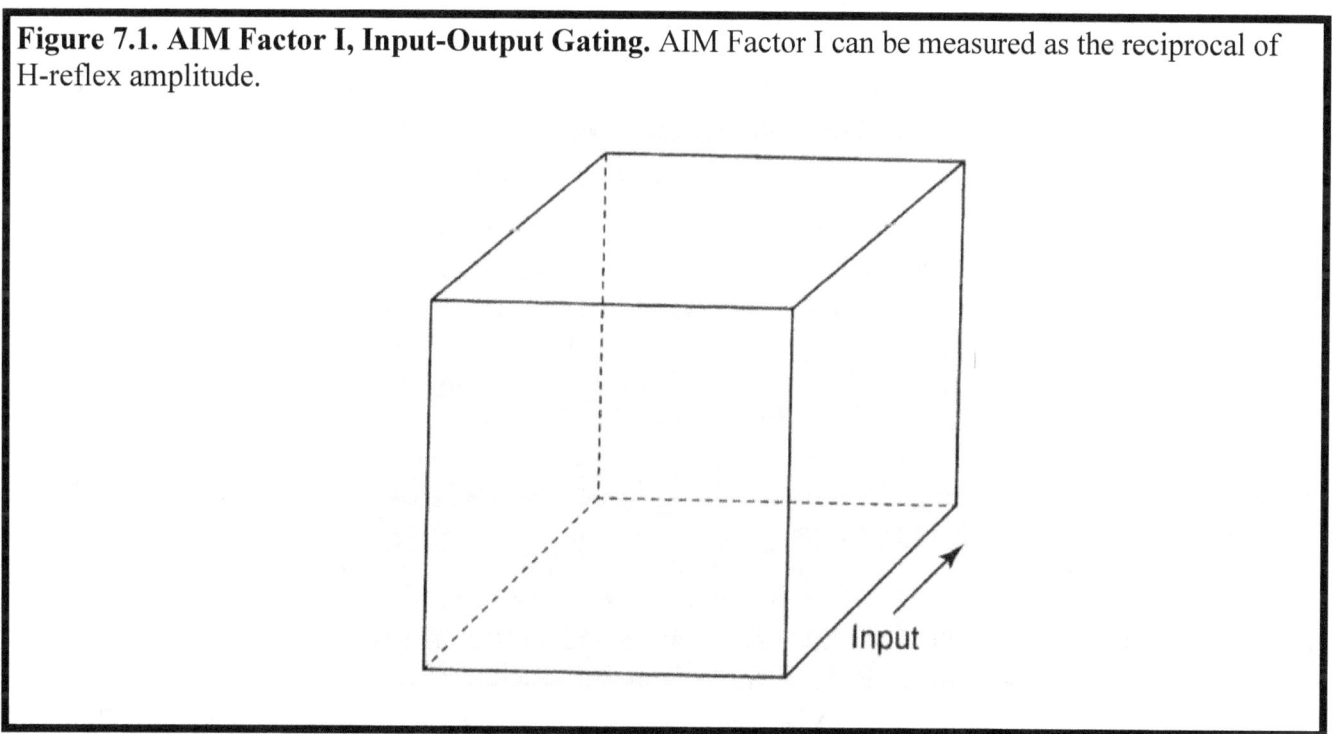

Input

The Reflex Doctrine

In order to eliminate the influences of the upper brain, Sherrington carried out most of his classical work on decerebrate animals created by severing the connections between the spinal cord and the upper brain. Hence, most of what we now know to be important in changing reflex activity in the mediation of conscious states was deliberately excluded from experimental study. Furthermore, Sherrington's animal subjects were not simply spinalized but also made spastic by his transections. In the interest of experimentation, normality was sidestepped and even subverted.

Sleep and dream science tries to restore normality while remaining experimental. This maneuver opens the door to the brain basis of consciousness.

The monosynaptic reflex is an integral part of consciousness science, however, and we owe to Sherrington its systematic investigation. He showed that sensory fibers from the skin and muscle tendon organs project directly into the posterior spinal cord, traverse it, and make excitatory synaptic contact with motor cells in the anterior cord. When stimulated, this reflex arc produces a contraction at very short latency (1.5 milliseconds). This rapid and reliable "knee jerk" reflex is adaptive in mediating both the resetting of posture and the motility of waking behavior.

Sherrington also showed that the excitation which resulted in the monosynaptic twitch of motor neurons on the same side of the body was mirrored by inhibition and compensatory relaxation of tension of the muscles on the other side of the body. This reciprocally balanced response is used by the brain at higher levels, resulting in conscious state changes governed by oscillators with much longer time constants. The time constants of Sherringtonian reflex responses also include longer delays above and below the level of stimulation, indicating what he called polysynaptic circuitry. These polysynaptic reflexes began to reveal the complexity of neuronal activity and anticipated the recognition of the spontaneity of brain activity that we now celebrate as the essence of conscious state change.

Integral to the paradigm shift occasioned by the discovery of the reticular activating system was the reflex inhibition of Sherringtonian reflex activity caused by Horace Magoun and Ruth Rhines' 1949 medullary brain stem electrical stimulation. A full decade before motor reflex response alteration was seen to be spontaneous in sleep as well as induced by experimental stimulation, it was clear that the brain stem mediated the intensity and type of change in spinal cord reactivity. That which Sherrington had purposely eliminated a half century ago was thus restored, enabling sleep and dream scientists to study what most interested Freud and Sherrington, the brain basis of consciousness.

Input-Output Gating

The most direct application of Sherrington's reflex model is found in the work of Michel Jouvet (in Lyon, France), Ottavio Pompeiano (in Pisa, Italy) and Michael Chase (in Los Angeles, USA). They showed that the REM sleep paralysis that prevents us from moving in our dreams was due to postsynaptic inhibition of the anterior horn motor neurons that Sherrington had identified at the turn of the 20th century as the output elements in the "knee jerk" spinal reflex.

Michel Jouvet and François Michel discovered REM sleep atonia inadvertently. They had implanted the posterior neck muscles of cats thinking that the EMG might be sensitive to lapses of attention in cats they were studying in learning experiments. Like Eugene Aserinsky, they wanted their subjects to remain awake throughout an experiment. Before they were aware of the existence of REM sleep, the French collaborators noticed that the EMG amplitude did indeed decline when the cats dozed and that it was completely suppressed if the cats fell deeply asleep. Because the EEG was simultaneously activated, they thought that the cats were awake. After they became aware of William Dement's work on REM sleep in the cat, they realized that their so-called "paradoxical" sleep was more interesting than the awake learning they set out to unravel. Thus, they switched their attention from one state of consciousness to another.

Michael Chase (1938–) was a student of Carmine Clemente and Barry Sterman at UCLA. In his own career, he investigated the mechanism of inhibition of the jaw-opener reflex of the cat in REM sleep (cf Factor I of REM). He went on to apply the carbachol induced REM effect (Factor M of REM) to facilitate intracellular recordings of neurons of interest. Chase's work combines the sophistication of experimental neurophysiology with sleep and dream research to illuminate the conscious state paradigm. We are entirely unaware of these exotic mechanisms when we lie in our theater beds at night. If it were not for motor inhibition, we would actually run instead of only dreaming that we do.

Jouvet's subsequent research was inspired by his meeting with David Rioch, a physiologist at Johns Hopkins Medical School in Baltimore, USA. Rioch was following up on Sherrington's studies on spinalized cats. He told Jouvet that the spastic rigidity of these animals was totally and spontaneously inhibited at intervals of about 30 minutes, the periodicity of REM in these animals. Jouvet then conducted experiments that proved that the brain stem was the site of the neuronal clock that timed REM and, with it, dreaming.

Having been trained by the Swedish Sherringtonian, Ragnar Granit, Ottavio Pompeiano was well versed in spinal reflex physiology. In a brilliant series of studies, he was able to show that post-synaptic inhibition was indeed the mechanism of REM sleep atonia and paralysis. He also confirmed Jouvet's theory that the spinal inhibition was the result of commands from the brain stem.

Pompeiano demonstrated that the sensory gating which isolated the activated brain from the outside world in REM was due to presynaptic inhibition of the Ia afferent fibers, which mediated Sherrington's knee jerk reflex. This effect was not only of brain stem origin but was associated with eye movements such that sensory input blockade was greatest when the internal activation of the visual brain was at its peak. Hence our dream visions are unlikely to wake us up.

That neuronal membrane hyper-polarization caused the spinal motor neuron inhibition and paralysis of REM sleep was demonstrated by Michael Chase using intracellular micropipette recording. Amazingly, Chase succeeded in impaling the neurons in immobilized but unanaesthetized cats, a technical feat replicated by Robert McCarley at Harvard, and possibly made feasible by the very immobility that they explained via their daring experiments. This achievement brought the Sherringtonian tradition into register with the modern science of consciousness.

Michel Jouvet (1925–). A neurosurgeon by training, Jouvet spent a year at UCLA in the early days of the Magoun–Moruzzi reticular activating system era (1949–1959). When he read of Dement's discovery of REM sleep in the cat, he abandoned his studies of classical conditioning and showed that REM sleep was actively orchestrated by the brain stem. He went on to investigate the role of aminergic and cholinergic brain stem mechanisms in REM sleep generation. Michel Jouvet was an amateur artist influenced by André Breton, whose interest in dreaming was surrealistic. Jouvet kept a dream journal and elaborated an imaginative dream theory, which combined his personal experience with the molecular biology of the gene.

Synchronization and Consciousness

Progress in EEG recording has revealed very high frequency (40 Hz), low voltage oscillations that are associated with waking and dreaming consciousness. Thus the activation of those states that are associated with subjective awareness was shown to be in a specific EEG power domain that could be characterized, localized, and quantified using statistical techniques. This power domain is often called gamma, as we have previously noted.

Working at the Max Planck Institut, first in Munich and later in Frankfurt, Germany, Wolf Singer showed that the 40 Hz EEG power range that is correlated with consciousness was the substrate of synchronized neuronal activity in widespread areas of the brain. According to Singer, there is no "center" of consciousness. Rather, consciousness is a distributed function of the human forebrain. This finding is related to the modern philosophical revision of Descartes' pineal localization theory, as in, for example, Daniel Dennett's "multiple drafts" hypothesis.

The diffuse synchronization idea fits with other evidence that consciousness is composed of the binding unification of multiple functional components (or information processing elements subserving sensation, attention, perception, emotion, and memory), each of which is relatively localized. There is no place where it "all comes together." Instead, consciousness is an integrated state of the human brain whose differential manifestations in waking and dreaming are otherwise determined (e.g. via input–output gating and neuromodulation).

Figure 7.2. When REM sleep begins, there is a progressive and simultaneous brain activation (orange) and muscle tone suppression (yellow). The activation of the brain induces eye movement (green) and PGO waves in the visual thalamus itself (purple). This complex but coordinated set of processes turns the brain on (orange) and turns motor output off (yellow) except for the eye (green). Information about the eye movements (purple) is generated and transmitted within the brain itself.

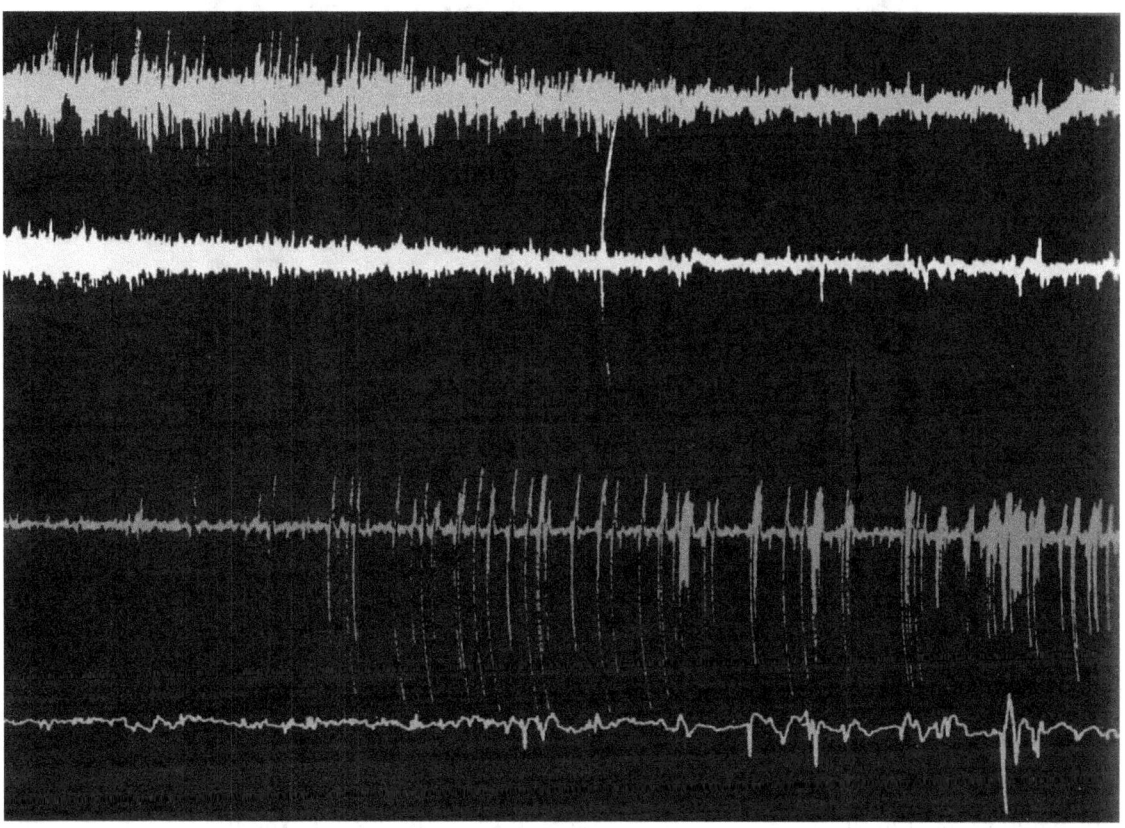

Other scientists, like the late Nobel laureate Francis Crick and Christof Koch, have championed the idea that 40 Hz synchronization and binding are abetted by the claustrum, a deep brain structure with widespread connections throughout the forebrain. The binding of information has also been advocated by António Damásio, a neurologist at the Salk Institute in San Diego, who has written eloquently on this and other aspects of the brain basis of consciousness. A positive feature of Damásio's writings is their philosophical orientation.

The voltage synchronization idea of Wolf Singer is consistent with the model that consciousness is a physical but immaterial manifestation of brain activity. According to the dual-aspect monism theory, consciousness is a state akin to gravity. Gravity exerts a force that can be shown to be related to mass even though gravity itself cannot be directly visualized. It is more widespread than EEG synchronization but the analogy is nonetheless valid.

For the first time, we have a rational theory unifying subjective first-person data with objective third-person observations. What is more, the new theory is consonant with quantum physics and can be mathematized. These innovations make a science of consciousness possible.
Neuromodulation

Anticipating the discussion of AIM factor M in the following Chapter 8 and extending Singer's synchronization hypothesis, neuromodulation may be defined as a process determining the chemical mode of the brain. Whereas neurotransmitters act very locally (between specific neurons), neuromodulators affect large areas of the brain.

Wolf Singer (1943–) is a German neurophysiologist who works in Frankfurt. He first pioneered confirmation and extension of the Hubel and Wiesel studies of the visual system but, imbued with the work of Michel Jouvet, has included attention to sleep in his recent studies of consciousness. How does the brain integrate its multiple elements to account for the very conspicuous binding, the unity of subjective experience noted by William James, António Damásio and Giulio Tononi? EEG synchronization is Singer's answer to this question. His experimental work has shown that gamma frequency EEG oscillations are rhythmic voltage changes that occur simultaneously throughout the brain in both waking and REM sleep dreaming, two intensely conscious states.

To prepare yourself for our later development of this idea, think of such contrasting modes as "remember vs. do not remember," "think logically and tightly vs. think associatively and loosely," and "look at the outside world vs. focus your attention on the internal world of your imagination." Which of these contrasting modes is predominant at any particular time is chemically determined by whether biogenic amine-containing neurons of the brainstem are excited or inhibited.

Michel Jouvet immediately adopted the discovery, in the early 1960s, of chemically specific neurons in the brain stem. Anica Dahlstrom and Kjell Fuxe, working at the Karolinska Institutet in Stockholm, Sweden, used fluorescent dye to stain brain cells containing the biogenic amines norepinephrine, serotonin, and histamine. Arvid Carlsson had previously done the same for the dopamine neurons of the substantia nigra and won the Nobel Prize for his efforts. Remarkably, the small-sized and numerically few cells studied by Dahlstrom and Fuxe had extensive axonal domains and, by means of branching, huge spatial reach. They projected to all corners of the CNS. There thus emerged the idea that these neurons could exert a ubiquitous chemical influence and this idea led to the theory that they might participate in the control of conscious state.

Subsequent studies led to the refinement of the neuromodulation concept. Instead of the precise, point-to-point targeting and rapid action of cells which use traditional neurotransmitters, like glutamic acid and gamma aminobutyric acid, the biogenic amine neurons were nonspecifically targeted, slowly discharging and slowly conducting as if they were an un-reticular formation system. They were also richly interconnected with each other, such that when one cell fired the whole network was excited, giving the ensemble the syncytial quality that the German reticularist histologists had lost to Santiago Ramón y Cajal.

Among the many findings which shored up a relationship of neuromodulation and consciousness was the potency of pharmacologic manipulation of aminergic neurons. For example, Michel Jouvet's early studies of drug suppression and enhancement of sleep and George Aghajanian's studies of suppression of aminergic cell discharge by the psychotogenic drug lysergic acid diethylamide (LSD), tried to come to grips with the Swedish neuroanatomy findings. The concept was more firmly established and made clinically relevant by the recognition that most of the drugs effective in the treatment of major mental illness exerted their effects via neuromodulation.

The importance of neuromodulation had, in fact, a much earlier history. The discovery of Thorazine in the search for an antihistamine was made in 1955. Its antipsychotic efficacy was correlated with its antagonism of dopamine (another biogenic chemical secreted by neurons of the midbrain substantia nigra), which had already been shown to be operative in the control of movement. Deficits in dopamine resulted in Parkinsonism, a condition also associated with defective consciousness (psychosis, memory loss, and attentional lapses). Dopamine effects oppose the action of acetylcholine, another neuromodulator that we will show to be involved in the mediation of both waking and dreaming consciousness. The orchestration of brain chemical modulators is fundamental to the experimental study of conscious states.

Acetylcholine and the Chemistry of Consciousness

The first evidence demonstrating that any form of neurotransmission was chemical and not electrical was the crossed frog perfusion experiment of Otto Loewi, an Austrian pharmacologist friend of Sigmund Freud. Loewi's blood-borne agent turned out to be acetylcholine, a chemical at first thought to be active peripheral to the brain but later recognized to operate centrally as well. Acetylcholine is a very versatile molecule: besides slowing the heart, it mediates skeletal muscle contraction and the autonomic ganglion activation responsible for intestinal motility, blood pressure change, and erection. In the brain, it mediates EEG activation and the brain stem triggering of REM sleep, among many other functions. That acetylcholine might be a neuromodulator was suggested by the work of Marsel Mesulam, working first at Harvard Medical School in Boston and later at Northwestern Medical School in Chicago.

Mesulam identified, localized, characterized, and enumerated six clusters of acetylcholine-containing cells, called CH 1–6. They were located in parallel with the biogenic amine containing neurons in the brain stem (called CH 3, 4, 5, and 6) and as far forward as the basal forebrain (CH I and 2). Both aminergic and cholinergic neurons release their chemicals differentially and measurably in waking and sleep, allowing scientists to create testable models of how consciousness might be engineered at the molecular level. Because acetylcholine has been so exhaustively studied by pharmacologists and physiologists, countless experimental options are available. We will summarize a few of them later.

Marsel Mesulam (1945–) was the heir apparent of the Behavioral Neurology movement at Harvard but has pursued his research and theorizing about consciousness at Northwestern. Mesulam used specific enzymatic markers to identify acetylcholine producing neurons in the subcortical brain. Other scientists have shown that both waking and REM sleep dreaming are associated with acetylcholine release from the cortex. For this reason, acetylcholine is reasonably considered to be a chemical mediator of consciousness. Mesulam has voiced more specific ideas of how acetylcholine may contribute to the cognitive features of waking. Allan Hobson had the honor of helping Mesulam choose neurology over psychiatry so it is a particular pleasure to welcome him into the unified field of psychodynamic neurology.

In honor of their respective contributions, Otto Loewi won the 1936 Nobel Prize for Physiology and Medicine while Sigmund Freud was awarded the 1931 Goethe Prize for Literature. The third-person work of Otto Loewi and the first-person approach of Sigmund Freud are now united in the new science of consciousness. An irony is the now famous dream of Otto Loewi that led to his crucial crossed frog heart action experiment. Sigmund Freud could not have interpreted this dream without derailing the science he thought he was creating. Freud's great literary talent got in his scientific way. Imagination is important to both art and science but critical thinking and experimentation distinguishes science from art.

Dialogue 7. The Marriage of Mind and Brain

TH: Do you, Sigmund Freud, take Otto Loewi, to be your lawfully wedded spouse?

AH: Metaphorically speaking for Sigmund Freud, I certainly do. Having lusted for one another for over a century, it is a relief finally to consummate our relationship.

TH: You guys are both Viennese with shared mutual interests and a passion for the relationship of the brain to dreaming.

AH: Freud and Loewi were friendly colleagues who just didn't see how they could get together in a scientific union. Psychology and physiology seemed worlds apart so they went their separate ways.

TH: As a psychologist, I must say that I am not yet sure this marriage will work but I am willing to give it a try. What should I do?

AH: Give up the focus on dream prophecy. Shift your attention from dream content to dream form.

TH: Can you draw me a picture by which to understand the difference between content and form?

AH; I *see* in my dreams (form) and I see *you* in my dreams (content) but you don't look like you (form) and your image challenges my theories (formal analysis) because you think I am your father (content analysis).

TH: Are you saying that I must stop trying to be a Sigmund Freud by interpreting the supposedly hidden meaning of dreams?

AH: Yes. And don't be an Otto Loewi waiting for dreams to dictate your experimental program.

TH: It is difficult to turn away from what still permeates my interest in the mind. The formal approach to mentation just does not cut it for me.

AH: Try harder, you might like it. In fact, if you learn a little physiology, you might make a marriage that works and is exciting.

TH: Brain science is crowding in on psychology. I am afraid that psychology may go down the tubes with academic theology. I want to avoid that fate. Meanwhile, I want to have fun.

AH: Without brain science, psychology departments may indeed be an endangered species. Cognitive science has recognized this danger and is flourishing today but dreaming has not yet made it to the curriculum of most college courses.

TH: Sigmund Freud was frustrated by both neuroscience and descriptive psychology. That's why he got a divorce from both and tried to take up with a mistress which he found more exciting: psychoanalysis.

AH: After Loewi dreamed of the crossed frog heart perfusion experiment, he contented himself with demonstrating humoral neurotransmission.

TH: Are you saying we can now have it all? Psychology, physiology and even philosophy?

AH: Three in a bed sounds like a *liaison dangéreuse*, but a successful marriage may depend upon it.

Chapter 8. M = Modulation

The Graz, Austrian physiologist Otto Loewi hoped to show that neurotransmission was occasioned by the ferrying of a chemical message across the synaptic cleft. In other words, he hoped to demonstrate a humoral mechanism in neuron-to-neuron communication. While pondering this problem, he awakened from a dream that revealed a crucial experiment but he was frustrated by his failure to remember the dream. He claimed to have repeated the dream the following night, to have awakened from it, and to have recorded its content. The result was the famous crossed frog heart perfusion experiment that won Loewi the Nobel Prize for Physiology or Medicine in 1936.

Upon awakening from his dream the second night, Loewi went to his laboratory and connected the circulatory system of one frog to that of another. Then, he electrically stimulated the vagus nerve, causing the heart of Frog 1 to slow. The heart of Frog 2 followed suit. Loewi concluded that the only possible explanation was the release from the heart of Frog 1 of a chemical which caused the heart of Frog 2 to slow.

Loewi called the hypothetical chemical the "Vagusstoff" in honor of its nerve of origin. German chemists later identified the "Vagusstoff" as what we now know to be the neurotransmitter *acetylcholine*. Acetylcholine was then shown to be the chemical that mediated many functions throughout the body, including the brain and the heart. In the following sections of this chapter, we will see that the brain uses acetylcholine to fabricate REM sleep. Thus, it is reasonable to suppose that Loewi's dream was caused by the very molecule that his experiment was designed to demonstrate. This anecdote shows how the brain and the mind may conspire with each other to enrich understanding.

Figure 8.1. AIM Factor M, Modulation. AIM Factor M can be computed as the rate of aminergic neuronal discharge.

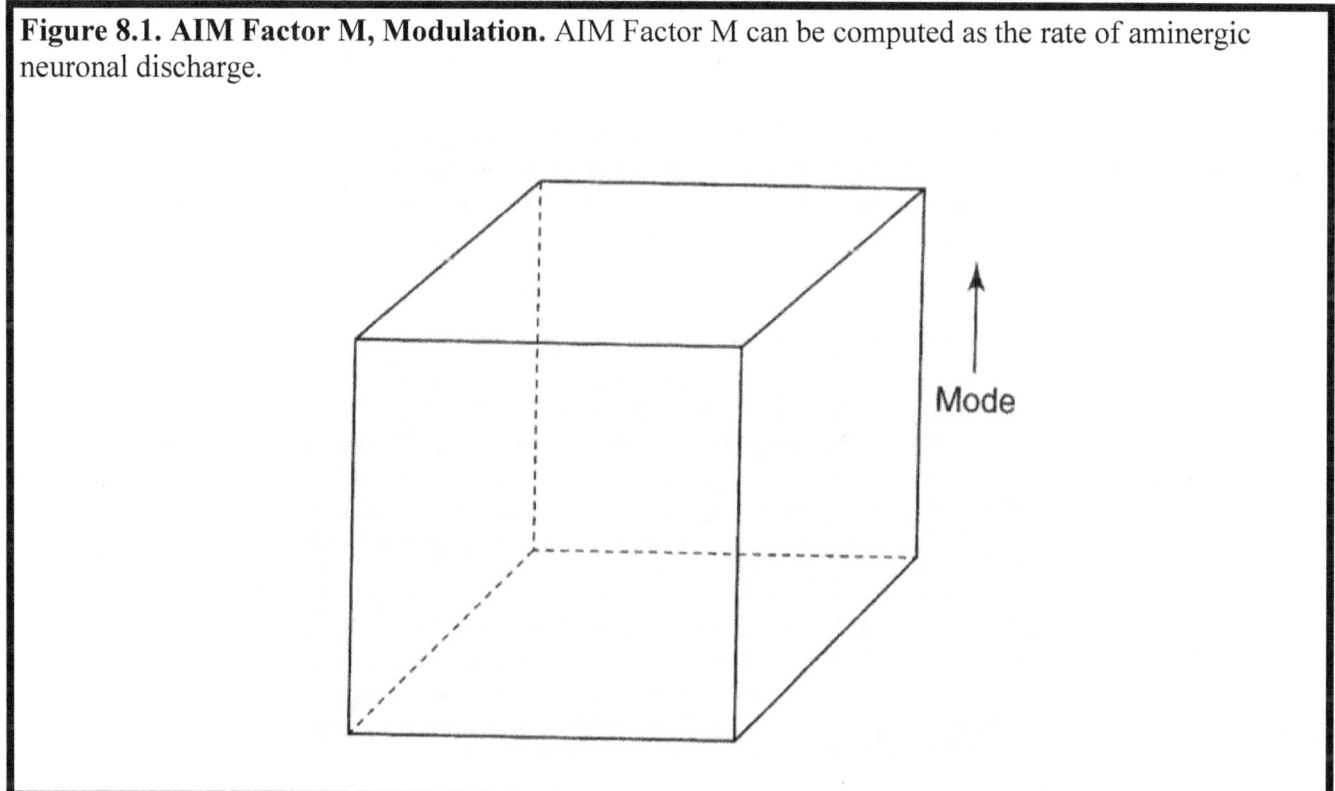

Mode

Activation and Modulation Revisited

The AIM model, which is derived from molecular neurobiology and discussed in more detail in the subsequent chapter, predicts that any chemical that impedes brain activation will induce sleep. Great progress has been made in discovering how the brain changes state in the production of waking and dreaming consciousness. We will emphasize this experimental story by way of exposing the science of dreaming, a state of consciousness now perhaps better understood than either waking or non-dreaming sleep.

Otto Loewi (1873–1961) was an Austrian pharmacologist who won the 1936 Nobel Prize for Physiology or Medicine by demonstrating humoral neurotransmission. His discovery complemented and enriched the mechanism of synaptic transmission by which nerve and muscle cells communicate with each other. The chemical nature of AIM factor M derives from Loewi's experiment, the design of which is said to have occurred to him in a dream. Thus the acetylcholine molecule which caused his dream was elucidated by it. The mind is a function of the chemical brain which guides science and other products of consciousness.

Later chapters will detail the neuronal recording studies that were used to create the reciprocal interaction theory of sleep cycle control. The pontine oscillator which governs the cyclic alternation of the NREM and REM phases of sleep is a push-pull device. When it pushes the neurotransmitters norepinephrine and serotonin, the brain is activated, with waking as the result. When it pulls these chemicals away, the brain reactivates cholinergically and dreaming is the result. Reciprocal interaction theory predicts that REM sleep and dreaming results by either reducing aminergic or increasing cholinergic drive on the oscillator. As we will show below, the cholinergic prediction has been amply confirmed.

Cholinergic REM Sleep Induction

The neurotransmitter acetylcholine is used throughout the body and brain to affect a wide variety of functions. As Loewi showed, the heart beats more slowly because of it. Other work shows that muscles contract because of it; the neurons of the brain talk with it: speed up, slow down, remember, forget, or hide your emotions — all are examples of acetylcholine talk. The acetylcholine instructions that interest the consciousness scientist the most are "now dream" and "now wake up."

Figure 8.2.

A. Otto Loewi Dream Experiment.

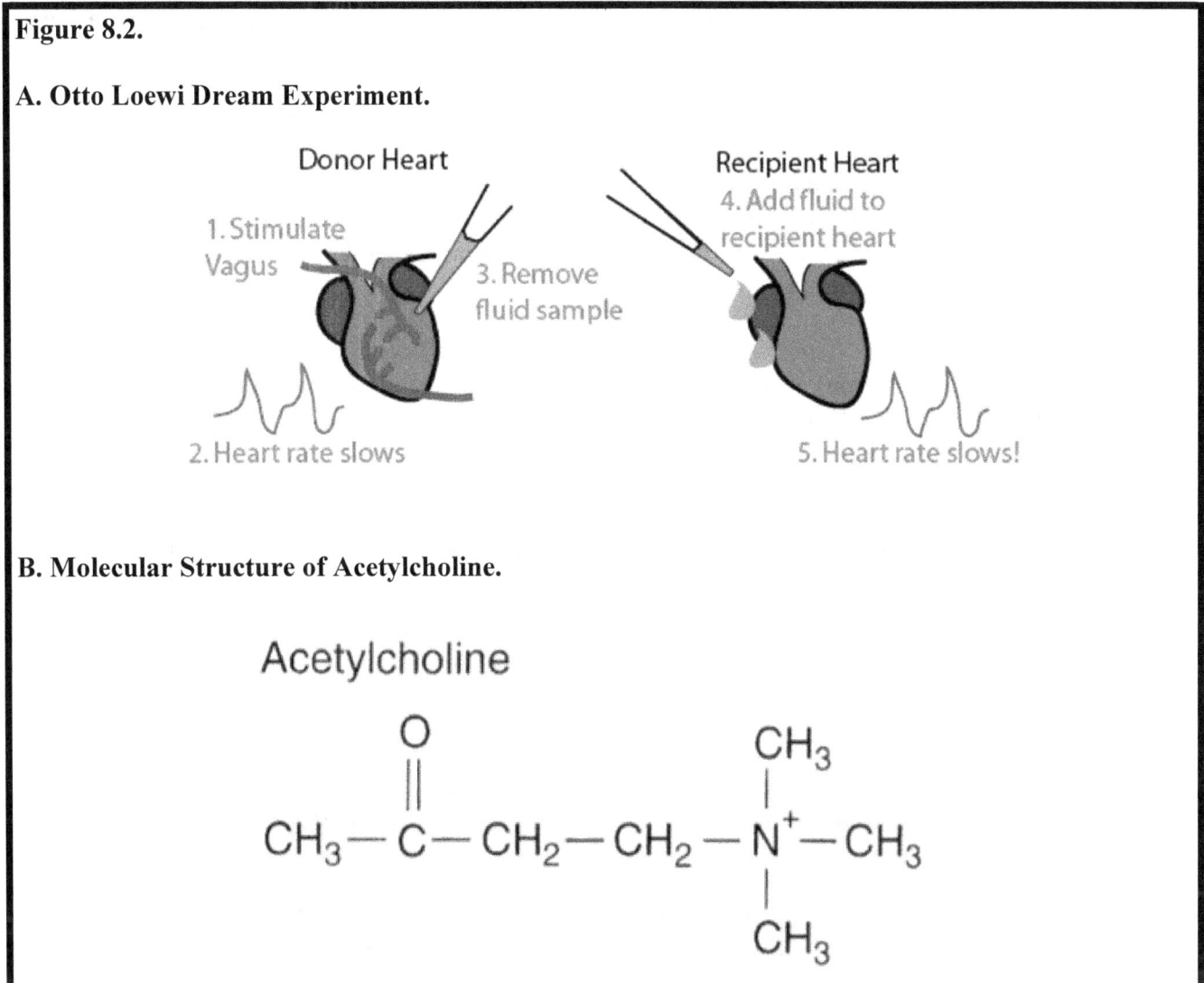

B. Molecular Structure of Acetylcholine.

To be efficient and effective, acetylcholine must be broken down as soon as it is released; otherwise the chemical message would be blurred and the neurons instructed by it would not understand what to do. The chemical enzyme that gets rid of the acetylcholine message as fast as it is delivered is called acetylcholinesterase. It was reasonable to assume that the REM-on cells might use acetylcholine to turn on other REM neurons. This original idea was a mistake. Instead, it was shown that the REM-on cells were activated by acetylcholine that was manufactured locally in a highly restricted remote zone in the lateral brain stem.

Great anatomical precision is necessary in carrying out this type of research. The delivery of drugs to the smallest possible locale in the brain is called *microinjection*; less is more because the spread of a drug activates REM-off as well as REM-on neurons. It is very difficult to limit spread because fluids diffuse widely. No wonder physicians and street drug users have so much trouble with specificity — they are both shooting fish in a barrel. It is a miracle if they ever kill a fish and not surprising that they often kill the fisherman.

Because acetylcholine is so rapidly destroyed, it is essential for consciousness scientists to use a chemical that fools receptive neurons into thinking that acetylcholine is talking to them and keeps on doing so because it is not broken down by acetylcholinesterase, the chemical that normally breaks down excess acetylcholine in the body. One such chemical is the synthetic drug carbachol. Carbachol can be microinjected into the pontine brain stem center of REM-on cells to artificially trigger REM. Another trick is to tie up the enzyme, acetylcholinesterase, so that spontaneously released acetylcholine is not destroyed and can act unimpeded for a longer time than is usual. Using these and other tricks, it has been possible to bring REM sleep under experimental control.

These studies were carried out on experimental animals (cats and rats), who cannot tell us whether or not they dream. Less neurobiologically precise but more psychologically informative experiments with acetylcholine-enhancing drugs cause human subjects to have earlier and longer REM periods that are associated with reports of dreaming, confirming the predictions of both the reciprocal interaction and activation-synthesis theories. These results have been replicated and other chemicals have been shown to participate in REM sleep production.

Follow-up studies indicate the neurobiological precision by which conscious states are regulated. Acetylcholine-containing neurons must be identified, as well as the synthetic enzyme, choline acetyltransferase (ChAT), not the degradative enzyme, acetylcholinesterase.

Where the carbachol is placed in the brainstem makes a crucial difference even with well controlled microinjection. Injected anterior to the pons, the increased waking that is produced reflects activation of the midbrain reticular formation. Clockwise and counter-clockwise turning may be caused by substantia nigra stimulation. Injected into the medulla, carbachol may induce abnormal trunk rotation, as well as arousal. Credit for this work is due to Helen Baghdoyan.

Using Carbachol-Induced REM as an Investigative Tool

The conversion of consciousness science from correlation to causation is epochal. Instead of waiting for a whimsical conscious state to occur spontaneously, it is now possible to cause it to occur at will. This experimental power has been used to explore brain physiology in the context of psychology.

One cogent example is the muscle inhibition component of AIM, Factor I. The question is: how does the brain stem control spinal reflex excitability? The answer could benefit sleepwalkers, narcoleptics, and REM sleep behavior disorder victims, as well as enlightening all of us normally paralyzed dreamers.

Sir Charles Sherrington would be pleased to know that his monosynaptic spinal reflex (MSR) was modified from excitation (in waking) to inhibition (in REM) but he might ask, "How is this reflex reversal accomplished? Are you telling me that a reflex spontaneously reverses its sign, from positive to negative? I don't believe it." Even brilliant scientists like Sherrington have blindered vision. Our current views of consciousness are likely to be no less limited. As long as scientific progress is made, we will be happy to be proved wrong.

Michael Chase, a neurophysiologist at UCLA, might respond, "I showed this to be true in the suppression of the jaw-opener reflex of the brain stem but now that carbachol REM and intracellular recording have become available, I have also shown it to be true of the spinal cord, as well. I suspect that it is a universal truth which makes REM sleep and dreaming possible." Neurons in the brain stem send inhibitory commands to every motor neuron in the brain. In REM sleep, these neurons are selectively activated and turn off movement by squirting an inhibitory amino acid called glycine onto motor neurons.

What is the network structure of the REM sleep-dream generator in the pontine brain stem? Carbachol-induced REM can be used to map the REM-on, REM-off neuronal populations and their connectivity. By conjugating carbachol to fluorescent microspheres, Dr. James Quattrochi, then a research scientist in the Laboratory of Neurophysiology at Harvard Medical School, was able to trigger REM and identify the generator network (see again figure 8.3). It consisted of a vast number of REM-on cells in the pontine reticular formation which received modulatory inputs from cholinergic and aminergic nuclei.

Fluorescent microspheres are microscopic blobs of paint which neuronal endings engulf, pump up their axons, and light up the cell bodies for visualization in the fluorescent microscope.

Carbachol-induced REM can be used to dissect the chemistry of the REM-off system. Ralph Lydic and Helen Baghdoyan, now Professors of Psychology and Neuroscience at the University of Tennessee in Knoxville, use the cholinergic model to learn about the interaction of serotonin with the inhibitory modulator GABA, a chemical known to turn off the aminergic neurons of the pons by local signals and messages from the circadian oscillator in the hypothalamus. When that system flops, dreaming flips and we go crazy in the safety of our beds. Carbachol-REM is now a widely used tool; consciousness science has come of age.

Helen Baghdoyan now investigates the molecular basis of REM sleep generation (AIM, Factor M) at the University of Tennessee in Knoxville. She uses mass spectroscopy to measure spontaneously released brain chemicals that result in our nightly dreams. She and her colleagues have shown that transmitter release changes in a behavioral-state-specific and brain-region-specific manner. For example, acetylcholine release in REM sleep generating regions of the brainstem increases during REM sleep while GABA levels simultaneously decrease. In contrast, acetylcholine levels decrease in the cortex during non-REM sleep, while GABA levels simultaneously increase. These findings help us derive a more nuanced understanding of the neurochemical basis of the dreaming and non-dreaming phases of sleep.

Cholinergic Activation of the Gamma Rhythm

A major source of cholinergic activation of the REM-on cells in the pontine reticular formation was shown by James Quattrocchi to be the pedunculopontine nucleus of the far lateral brain stem. By microinjecting carbachol, José Calvo and Subimal Datta, working together in the Laboratory of

Neurophysiology at Harvard Medical School, were able to induce immediate PGO waves and delayed, but long-lasting, increases in REM sleep.

Figure 8.3. Graphs showing results of carbachol and neostigmine injections.

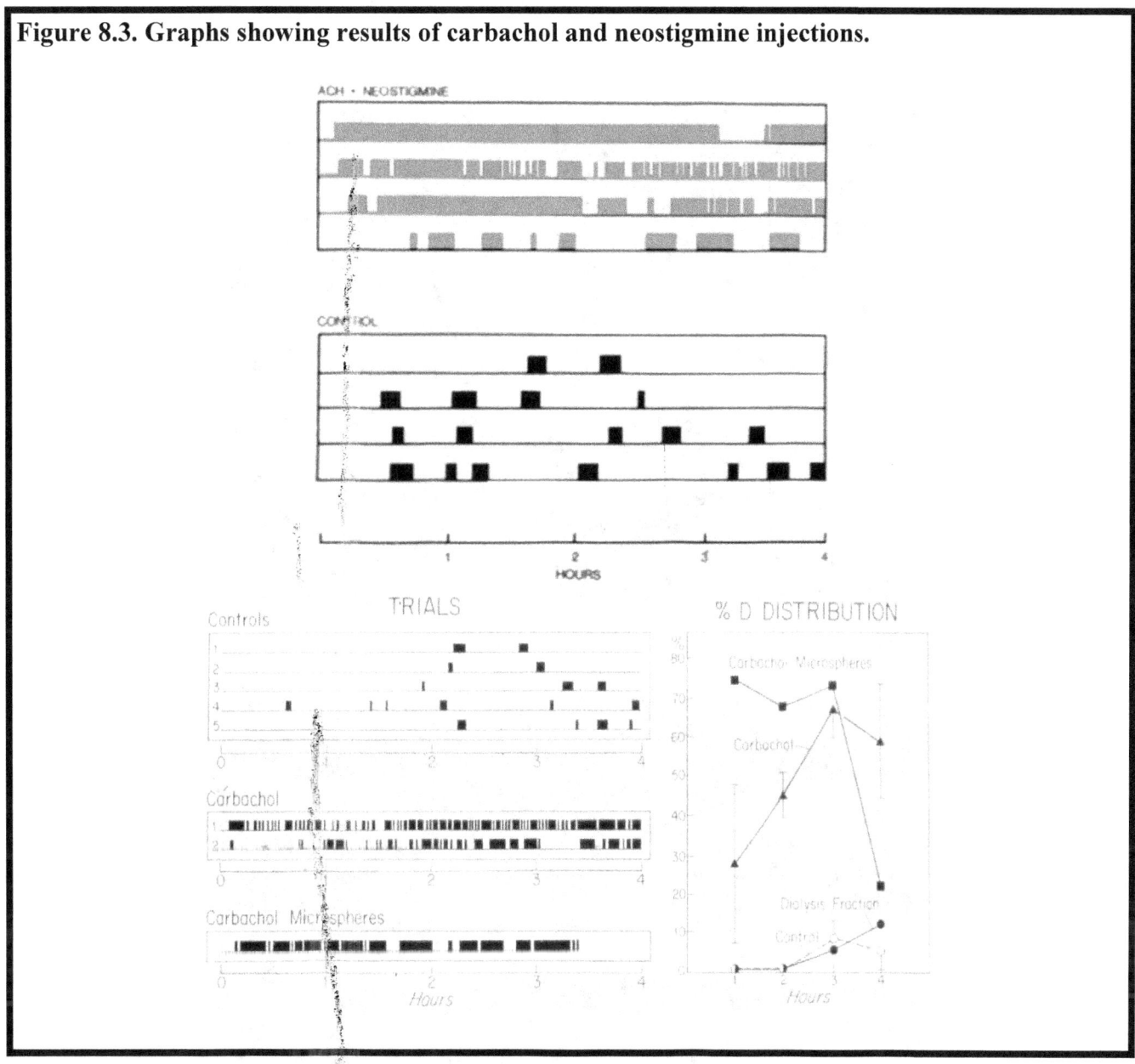

Each of the carbachol-induced PGO waves was associated with an eye movement directed toward the side of the injection. The waves were recorded in the lateral geniculate body of the thalamus, again predominately on the same side of the brain as the injection. Because waking persisted, Datta and Calvo concluded that they had triggered a local circuit usually only activated in REM. They naturally wondered if the cats were experiencing seizure-like visual hallucinations instead of full-blown cat dreams. To their surprise, as the PGO waves declined at six days post injection, REM sleep increased to a peak that was four times its normal height.

The carbachol had provoked not only a local PGO circuit activation but a widespread and sustained excitatory effect on the REM generator network. Calvo and Datta hypothesized that the carbachol had initiated a two-stage process: the first stage was a pharmacological incubation of PGO waves and the second stage (which appeared long after the carbachol had become inactive) was a shift in the excitability of the REM sleep generator network. Not only could REM sleep be increased in the short term but long-term regulation could also be experimentally effected.

Figure 8.4. Quattrocchi photos of the computerized neurons of the REM generator network (a) and the cholinergic input (b).

a.

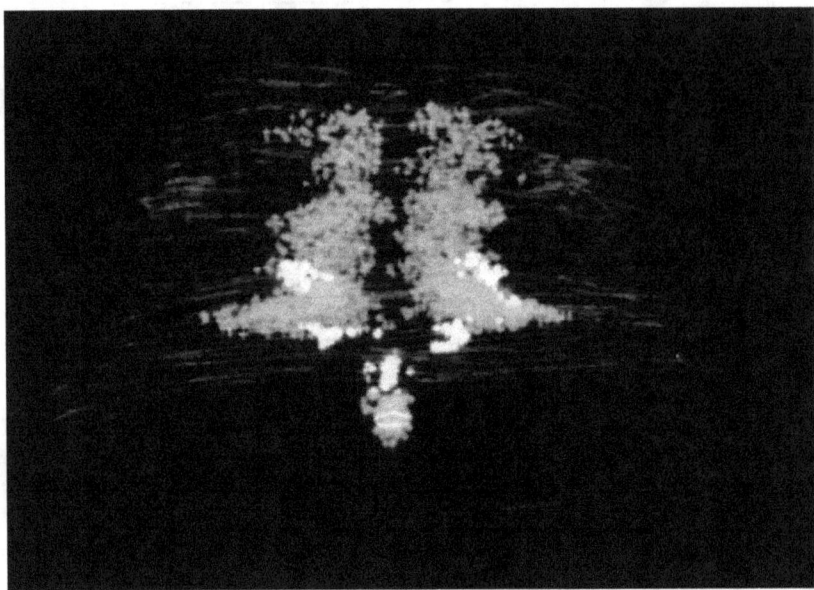

b.

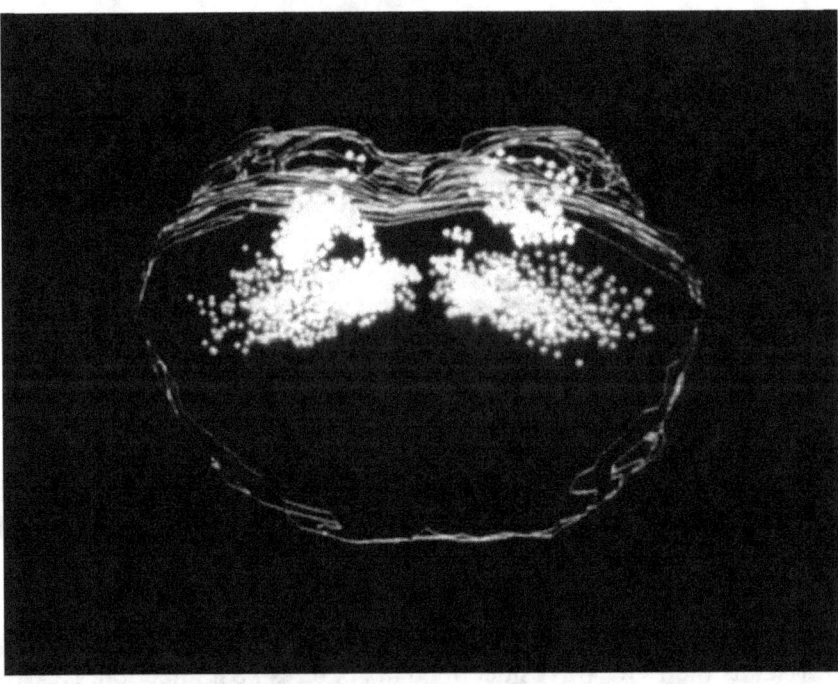

A short-term trigger zone in the paramedian reticular formation was linked to a long-term regulatory region in the lateral pons. It was the latter region that was naturally cholinergic. The paramedian trigger zone was cholinoceptive, that is, responsive to the acetylcholine coming from the lateral regulatory region. We recount these details in order to illustrate the complexity of brain physiology and its progressive elucidation.

Further study suggested the pedunculopontine nucleus was a possible source of the 40 Hz gamma rhythm that other neuroscientists had identified as essential to the synchronization of forebrain neurons underlying waking consciousness. The theory that synchronous activation underlay dreaming as well as consciousness fits well with the activation-synthesis model of dreaming and the AIM model of conscious state regulation. The brain is activated and synchronized in both states of consciousness; factor A is high in both waking and dreaming.

The vertical analysis of conscious state activation has now reached the subcellular and genetic levels at the hands of Subimal Datta and his colleagues in Knoxville, Tennessee, and Edgar Garcia-Rill, in Little Rock, Arkansas. Pedunculopontine cholinergic neurons can be studied in tissue culture, allowing individual brain cells to be recorded and manipulated in vitro. A 40 Hz rhythm, which can be activated via specific membrane receptors (proteins whose genetic origin can be determined), is recordable from these cells. These authors relate their findings to the Freudian preconscious, a concept akin to the background process of waking fantasy in adults and to protoconsciousness in immature animals.

Mad as a Hatter

The actions of acetylcholine can be blocked by occupying its neuronal receptors with a chemical called *atropine*. Atropine is naturally occurring in plants such as deadly nightshade and Jimson weed, and has long been known as *belladonna* (Italian for beautiful woman) because it blocks acetylcholine-induced contraction of the pupil. The limpid, dark depth of the atropinized eye (when the pupil is dilated) has been regarded by beautiful women in the premodern era as integral to seductive charm. Atropine is now used by ophthalmologists to facilitate the visualization of the retina in clinical eye exams.

Figure 8.5. Atropine molecule.

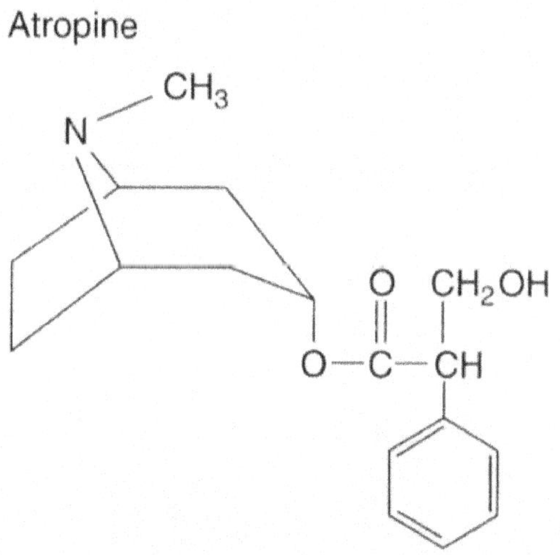

Atropine use is not restricted to cosmetics and ophthalmology. Among other desirable effects, it slows the bowel's movement and thereby counteracts diarrhea. It is available as an over-the-counter medication. Atropine poisoning, caused by intentional or accidental overdosing, is instructive to students of consciousness science for two reasons; it shows how different the effects of a drug taken by mouth are from a drug microinjected into the brain, and it reveals how potent a drug is atropine regardless of the route of administration. When acetylcholine is blocked throughout the body, all hell breaks loose. More than the control of dreaming is deranged. Consciousness of both varieties, waking and dreaming, is affected.

Atropine poisoning causes a panoply of symptoms remembered by medical students via the mnemonic: mad as a hatter; red as a beat; hot as a fire; blind as a bat; and dry as a bone. We will focus here on atropine madness because it bears the closest relation to the science of consciousness. What was wrong with the mad hatter of Lewis Carroll's *Alice in Wonderland*? He was as crazy as a dreamer and a victim of atropine poisoning because the mercury he used to shape hats was neurotoxic. The organic psychosis of mad hatters consisted of the same cognitive defects of dreams: memory loss, disorientation, confabulation, delirium, delusions, and visual hallucinations.

Acetylcholine must be tempered by harmonious balance with other neuromodulators, including those specified by AIM (see Chapter 9) and, certainly, many, many more (See Chapter 21 on Psychopathology). When that balance is upset (as it is by atropine), the brain-mind cannot function properly and consciousness suffers. The pharmacology of consciousness is in its infancy and, although it is growing painfully slowly, the baby appears to be healthy.

Dialogue 8. Discover Dreaming for Yourself

TH: What does the Otto Loewi dream story tell us?

AH: Two things: one is that dreaming is not only meaningful but sometimes useful; the other is that we can teach ourselves to wake up and remember our dreams.

TH: Although I haven't yet made an earth-shaking scientific discovery, I have enjoyed the dream fruits of self-awakening and enjoy sharing edited versions of them with others.

AH: We cannot be sure that Otto Loewi really had the same dream twice, but he certainly did lay out the self-awakening technique.

TH: He put a pen and pencil on his bed table and gave himself a pre-sleep autosuggestion: if you dream, you will notice that it is a dream and not waking because of its bizarreness. You then wake yourself up to record its content.

AH: Noticing dreaming is now even easier because we now know more precisely how to distinguish dream subjective experience from that of waking.

TH: The essence of dream bizarreness is inconstancy of times, places, and persons. I use these formal features to tip myself off that I am not awake but dreaming when I sleep.

AH: How often are you successful and are you still good at self-awakening?

TH: Like most other people, I was better at this when I was in my twenties or thirties, but I have learned, over the years, to capture hundreds of dreams.

AH: What is the secret of your success?

TH: Whether or not I have given myself a pre-sleep autosuggestion, I wake up from dreams fairly frequently. Most people wake up several times during the night; the secret is to resolve to record your dreams then. I have used pen and paper in the past, but that is often too much effort, and besides, my writing is often illegible the following morning! I use a voice recording app on my smart phone sometimes. Easiest of all, though, is a digital voice-activated recorder; I just put it beside my head and speak into every time I wake up in the night. As is normal, this approach captures some fairly frightening things that I would prefer to ignore.

AH: Frightening?

TH: I am surprised how many times I wake up screaming in the night. On one occasion, I let out this blood-curdling yell, only to say something like "I was dreaming I was making a cup of tea." On other occasions, I have spoken in peculiar voices, like something out of a movie such as *The Exorcist*.

AH: You're doing well then. As we age, our ability to program dreaming normally declines. That's the bad news. The good news is that sleep is increasingly fragmented and spontaneous awakening is more frequent. Maybe that's why you've gotten better at dream recall over the years.

TH: It is said that awakening conditions are the most potent determinants of dream recall.

AH: On awakening from a dream, it is important not to move and to use waking memory to fix the recall for later transcription.

TH: Sometimes this works so well that I have more recall than I can transcribe.

AH: The same rules apply to sleep lab dream collection. Subjects who never before recalled a dream are flooded with memories when awakened from REM. They are often surprised and delighted by their self-discovery.

TH: How much dreaming is normally unremembered?

AH: It is impossible to be sure but 90 minutes a night is a modest estimate.

TH: Doesn't that loss of recall make you think that dreaming is a waste of time?

AH: Some experts hold that it is REM itself, not dreaming or its recall, which is important. But others, and I am one of them, hold that dreaming and its recall are at least an important clue to how the mind works.

TH: I have kept my dream transcripts and also recorded other information about each dream, and kept a spreadsheet of all the information for one calendar year.

AH: And what have you learnt?

TH: You must be careful about generalizing from just one person. But my dreams are full of fear and anxiety. There are also several recurring themes. The complexity and originality of the dream world can be amazing.

AH: Anxiety is the most common dream emotion. Frightening dreams are normal. Like riveting films, dreaming is great show.

Chapter 9. AIM

Although the brain-mind is constantly changing, it passes through the relatively enduring states of waking, sleeping, and dreaming. Consciousness changes state accordingly. In this chapter, we present a model framework for understanding these three states of consciousness; we call it **AIM**. It is physiologically based but helps us think about the psychology of the normal and abnormal mind.

The State-Space Concept

Each brain-mind state can be specified by quantifying the intensity of activation (A), input-output gating (I), and modulation (M). These three functions yield scores on three orthogonal variables, which together constitute the x, y, and z axes of the virtual space that states of consciousness inhabit. The variables represented are activation (A) on axis x, input-output gating (I) on axis z, and modulation (M) on axis y. The result is a three-dimensional state-space plot. Each of the three states (waking, NREM, and REM sleep) can be expected to occupy a specific zone of the state-space diagram and, indeed, this is the case. It is possible to represent the states as a cluster of values and to track changes within and between states in real time.

Figure 9.1. Normal domains of AIM state space. The domains of waking, NREM, and REM sleep are shown as cloudlike clusters of points, each of which represents the strength of A, I, and M at any instant in time. The waking domain is in the right upper back corner because activation is high, input-output gates are open, and aminergic modulation is high. The NREM sleep domain is in the center of the space because the values of all three parameters have fallen by about 50% of their range. The REM domain is again at the right (because as in waking, activation is high) but it is on the front wall (because the input-output gates are closed) and at the floor (because aminergic modulation has fallen to zero).

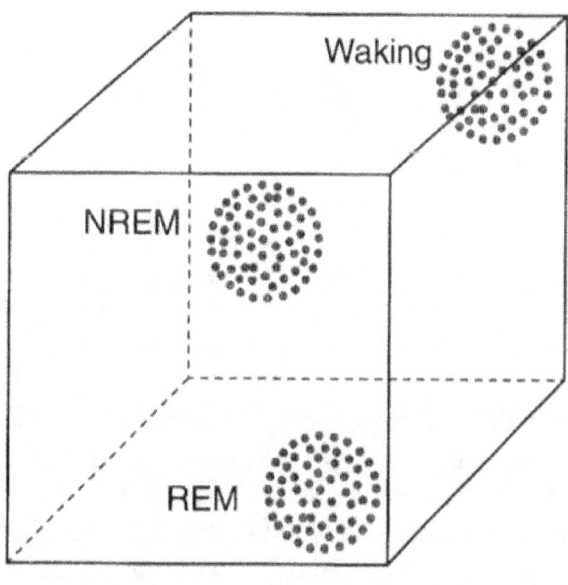

A unique advantage of state space graphing is the ability to specify unusual and abnormal states of mind. Examples of unusual states are lucid dreaming, hypnosis, and meditation. Examples of abnormal states are depression, narcolepsy, and coma, each of which might be expected to occupy a unique part of the state-space and/or to exhibit a distinctive trajectory as the brain-mind changes from a normal to an abnormal state or from one normal state to another. Preliminary mapping of unusual and abnormal states has been accomplished but there is much more to be done. For example, the neuropsychological syndromes discussed in Chapter 3 could be better understood in this way insofar as their anatomical substrate overlaps with state control systems.

The shift from the analysis of consciousness content to the analysis of its form initially seems like a regrettable loss of information. What could be more valuable that the niceties of individual history? Individual history is what most people want to talk about. But science is not conversation. In its reductionistic goal of explaining the greatest number of variables with the fewest assumptions, science is the very opposite of individuality. Science is generic and not yet individual. The state concept and the state-space model are dedicated to the universal generic nature of things, in this case things human. AIM wants to be a theory of everything, of all states of all people. The state-space model is potentially infinite and certainly needs to be more that three-dimensional but it is a marked improvement on the current uni-dimensional paradigm of sleep and dream lab science.

What besides the loss of individuality are the shortcomings of the three-dimensional AIM state-space model? The major weakness is its assumption that the whole brain is always in one and only one state at a time whereas it is already abundantly clear that while one part of the brain may be in one state, another may be in another. Examples are sleep talking and lucid dreaming, but even normal states cannot be done justice in only three dimensions. How many dimensions would a realistic state-space model require? How many discrete brain regions are there? How many different thoughts, feelings, imaginings do you have?

Activation: AIM Factor A

If the brain and the mind are co-activated, how can the activation of each be measured? It is easier to specify the activation for the brain than for the mind because it is direct, instantaneous, and objective. In humans, we can easily obtain EEG measures and subject them to a variety of analyses. The simplest measures are the dominant frequency and its inverse, the amplitude. Certain patterns, such as sleep spindles, are easily identified.

Scoring of all–night sleep recordings is a good case in point: waking and REM sleep are identified as alpha (8–12 cycles/second or beta 12–15 c/s); drowsiness and sleep onset Stage I as theta (4–7 c/s); Stage II NREM as theta with distinctive spindles; Stage III NREM as spindles and slow waves (1–4 c/s); and Stage IV NREM (1–4 c/s). This sequence proceeds from low-voltage fast for the most activated states of waking and REM to the most inactivated state, Stage IV NREM. Automated sleep-stage scoring makes these distinctions with absolute objectivity and reliability. Quantitative EEG analysis achieves objectivity by computing power spectra instead of visual pattern analysis.

A further technical advance in quantifying AIM factor A is provided by PET and fMRI imaging. Activation is determined by the automatic assessment of blood flow (a proxy for neuronal activation) and depicted as color-coded pixels in the display of the electronic data. Red is activated and blue-violet is deactivated with the intervening values symbolized by intermediate colors on the visible spectrum of light. The absolute values of A from state to state and brain region to brain region can be statistically compared thanks to the work of the University College London psychiatrist, Karl Friston. Friston is a leading mathematical theorist of consciousness.

In experimental animals (who are not reportedly conscious), the rates of firing of neurons can be taken as the most direct quantitative assessment of AIM factor A. This measure can be made with three desirable features: spatial precision, anatomical identification, and discrimination between inhibition and excitation. These physiological desiderata more than make up for the absence of subjective values. The missing link for a tight chain of logic is the absence of imaging studies of the animal subjects' brains. Until those are obtained, we will be unable to use the cellular-level data confidently in place of the missing human cellular physiology. The recent analysis by Giulio Tononi of the neuronal activity in the temporal lobe of human epileptics confirms the work of Charles Hong regarding dramatic phasic excitation of neurons in conjunction with the eye movements of REM.

To date, instead of more precise measures, we have only realistic estimates of A and the three-dimensional graphs of AIM must therefore be regarded as feasibility demonstrations only. The science of consciousness is only beginning and will grow as fast as the scientists of tomorrow rise to the challenges and opportunities of today.

Input-Output Gating: AIM Factor I

Consciousness has access to the outside world in waking but not in sleep, especially in REM sleep, when ambient information is actively excluded. In waking, the brain is privy to the environment as it must be to serve as a guide to behavioral action. In dreaming, the human subject creates an entirely imaginary virtual reality for reasons that are not yet entirely clear. In the waking state, we can act upon the world intentionally; in sleep, we are either effectively paralyzed or we move spontaneously without any intention to act upon the real world. So at some level the brain is deciding on how deeply we process inputs, and what actions we can effect on the world. We focus here on the difference between the quasi-real and the artificial as a way of better understanding both waking and dreaming consciousness. There are several easily understood ways to measure this difference.

Because sensory gating is tied to the state-dependent inhibition of motor output, the simplest measure, EMG amplitude, is a routine sleep lab datum. The EMG records unit activity of the posterior neck muscles (in the cat) and of the chin muscles (in humans). Each muscle fiber is a unitary structure akin to central neurons, and muscle fiber action potentials are easily recordable because they are subcutaneous rather than protected by the skull and dura mater. When chin muscle EMG is visible and well differentiated, it is useful. Since humans sleep in a recumbent position, however, chin muscle EMG may be of very low amplitude at sleep onset such that further suppression is impossible.

When the EMG amplitude is low (or when more precise measurement is desired), the so-called *H reflex* may be elicited. An electrical stimulus is delivered to the skin and the amplitude of the resulting muscle twitch is measured. This is a quantitative assessment of Sherringtonian monosynaptic reflex excitability of the spinal cord, and its measurement ties consciousness science to classical neurobiology.

Surprisingly, laboratory subjects are able to sleep despite H reflex sampling. This stimulus insensitivity is a result of the very process that we would like to measure.

The physiological studies of Ottavio Pompeiano elaborated the way that the brain isolates itself from the outside world in sleep (factor I) and coordinates that exclusion of external stimuli with the internal brain activation (factor A) of REM. Such a complex piece of biological engineering must have a vital adaptational function to which millions of years of evolution were dedicated. Yet only within the very recent past have we begun to understand how and why this was done.

Normal human subjects may be naturally anaesthetized and paralyzed so that they can run a simulation of waking programs in their sleep. Failure of input-output gating sometimes occurs as if to underline this point. Subjects who do not adequately control factor I are insomniac or very light sleepers (if their arousal threshold to sensory stimuli is low); others become motorically active in sleep (if their muscle tone is not sufficiently quenched by inhibition). Sleepwalking is the most dramatic example of a failure to control factor I. Sitting atop those slavish brain stem servants is the cortical master of masters, with its consciousness. When it commands movement, movement is executed.

Modulation: AIM Factor M

The most novel and important axis in AIM is unfortunately impossible to measure in humans. Modulation (AIM factor M) is a process of undoubted importance in man, as is made clear by the recognition that so many of the potent psychoactive drugs used in medicine and on the streets effect their changes via the same brain stem aminergic systems that control the normal conscious states of waking, sleeping, and dreaming. Fortunately, there are abundant quantitative data regarding factor M in animal models. The aminergic subsystems of the brain stem include the serotonergic nuclei of the midline, which Jouvet thought mediated NREM sleep.

As a function of the widespread distribution of their axons, aminergic neurons are capable of establishing the chemical microclimate of the brain and, by the philosophical principle of dual-aspect monism, change the mode of operation of the mind. In waking, the brain is maximally aminergic. Evidence suggests that attention and memory (described in Chapter 3) depend upon aminergic modulation. Information acquisition and information storage, two of the most critical features of waking consciousness, are thus tentatively explained biologically. We experience and we remember because our brains acquire and store data. Not surprisingly, the converse is also true. If, as in REM sleep, aminergic modulation is unavailable, our dreams are not attended to and they are forgotten.

Ralph Lydic is a professor of psychology at The University of Tennessee. Among his many contributions to dream science, none is more germane to the analysis of the brain's determination of conscious states than his long-term recording for neurons in the dorsal raphé nucleus of the cat. These data showed, for the first and only time, that brain modulation by putatively aminergic (AIM factor M) neurons was repetitively variable, with highest discharge levels in waking and lowest levels in REM sleep. This finding strongly supports the idea that the two activated states of waking and dreaming are strongly differentiated at the chemical level. These chemical differences help us to understand the psychological features of dreaming that have heretofore been a speculative guessing game.

We wish that we could quantify factor M in humans because only words can convey conscious experience. But humans do not welcome microelectrodes in their brainstems and current neuroimaging methods lack the spatial resolving power necessary to track the activation and inactivation of the very small and very few aminergic neurons. We are at our scientific limit here and can only transpose data from other mammals with the reasonable conviction that humans share these phylogenetically old systems.

Figure 9.2. AIM model of brain-mind state control. The three-dimensional AIM state-space model, showing normal transitions within the AIM state space from waking to non-rapid eye movement (NREM) and then to rapid eye movement (REM) sleep. The x axis represents A (for activation), the y axis represents M (for modulation) and the z axis represents I (for input–output gating). The values of A, I, and M can be derived from the neuronal data of animal experiments; factors A and I can also be estimated in human sleep laboratory data but, as yet, there is no way of measuring factor M in humans. Waking, NREM sleep, and REM sleep occupy distinct loci of this space. Waking and REM sleep are both in the right-hand segment of the space, owing to their high activation levels, but they have different I and M values. Thus, the activated, REM-sleeping brain-mind is both off-line and chemically differentiated compared with the waking brain-mind. NREM sleep is positioned in the center of the space because it is intermediate in all quantitative respects between waking and REM sleep. The values of A, I, and M change constantly, but the changes are constrained. Sleep and waking states alternate owing to circadian influences (not shown). During sleep, AIM values tend to follow elliptical trajectories through the space. As sleep advances in time, AIM values go less deeply into the NREM sleep domain and more deeply into the REM sleep domain. The normal, cardinal domains of waking, NREM and REM sleep occupy relatively limited zones of the space. Diseases, such as those neurological conditions that produce coma and minimally conscious states, can be arrayed in the left-hand segment of the space, owing to their low activation values. Lucid dreaming, which is a hybrid state with features of both waking and dreaming, may be situated in the middle of the extreme right-hand side of the AIM state space between waking and REM, towards either of which lucid dreamers are drawn. Sleep and psychiatric disorders can ultimately be placed in such a schema. ACh, acetylcholine; NA, noradrenaline; 5-HT, serotonin.

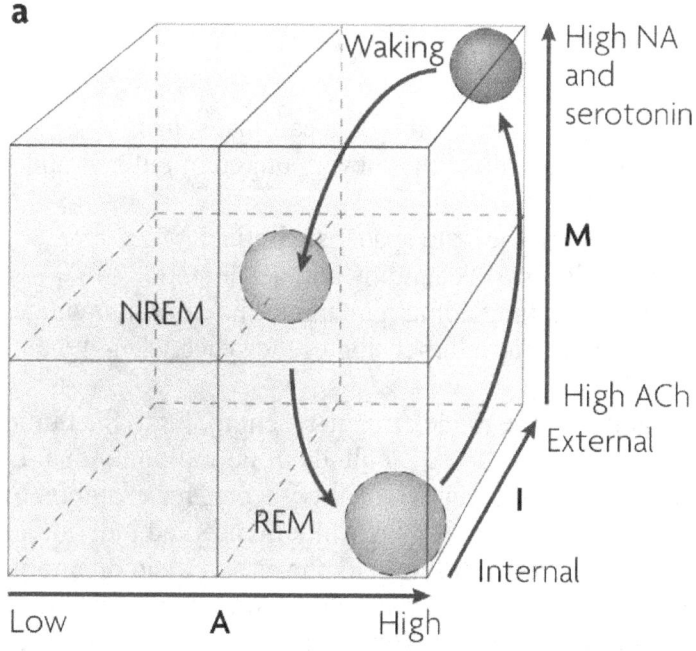

If these assumptions are correct, then we are justified in using the firing rates of cat and rat aminergic neurons in AIM as proxies for their human equivalents. Although the waking rates are low (2–4 Hz), the selectivity ratios are very high for waking because the rates go to zero (or very near zero) in REM sleep. Statistics are not necessary for these almost infinite differences. These are the sort of findings that David Hubel hoped to see upstairs in the cortical reception rooms but they lay deeply buried in the wine cellar.

The difference in psychological measures is not infinite because waking and dreaming consciousness share enough activation to be simulacra of each other. This fact has important functional implications. Because factor M reflects the activity of both aminergic and cholinergic elements, the values can be expressed as ratios of the two modulatory systems. An important reason for doing this is to recognize the fact that the cholinergic system is much less state specific than the aminergic system and more inconsistent throughout its extent from the basal forebrain to the pontine brainstem. When released, acetylcholine concentration is measured in the forebrain. It tracks activation quite well, indicating that both waking and REM sleep are cholinergically driven states of consciousness. The critical difference between the two states must be aminergic.

Structural and Dynamic Aspects of AIM

As a three-dimensional model structure, the AIM cube has x, y, and z axes to which activation (A), input-output gating (I), and modulation (M) values are assigned. The result is that the state space is bounded by values for activation (front and back walls), input-output gating (ceiling and floor), and modulation (side walls). Values of the three dimensions run from left to right (for activation), front to back (for input-output gating), and top to bottom (for cholinergic to aminergic ratio of modulation).

Each point in the state-space is determined by these three values; we may say that our consciousness is continuously moving through the space. The points cluster as clouds, however, one for each of the three cardinal states of waking, NREM, and REM sleep. The waking domain is at the rear, near the ceiling and at the right of the space; NREM is near the center of the front wall; and REM is at the front right-hand corner, near the floor of the space. Like waking, REM is near the right wall of the cube. This treatment of values differentiates the two brain activated states, waking and REM, far better than the traditional two-dimensional graphs based on polysomnographic data.

Once the AIM data are displayed, several features become evident:

1. Very little of the AIM space is occupied by the cardinal states.
2. Vast domains of the AIM space are either never entered or entered under only pathological or exceptional conditions.
3. The normal trajectory through the state space is elliptical.
4. AIM is a cycle because it is always changing and never static.
5. AIM always returns to the waking domain no matter how far it navigates away into the NREM and REM regions. Time is a fourth dimension of the space when dynamics are considered.

The implications of these features of the model are far-reaching. First, the number of possible states is virtually infinite, in keeping with the observed plenitude of normal individual traits. AIM (or something like it) could thus be useful in reconstructing personality theory, for example. Exuberant, outgoing people might be expected to show high levels of A while in reserved individuals A might be lower. Second, the accommodation of unusual states, like hypnosis and lucid dreaming, is already evident. They are states of the brain-mind adjacent to waking but distinctly separate from waking. Third, and most important, the AIM model yields testable hypotheses: vary one factor or another and observe what happens, both in the virtual space of AIM and in the real space of life.

Figure 9.3. Overnight trajectory of AIM.

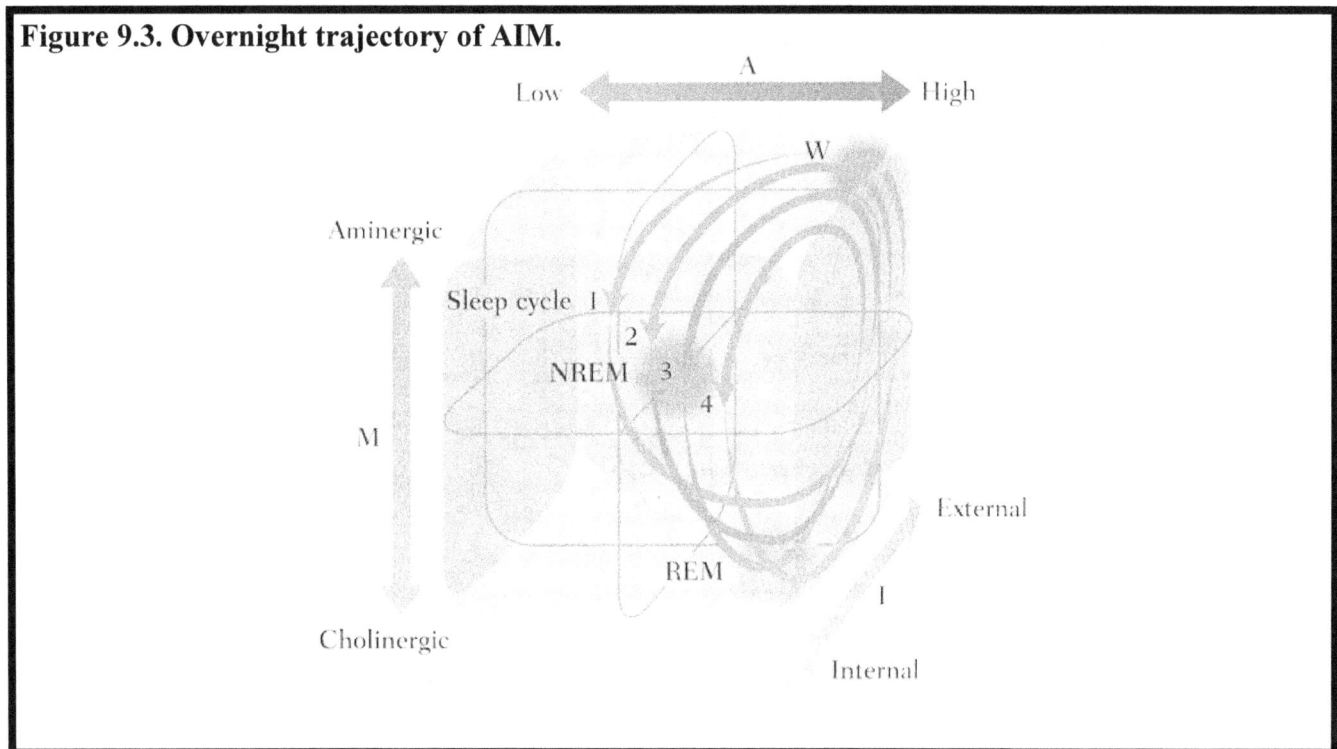

Clinical Implications of AIM

The diagnostic dilemma of the linear, all-or-none, state paradigm is resolved by the state-space model. The question is not who is clinically depressed but how depressed is each of us? Neither are we ever entirely insomniac. Each of us has a bad night now and then so the real question is how bad are our bad nights and how often do we experience them and with what consequences? The answers to these questions are quantitative, not qualitative.

One important difficulty that is constantly present in the history of modern psychiatry is the flux between the linear and discontinuous diagnoses of schizophrenia and major affective (mood) disorder. A psychotic schizophrenic this month is a depressed but no longer psychotic depressive three months after the schizophrenia diagnosis is made. Medication (with aminergic blockers or boosters) is partially responsible for effecting this dramatic transformation. Drug treatment alters AIM position dramatically.

Millions and millions of US$ are spent each year in the laudable, but futile, attempt to make new versions of the Diagnostic and Statistical Manual more reliable. Meanwhile the patient, at one instant meticulously classified, has changed beyond recognition in the next instant. Just as normal consciousness changes almost qualitatively over the course of a single day, so does pathological consciousness change dramatically over time. It is this dynamism that we aim to capture with our new way of thinking about human nature. We are awake now but within hours we will be hallucinating and deluding ourselves that we are flying, spinning, and running away from imaginary pursuers.

Doctors see patients by day, when both the observer and the observed subjects are awake. Only within our lifetime have sleep labs sprung up to counter the belief that waking is the only conscious state. The sleep disorders movement is salutary but we still have a long way to go to achieve anything like adequate assessment and treatment of most people's sleep and dream disorders, especially if we realize that so-called mental illness is at once a functional brain disorder and a disorder of consciousness.

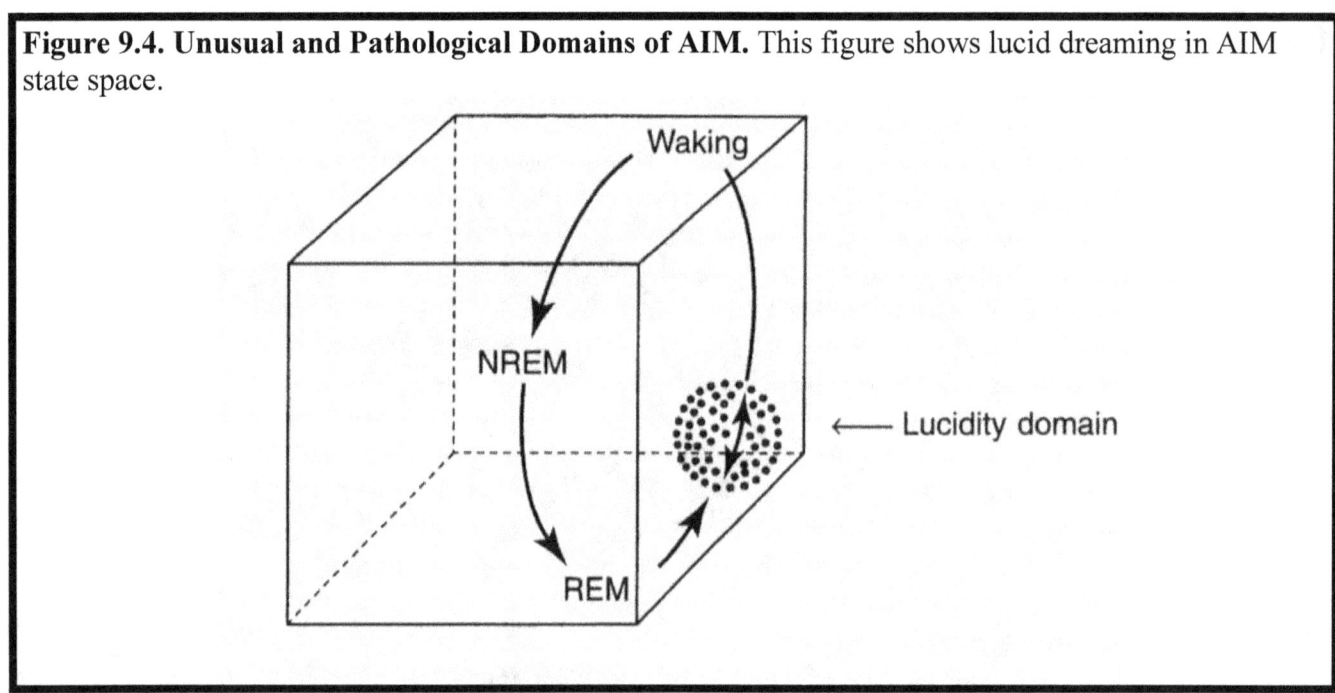

Figure 9.4. Unusual and Pathological Domains of AIM. This figure shows lucid dreaming in AIM state space.

Sleep and dream science is not governmental bureaucracy, insurance reimbursement, or even clinical medicine. But science is financed by the same government and business agencies that must make practical decisions. Hence, there is a great tendency for institutional convention to resist change. But change is what our state-dependent focus on consciousness is all about. We hope to imbue our students with this now obvious fact and to inspire them to explore the new scientific horizons opening up before them. Consciousness is a moving edge, both phenomenologically and biologically. AIM is clearly inadequate and must be improved immediately, if not sooner.

Dialogue 9. The AIM Model

TH: Your cute little AIM cube has attracted a good deal of attention.

AH: People like to be able to visualize concepts and scientists want their pictures to be quantitatively accurate.

TH: You have succeeded in the first goal. What about the second?

AH: The quantitative assumptions are reasonable but arbitrary and not yet precise.

TH: You say the model needs many more dimensions. One of the advantages of three-dimensional space is that it's easy to visualize. Why do you want to complicate things more?

AH: I don't. In empirical science, you have to look where the light is. Theoretical types, like Albert Einstein, want to understand what the light is in physical terms. I share that ambition even though I am not a math genius.

TH: In that case, your model should be driven by empirical considerations. Empirically, how many dimensions does the model need to capture the space of consciousness states?

AH: We don't know that yet.

TH: Hmm. It sounds a bit unconstrained then. Is AIM a neurological or a psychological model?

AH: It tries to be both. It succeeds best when one conceives of the factors as dual-aspect monism axes.

TH: Factor A, the x axis, estimates the strength of both psychological and physiological activation. The two functions may not be identical but they are highly intercorrelated and my hunch is that they are two aspects of the same underlying physical process.

AH: I think of factor I, the z axis of the AIM cube, as input-output source; it tracks the direction of information flow: Is the information partially external (as in waking) or predominantly internal (as in dreaming)?

TH: Factor M tries to quantify what is done with the information being processed. Is it being remembered or forgotten?

AH: M is for Memory and Modulation. That's Mnemonic (meaning easy to remember).

TH: What else do we need to consider?

AH: The imaging data already show that a virtually infinite number of brain regions differ from each other. AIM treats the state space as a whole. This assumption assumes a vast number of possible microstates, a formal feature which leads many to despair of ever understanding consciousness at all.

TH: All imaging data gives us are pretty pictures. We then need to interpret them. Often that proves beyond us. No wonder there are so many mysterians.

AH: I am so glad you are not a perfectionist. Try to be patient and a little more optimistic. We have just been working on this problem for half a century. Nature spent hundreds of millions of years cooking it up for us.

PART II. BIOLOGY

Introduction and Summary, Part II.

AIM is based not only upon the neurophysiological facts outlined in Part I but also upon more general biological considerations.

Chapter 10 discusses its developmental origins which show that the physical mediation of AIM is lifelong, beginning in utero and continuing to change throughout life. AIM is not only a human phenomenon but is shared by most mammals and has an evolutionary history in avian and even reptilian animals. This is the subject of Chapter 11.

Much of Part I focuses on the brain mechanisms of AIM, but what survival purpose do these exquisite mechanisms serve? Chapter 12 begins to answer this question by reviewing the deleterious effects of sleep deprivation, which indicate that AIM benefits not only cognition but also the energetic basis of life itself. AIM is much more than a model of the niceties of consciousness. It also helps us understand the biological underpinnings of survival. AIM Factors A and I can be measured in the field as well as in the laboratory. Chapter 13 discusses the way in which the new science can be pursued under the more natural conditions of the wild, that is to say in the human home.

Chapter 14 returns to basic neurobiology by presenting the cellular and mathematical reciprocal interaction model on which AIM is built. This model accounts for the alternation of conscious states in terms of the brain cells and molecules that determine how we wake and dream. An understanding of this model is crucial to the appreciation of AIM factor M and is an essential prelude to the clinical considerations of Part III.

Chapter 10. Development

Until about age 30, human wake state consciousness increases at the expense of sleep. Put the other way around, sleep is lengthier in younger people than middle aged ones. This reciprocal relation between age and waking consciousness suggests that sleep may play a functional role in the development of the brain-mind. On this view, the frequent observation of increased sleep in old age would be due to the degeneration of the brain as well as to social factors.

Sleep in Utero

Jason Birnholz is a radiologist who used echographic imaging to demonstrate that human fetuses have eye movements in a REM-like state as they float in the amniotic fluid of the womb. Echographic imaging is a standard clinical tool that enables obstetricians to advise expectant mothers about the health of their babies-to-be. Fetuses have long been known to be mobile but the Birnholz movies revealed a dramatic facial and ocular dynamism after thirty weeks of gestation. This means that facial and ocular motility is present in abundance during most of the last trimester of intrauterine life and possibly earlier. It is irresistible to speculate that the brain development needed to support consciousness is well advanced by the time of birth.

Figure 10.1. Birnholz fetal faces.

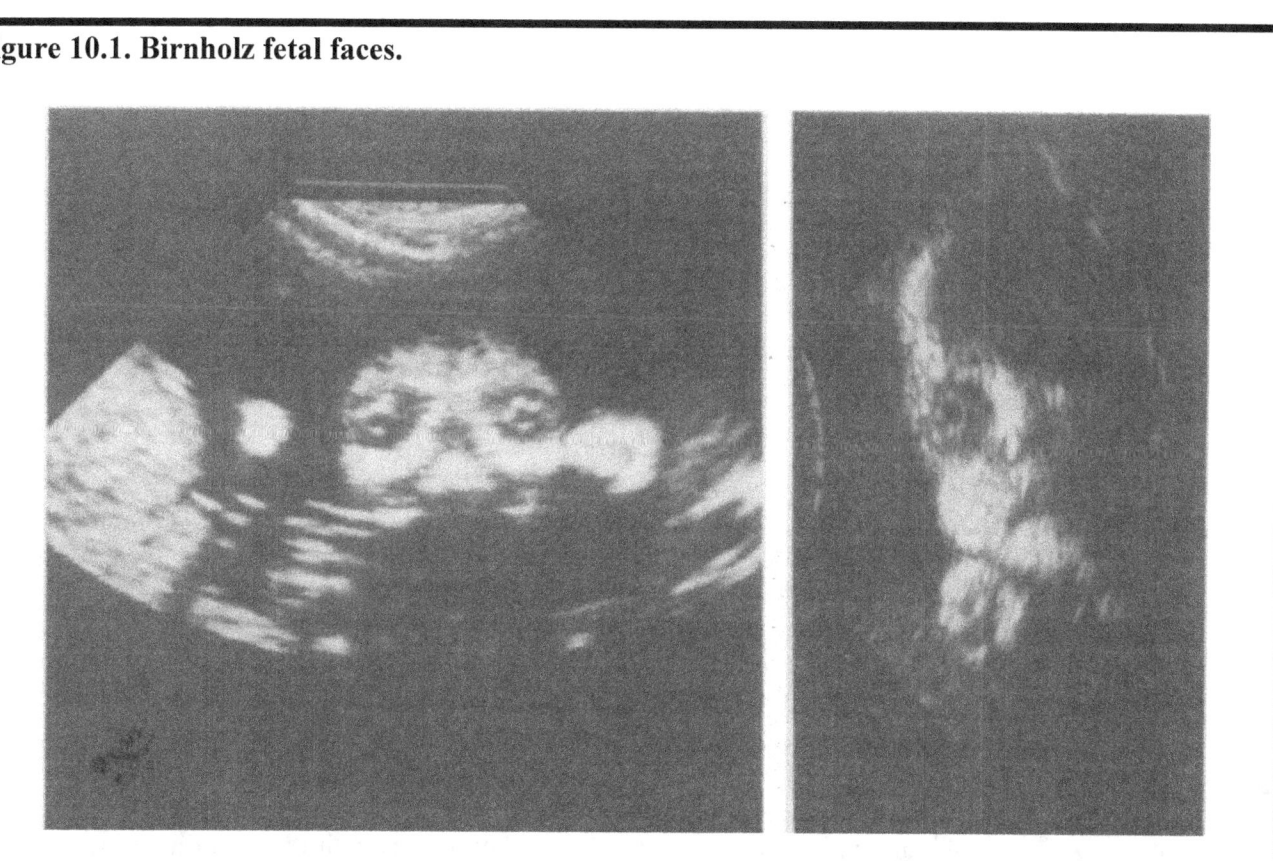

The implication of these findings is that the physical basis of our consciousness antedates our birth. We speculate that prenatal brain activation supports a protoconscious subjective state. This speculation is fortunately not logically relevant to the developmental hypothesis but it does express the likelihood that consciousness has a prenatal *Anlage* just as it has infra-human mammalian roots. In other words, consciousness is gradual over ontogenetic and phylogenetic time just as it is partial and fragmented over diurnal time in adult human life. It is the task of science to explain why this is so.

Not surprisingly, the human prenatal sleep pattern is present in many other mammals. This fact has permitted scientists to confirm that prenatal brain activation does constitute primordial REM sleep. The English physiologist G.S. Dawes exteriorized the uterus of pregnant sheep ewes and implanted a Plexiglas window to observe the fetal lambs. He also recorded brain waves and muscle tone from sheep fetuses, enabling him to identify REM sleep positively. This result strengthens the hypothesis that REM plays a developmental role in setting the sensorimotor stage for consciousness, whether or not subjective awareness is present before birth.

Figure 10.2. Operative exposure of the fetal neck (Dawes et al.).

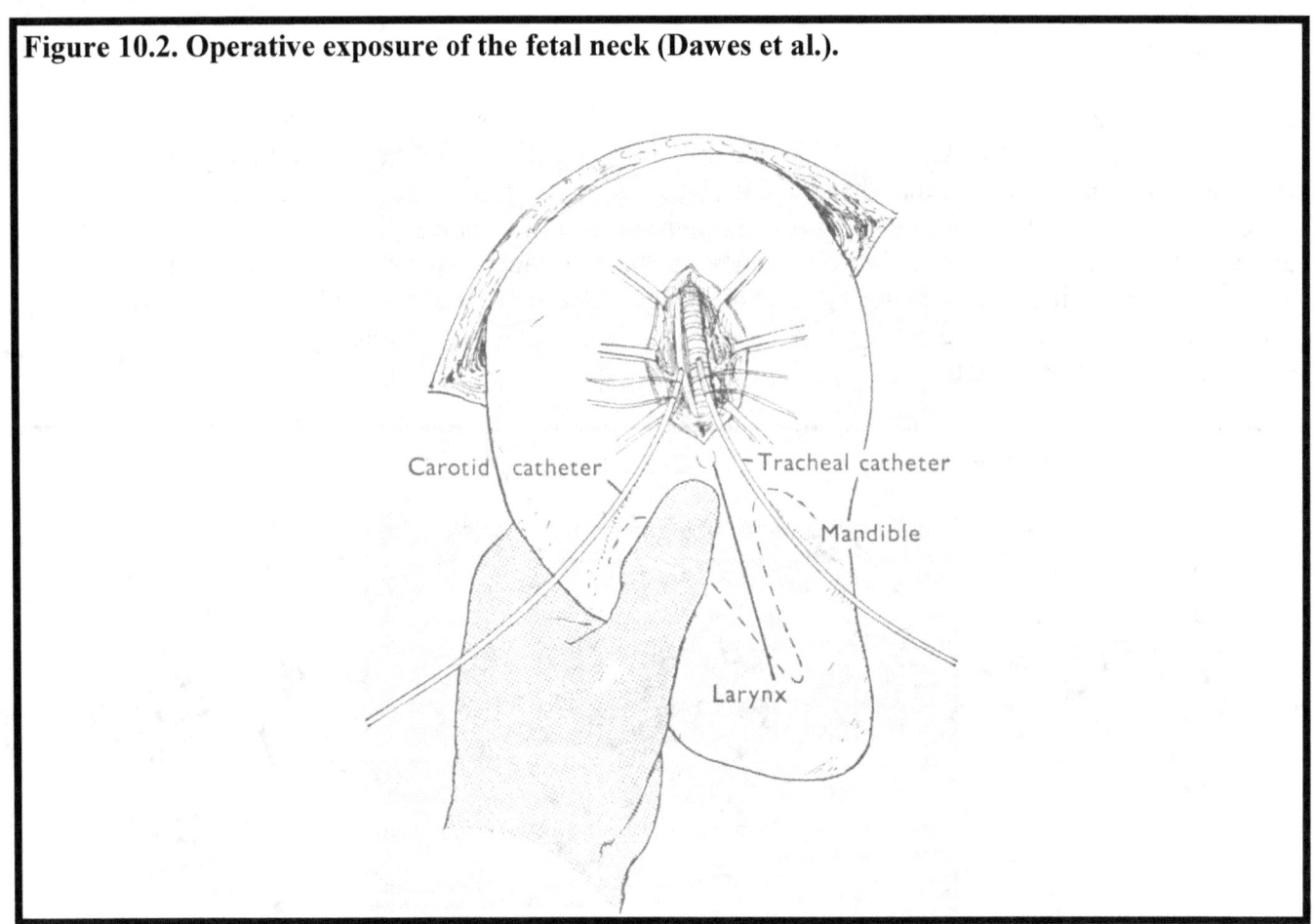

Sleep in Infancy

The full-term human newborn sleeps 16 hours a day, and 50% of that time is REM, meaning that 8 hours a day is devoted to REM. This quota is more than four times as much REM time as occurs at maturity. The immature brain generates brain waves that are less well defined than adults, making electrographic state identification more difficult, but the muscle twitches, facial expressions, and eye movements are directly observable. There is thus no possible doubt that REM is a very early, if not the earliest, organized state of mammalian existence. REM may therefore constitute a building block for waking consciousness. In this connection, it is remarkable that NREM (or SWS) actually develops (along with waking) later than REM. In adulthood, its precedence in a night of sleep suggests that its energetic and hormonal roles may favor higher-order functional adaptation with REM supporting the substrates of survival.

That this infant propensity for REM is a continuation of prenatal brain activation is further guaranteed by the still greater REM propensity in premature babies. REM % climbs toward the 100 percent level estimated at 30 weeks gestation by Birnholz. 30 weeks is near the limit of viability in premature humans. The earlier a baby is born, the more time it spends asleep and the more of that sleep is REM.

Realistic speculation estimates that human fetuses devote 24 hours a day to REM sleep in utero at age 30 weeks.

Figure 10.3. Ted Spagna photo of sleeping infant.

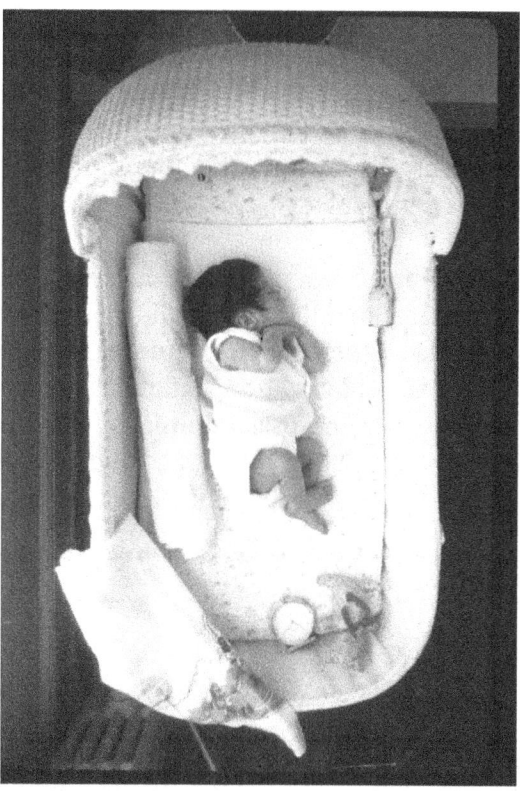

So far, we have focused on REM propensity but NREM-REM cycle length is also age dependent. Smaller brains cycle more rapidly in keeping with most other rhythmic physiological functions. The newborn human has a cycle period of 50 minutes, slightly more than half the 90-minute cycle of adult humans. That brain size is the basis of cycle length is suggested by its correlation within and across species. Mice, which are tiny, have REM every 6–8 minutes, and baby mice cycle even more rapidly. This correlation may subserve metabolic needs and determine the frequency of feeding as well as lay the foundation for consciousness. Consciousness is a relatively luxurious commodity, after all.

Mothers rejoice as their infants consolidate their sleep cycles and spend more of their time asleep in the night, when the mothers can also sleep. Meanwhile the infant's time awake is progressively more socially rewarding. This slow shift is associated with instantiation of the circadian rhythm, one of the few exceptions to the size/cycle length rule. The circadian rhythm, which brings us into harmony with cosmic forces, is temperature and size *independent*. Organisms activate and rest once a day year in and year out. Brain state differentiation and consciousness come to ride on top of this rhythm as its postnatal installation is slowly made.

Sleep and Dreams in Early Childhood

Most modern scientists agree with Freud, who felt that dream reporting was unreliable before age 5. Children are unreliable informants because they are so poorly rooted in reality. Fantasy rules the child almost to the point of dreaming while awake. The childhood love of play is a manifestation of the mixing of dreamlike imagination with waking consciousness. This fact is further support for the idea that REM physiology is irrepressible in early life. No autosuggestion is required to release the brain-mind's imaginativeness in the young. It makes children at once charming and difficult to manage by adults, who tend to assume that a child's consciousness is exactly like theirs. We don't really know any more about what its like to be a child than what its like to be a bat.

Sleep declines in duration but not in depth during childhood. Slow-wave sleep dominates the early night and arousal may be difficult or impossible to stimulate. Bed-wetting and spontaneous confusional awakenings are not uncommon. In *night terrors*, children may literally hallucinate as they sit, ostensibly awake, with a concerned parent at the bedside. The child is, literally, psychotic, and parents are understandably concerned until they are assured that the brain is asleep despite the outward signs of waking. Such porous state boundaries are also responsible for other so-called parasomnias, such as sleep walking or sleep talking, to which we return later.

Naps are still common until school commences at about age 6. It may be no coincidence that persons begin to become educable at the same time that their sleep allows their waking consciousness to be relatively free of dreamlike intrusion. The prolonged admixture of dreaming and waking mental content is testimony to the slow pace of brain state stabilization, and it is little wonder that such maturational changes are sometimes incomplete. The still impressive formal similarity of dream consciousness and major mental illness has been noted since the heyday of Freud and Jung, but our understanding of underlying mechanisms is still, pardon the expression, immature.

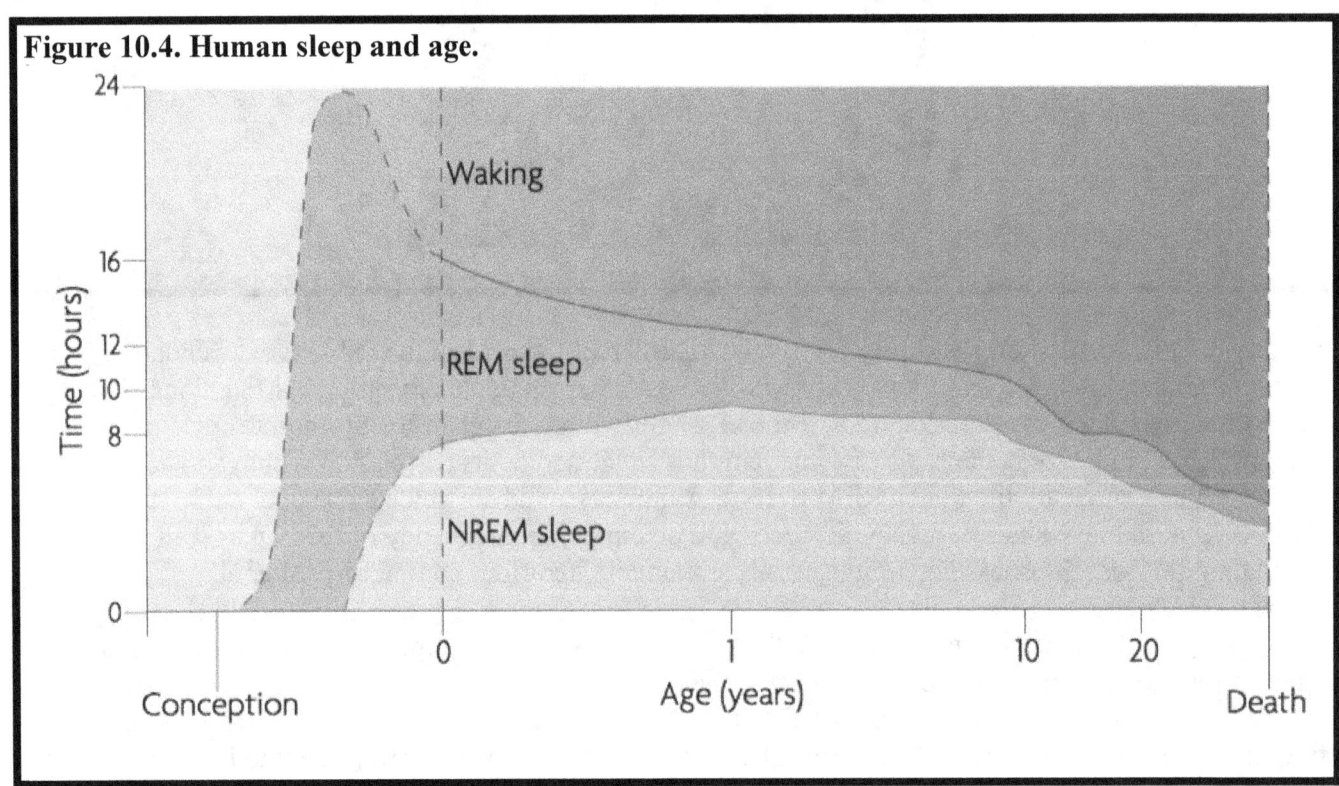

Figure 10.4. Human sleep and age.

Sleep and Dreaming in Late Childhood and Early Adolescence

A very curious inversion of consciousness occurs at age 9, when American youths begin to take intellectual hold in the third grade of school. This is the peak year for the occurrence of spontaneous dream lucidity, the paradoxical admixture in sleep of waking and dreaming. Instead of dreamlike fantasy invading wakefulness, as seen in earlier childhood, wakeful consciousness invades dreaming, so that subjects become aware that they are dreaming when they are dreaming, rather than erroneously supposing that they are awake. We will later describe what is known about the mechanism of lucid dreaming in young adults but here only emphasize that a shift in balance between waking and dreaming consciousness has occurred together with the shift in cognitive scholastic ability.

Marie-Jean-Léon, Marquis d'Hervey de Saint-Denys (1822–1892) was a scholar expert in Chinese culture who lived in late nineteenth-century Paris, where an interest in sleep and dreams was already nascent. In an effort to voluntarily control his own dreams, he pioneered the study of what is now called lucid dreaming. Saint-Denys is justifiably famous for his use of the subjective self as both an instrument and object of scientific investigation. His life affirms the paradox that aristocracy and privilege are not necessarily obstacles to scientific progress and that concern about morality may be a goad to important discovery. How can we reconcile our oneiric impulses with waking law? We cannot control most of our dreams and, pace Saint-Denys, they often convey deep truths about ourselves that we would prefer to deny.

This observation corroborates the cognitive science view that holds to a brain maturation paradigm for sleep and dreaming. By age 9, youngsters are capable of reliable reports of their inner life and this, of course, is dependent upon the language mastery that contributes to their improved scholastic aptitude. If we add in the 9 months of intrauterine time, we recognize that ten years are spent in developing something like an adult brain. That comes to 4,000 days and nights and roughly 40,000 hours of sleep. Building a brain-mind that is capable of consciousness requires a great deal of fine-tuning of a highly complex structure.

Sleep in Adolescence and Young Adulthood

Social demands, already great in elementary school, increase in adolescence and young adulthood. At the same time hormones dictate sexual maturation with the release, in sleep, of these chemical mediators of bodily change. The net result is often a standoff between internal and external forces. This conflict results in the paradox of chronic sleep deprivation, as study and courtship exert their demands, and physiology commands orgiastic sleep bouts of up to 12 hours in length, especially on weekends. Parents despair of rousing their kids even on weekdays. Dreaming is a luxury, which few young people can enjoy; this is unfortunate because dreaming is so psychologically informative from both the formal and content analytic points of view. Innovative educational programs might well consider taking advantage of these universals.

Sleep and Dreaming in Middle Age

We may define the beginning of life as conception to age 30 and the end as age 60 until death. The years between 30 and 60 are thus precisely in the middle, and this period is often associated with employment, marriage, and creative activities. We are lucky if we can sleep soundly during these years. Exercise may be helpful especially if it is timed properly (before supper is ideal). Anxiety, marital discord, and occupational failure often get in the way, however, and many insomniacs turn to alcohol or sleep medication to help them relax and rest. Neither popular remedy is without cost and sleep hygiene is not a popular topic. An additional problem is that the deepest sleep, EEG Stage IV, declines in the decade between 30 and 40 and is gone long before the climacteric. The bad news is that sleep is normally less deep and more fragmented but dream recall and book reading are facilitated, as if in reciprocal reaction.

Sleep and Dreaming in Old Age

The bad news about sleep in middle age is that it gets worse as time goes on. Our ability to sleep diminishes as we age. We often have more difficulty getting to sleep and more difficulty staying asleep than when we were younger, although it is a common misconception that our need for sleep decreases as we age — it does not. Our need to sleep remains pretty much constant throughout adulthood — we just have more difficulty getting it when we're older. Older people spend more time in the lighter sleep stages than the deeper ones. Older adults also show advanced sleep phase syndrome: they fall asleep earlier and consequently wake earlier.

Sleeping pills are tempting but there are none without side effects. We just don't know enough yet to preserve eternal youth. Philosophical contemplation may seem an odd prescription but lying awake in thought may be preferable to chemical obliteration. Our lives go unexamined all too often and while obsessive rumination is neither pleasant nor productive, mind travel may be quite entertaining and edifying. With the death of Freudian psychoanalysis, we can welcome an age of self-analysis. If it was good enough for Sigmund, it may be good enough for you.

Consciousness and Protoconsciousness

The prolonged time course of human brain-mind development, including its intrauterine, primordial aspects, strongly suggests that consciousness develops gradually and in a piecemeal manner. On this view, consciousness does not suddenly spring full-grown, from the brow of Zeus, as it were. We suggest that it may begin as *protoawareness* or *protoconsciousness*. Protoconsciousness is a primary or preconscious instantiation of self-as-agent which acts and senses in a virtual space. The emerging self-agent integrates emotion but, having no linguistic structure, cannot create the abstractions necessary to represent itself to itself or report to others its constructions.

One reason for making this suggestion now is the prior occurrence of REM sleep in the development of man and his fellow mammals that we have just reviewed. As we will further stress in the following chapter, infra-mammalian and infra-avian species do not have REM (and we theorize do not have protoconsciousness). We propose that protoconsciousness is a primary state of consciousness that we experience and report in secondary consciousness when we humans (and only we humans) acquire a narrative language capacity.

Protoconsciousness is thus early in an individual's lifetime but develops late in evolutionary time. We will reiterate this point when we discuss the phylogenetic evidence in the following chapter.

Anticipating that discussion, we will now assert that secondary consciousness appears late both in an individual's life and in evolution.

Why make so speculative a hypothesis? Because we must recognize the possibly adaptive function of brain activation in sleep — the brain activation that becomes REM sleep. Until now, scientists have regarded dreaming as a response to previous events and even as an unconscious mental process. But dreaming is a conscious mental state that usually has no access to waking consciousness (and vice versa). Why should nature have created two states of consciousness if not to model waking in advance, thus permitting an anticipatory, virtual reality by which to predict general properties by means of which it can prepare and repair off-line in sleep?

An unfortunate feature of this idea is that it cannot be directly tested. But can Darwin's theory of evolution be directly tested? No. What's past is past and we can only perform indirect tests of the implications of Darwin's paradigm. Against this glaring weakness is the explanatory power of evolution and protoconsciousness theory. Protoconsciousness theory is really a theory of the evolution of consciousness. What will be the DNA of consciousness? Might it be brain activation? Might it be the action potential? What about ion flow across the neuronal membrane? Perhaps the Ionic Greeks were right after all.

The Function of Dreaming

Dreaming has many features which, without a virtual reality model, are otherwise inexplicable: Q: Why are dreams so unpredictable? A: Because they are so creatively predictive. Q: Why does REM sleep occur in infra-human mammals? A: Because those animals are making up such minds as they have. Q: Do infra-human mammals dream? A: Almost certainly not as we adult humans do, but it is nonetheless to their advantage to create an internal model allowing them to predict and recognize future environmental challenges. Q: Why are dreams so difficult to remember? A: Because it is not the dreams as remembered stories which are important. It is rather the running of the developmental and predictive program that permits an animal to reiterate its novel genetic expectations. An animal can then alter its model of the world in the face of unexpected environmental data.

This argument, to which we will return repeatedly in this book, turns the stimulus-response paradigm on its head and enables us to regard consciousness in an entirely new light.

Dreaming is Lifelong

There is no evidence that dreaming is much affected by age, other than that older people often find their dreams more difficult to recall than younger people. This change could be driven by changes in memory rather than changes in dreaming itself. Senile brain degeneration almost universally diminishes memory, most typically for names and recent events. When older people keep a journal, their memory for dreams and other conscious life events are enhanced. This can be a part of self-analysis.

We close this chapter with an exhortation; the study of consciousness via the states of waking, sleeping, and dreaming is a field of enormous importance and, in our opinion, a neglected one. Pediatricians, child psychologists, and mothers take heed: the future is in the egg.

Dialogue 10. Consciousness from Womb to Tomb

TH: I wonder what it is like to be a fetus.

AH: I am afraid we can never know because memory does not develop until much later in life and even then, it is notoriously fallible.

TH: But fetuses have brains and can activate them at thirty weeks of life. Doesn't this prove that they may be conscious?

AH: I hedge my bets on that one and introduce the word "protoconscious" to signal my reserve about the possibility that fetuses dream (or are subjectively aware of anything).

TH: So must you flush down the toilet such psychoanalytic concepts as birth trauma and advise expectant mothers to dispense with Mozart entertainment of their offspring?

AH: I love Mozart and assume many pregnant women (who are uncomfortable and quite understandably fear the pain of delivery) do, too. It relaxes them.

TH: Why do you admit the possibility of "protoconsciousness"?

AH: To keep my mind open to the strong possibility that some aspects of consciousness are being prepared, even in utero. Sensorimotor integration, for example, is an essential element of the bodily awareness that constitutes adult consciousness. The evidence strongly suggests that it develops quite early in life.

TH: We think of newborn babies as entirely helpless.

AH: Yet they can sleep, feed and make their other bodily needs known to caregivers.

TH: You are referring to the mechanics of eye/hand/and trunk coordination that are such an important part of behavior. But surely, even these important mechanisms are mostly unconsciousness?

AH: Mostly, yes. But entirely, no. Parents can become quite aware of them and even consciously train their use when their own consciousness is more fully developed.

TH: You seem even committed to the idea that consciousness has its origins in motor acts. How can consciousness be motoric?

AH: Consciousness is virtual reality, which begins and ends in motility. I know this sounds behavioristic, and it is, but it is more than mere Skinnerism. As Karl Friston points out, the mind-as-force concept implies movement. Ions move as dramatically as limbs.

TH: So Skinner and Freud were both psychologists upon whose work consciousness science builds.

AH: Both Freud and Skinner had good historical reasons for restricting their attention to the mind and the body respectively. Thanks to them, we can now consider the two parts of the story together.

TH: Cognitive neuroscience has a strong developmental orientation of which neither Freud nor Skinner could avail themselves. Developmental neurobiology scarcely existed and what little there was had no relationship to psychology.

AH: Now there are college courses and textbooks written on the subject. That's what we hope for consciousness science: by bootstrapping it to sleep and dream research, we can point out links to other parts of the cognitive neuroscience enterprise.

TH: Does consciousness run downhill as we age?

AH: Some modules do decline with age. The senescent impairment of memory is a good example. But, as if in compensation, emotional expression often remains intact and may even be refined with what we call the wisdom of old age. Experience is a very good teacher.

Chapter 11. Evolution

In the 1950s, the then-young William Dement applied human sleep lab technology to cats and discovered brain activation with REM sleep in 5- to 10-minute-long phases which punctuated sleep at intervals of 30 minutes. This finding opened Michel Jouvet's eyes to new experimental possibilities. Dement's discovery opened a floodgate of comparative physiology studies with strong implications for consciousness science.

Cats are domestic pet favorites because they decorate furniture with their languorous indolence. Cat worship depends in part on the felines' high resting body temperature, which renders them exercise-intolerant. We never see cats on leashes. They stalk and pounce but do not, for long, chase prey. They sleep a lot because they are naturally warm; indeed, they sleep up to 18 hours a day in heated houses. Cat lovers are sure their pets are conscious and converse with them lovingly.

As Michel Jouvet has pointed out, the REM sleep scientifically described by Dement is directly observable (and has been easily visible since at least since the time of Lucretius in 44 BC). Cats like to loll about in the sphinx position, forepaws curled under the chest, head held high and eyes closed as they doze. If left undisturbed, the cats' erect posture melts and they fall over on their sides. After ten or fifteen minutes, what is left of their postural tone is lost but sleep persists and may even deepen. Fine muscle twitches of arms and legs begin; if we look carefully, we can observe piano playing flexion and extension of the clawed digits. Now that we know that these are the outward signs of REM, we can open the closed lids with a pencil eraser and see the cats' saccadic eye movements with our own waking eyes.

Because the cats' pupils are closed up tight, the prying does not awaken our pet. Unwanted intrusions are actively excluded. If cats were literate, they might hang out signs saying, "Genius at work. Do not disturb." Are the REM sleeping cats dreaming? Are they playing? Are they stalking and pouncing? Your guess is as good as ours but one thing is clear: their motor behavior is muted but dynamic. In addition to the twitches of the limbs and the REMs, whisker twitches and facial expressions are visible. Underneath the cloak of paralysis, this means, for sure, that the brain is not only tonically (the EEG) but phasically active (it generates the eye movements and the muscle twitches).

François Michel and Michel Jouvet showed that the collapse of cat REM muscle tone was actively engendered as if to allow central motor programs to be run in sleep. REM may thus be a built-in gym for the practice of behavior. Not only that, but it was clear that the motor commands that triggered the muscle twitches and the REMs were tied to large phasic EEG signals (the PGO waves) in the sensory systems of the thalamus and cortex, as if to coordinate the muted movement with internally-generated sensation. So whether or not cats dream, they certainly do have internally evoked patterns of behavior. This phenomenon has been extensively investigated by Charles Chong-Hwa Hong at Johns Hopkins University.

The Sleeping Brain of Other Animals

All mammals need to sleep. Among mammals, carnivores sleep more than herbivores. Of course, the further away a species is from a mammal in evolutionary terms, the more difficult it becomes to characterize sleep, but even small insects such as the fruit fly need to rest periodically, and show learning deficits if they are deprived of this rest period. The most primitive organism which sleeps, or at least manifests a sleep-like state of rest, is the nematode, *Caenorhabditis elegans*. This worm is scientifically popular with molecular neuroscientists because its entire genome has been characterized.

One of the mistaken ideas that drove early research on the natural history of sleep was that REM was archaic, phylogenetically (evolutionarily) ancient, and probably even prehistoric in evolutionary terms.

In fact, it was just the other way around: REM sleep is a modern, phylogenetically young, and therefore relatively recent adaptation in evolutionary terms. This reversal turned upon its head the theory that sleep, and especially REM, was a retreat into the instinctual past. Sleep was, rather, a great leap forward and was therefore associated with the highest levels of consciousness, first primary consciousness in birds and mammals, and later secondary consciousness in man.

Figure 11.1. Sleeping cat.

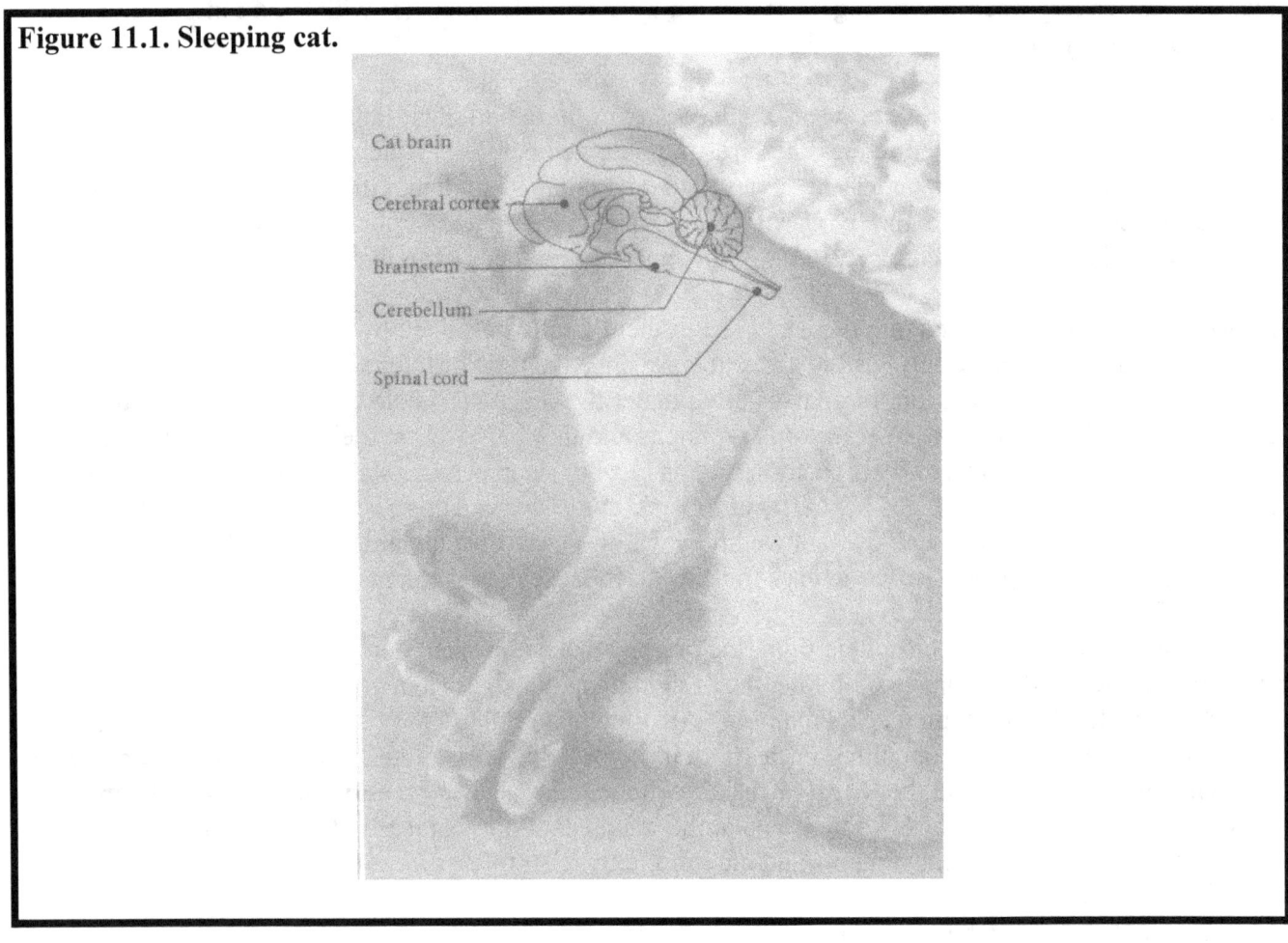

Sleep is widespread throughout the animal kingdom, and, unsurprisingly, there are several curious features with some animals. No bird spends more of its time on the wing than the swift, and the swift rises to great heights where it sleeps while still flying. Whales, dolphins, porpoises, and orcas appear to let only one hemisphere of their brain sleep at any one time, pausing or even swimming with the contralateral (opposite) eye open. They need at least one hemisphere to remain awake while the other sleeps. Sea otters hold hands while they're sleeping, perhaps to prevent themselves from getting separated.

Led by Michel Jouvet, sleep scientists kept looking for the primitive brass ring but always came up empty-handed. Researchers including Fred Snyder at the NIMH (who looked at opossums) and Yves Ruckebusch working in Toulouse, France (who looked at horses, cattle, and other hoofed beasts) found REM everywhere. It was soon clear that REM sleep was a behavior shared by all mammals. It was present in birds, too, but not seen in reptiles, amphibians, or fish. Of the latter three classes of animal, only the reptiles evinced anything like slow-wave sleep. They simply had not enough brain to exhibit analogues of mammalian sleep.

It was later realized that in REM sleep, birds and mammals shared an adaptive talent that gave them energetic advantages to complement their greater motor and cognitive skills. Both classes of animal regulated body temperature, allowing them to function above ground even in cold weather when some of their competitor species dug deep, turned off, and hibernated.

Fred Snyder was an unusual research scientist. He studied the sleep of patients with affective disorders but his real love was natural history. Fred kept a macaque monkey in his house and tried, unsuccessfully, to civilize it. When he tried to run a sleep lab, he became so tired that he fell asleep in his nearby office as he tried, again unsuccessfully, to practice psychoanalysis in the late afternoon. Worse yet, he fell asleep the next morning on the couch of Harold Scarles, the famous Washington D.C. analyst. Searles kicked him out saying, "You are the most exasperating patient I ever knew." Snyder was thus liberated to pursue his real love, naturalism. He ended his days among the American Indians, a tribal shaman if there ever was one.

Studies of Hibernation Long Antedated Sleep Science

With sleep quantification techniques available, it was soon learned that the temperature lowering entries into hibernation were initiated in NREM sleep, the state in which even non-hibernators are coolest. It was furthermore discovered that hibernators maintained low levels of neuronal activity as well as low levels of body temperature as they became calorie efficient throughout the winter. In the spring the hibernators emerged, with the thaw, via the same NREM sleep door through which they had entered hibernation. These findings reinforced the general conclusion that sleep, like hibernation, was an energy-saving adaptation.

What about REM? It is suppressed in both hibernation and experimentally-induced body cooling, not because it has nothing to do with energetics, but because it inhibits temperature regulation. This is a paradox because REM sleep has since been shown to be essential to body temperature control in both mammals and birds. Thus we now realize that REM sleep is the only phase of mammalian life in which temperature control is abandoned.

Figure 11.2. The Evolution of REM sleep.

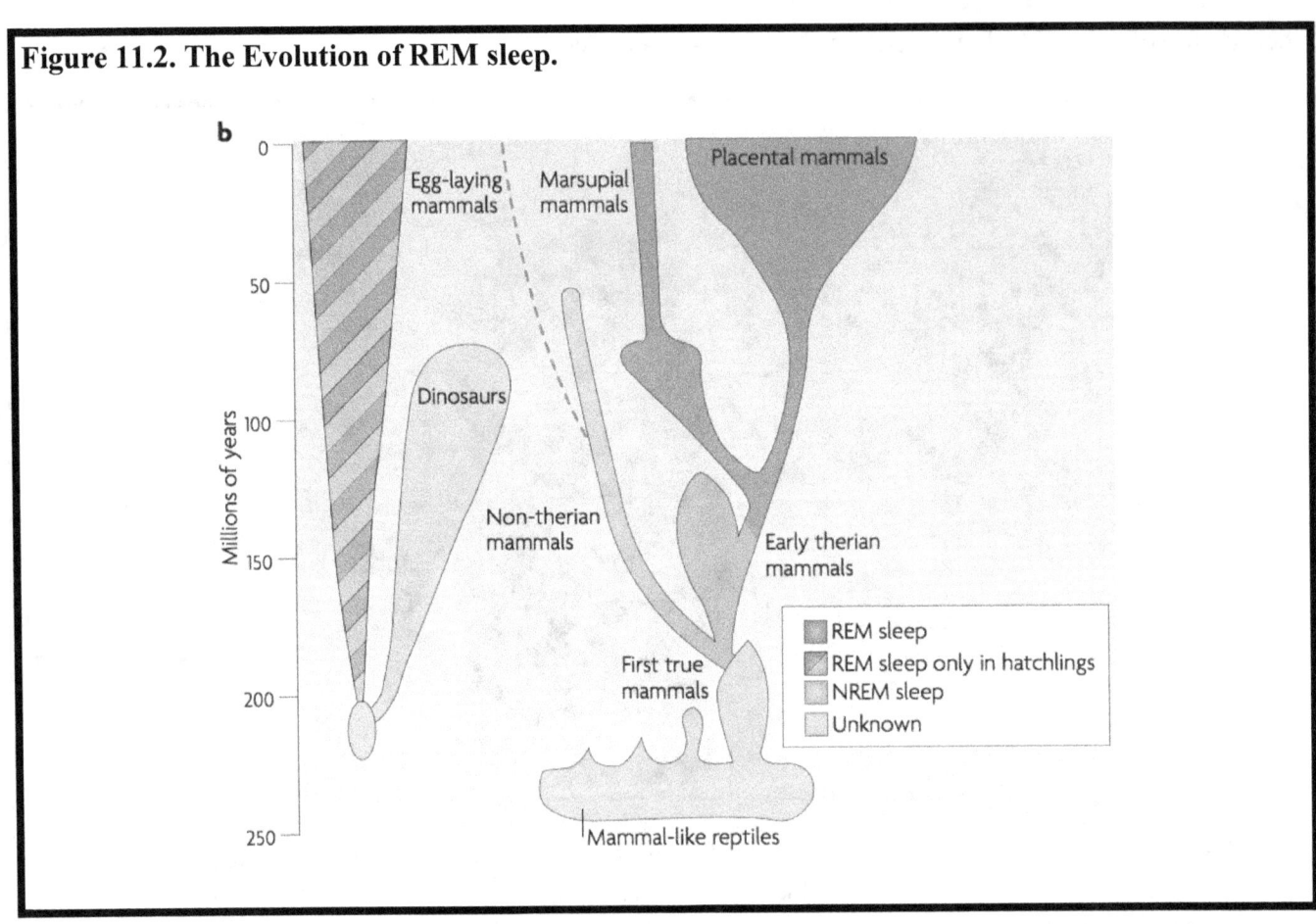

Why don't we freeze to death in the cold? Because we do not fall asleep or are awakened from sleep by unexpected temperature drops. This paradox may help explain the presence of REM in very primitive egg-laying mammals, such as the duck-billed platypus and the echidna; both of these beasts regulate body temperature at relatively lower levels and the presence of respectable mammalian levels of REM sleep may help them do so.

There is nothing primitive about REM. On the contrary, the evidence suggests that animals are conscious (and have REM) if and only if they have large, complex brains that can afford its energetic risks.

Phylogenetic Findings and Theories

Advances in thinking about the adaptive advantages of sleep were made in several steps. The first step was taken by Truett Allison and D.V. Cicchetti, who studied the relationship between several ecological variables and sleep (See Figure 11.2). They reported that sleep was longer and deeper in carnivores (like the lions) than in their herbivore prey (like the zebras). Sleep was therefore a function of prey-predator status: the hunters slept well in the open because they were safe from attack while the hunted slept fitfully, often in nests or in underground burrows. Large body size was likewise a predictor of greater sleep length and depth. This confirmed the hypothesis that sleep was a luxury, like good food, wine, and shelter.

Economic considerations like these were bought home by the startling finding of the REM sleep discoverer Eugene Aserinsky, that college students could double their EEG documented sleep length from 7–8 to 15–16 hours when paid five dollars an hour for the surplus. We sleep longer if motivated by environmental variables (including money). That's the good news. The bad news is that it is difficult to shorten sleep below what appears to be a genetically fixed set point. Short (4–6 hour) sleepers tend to be extraverted, upbeat wakers who get a lot of work done. They contrast with long (9–11 hour) sleepers, who tend to be introverted, feel gloomy, and be relatively unproductive. Sleep is not only environmentally sensitive: it marches to the beat of its own genetic drummer.

H. Craig Heller (1943–) Hibernation had been a subject of interest to scientists long before sleep burst on the scene. Craig Heller was quick to realize that sleep, too, was an energy-saving ploy. Moreover, he appreciated the shared function of thermoregulation, that is, the capacity to maintain body temperature that mammals have perfected. Working at Stanford University, Heller trained his students to record the sleep of animals as they entered, endured, and left hibernation showing the tight link between the two energy conserving behaviors. We owe Heller a thank you for demonstrating why we all hunker down for that long winter's nap.

Good dream recallers wake up often and return to sleep easily. As noted earlier, good dream recallers learn to enhance memory for their dream experience by the waking repetition of their antecedent sleep mentation. They tend to be psychologically minded and their REM propensity and the richness of their dreams appear to be positively correlated with IQ. Good dreamers may thus be born and not made but environmental conditions, not always easy to manipulate, do play a part. If you want to better recall (or control) your dreams, pay close attention to the later chapter on lucid dreaming.

Allison and Cicchetti found a correlation between NREM-REM cycle length and brain size. This led them to the more precise finding that it is the width of the brain stem that best correlates with the time constant of intra-sleep cycle length. These data set the stage for the hypothesis that intra-sleep cycle length was determined by the protein transport time between two half-centers in the brainstem REM sleep generator. The ecological implication of all this data is that waking, dreaming, and waking consciousness are physiologically enhanced via sleep when that commodity is genetically mediated and assured by positive energy economy. "Choose your parents carefully" could be the take-home message.

Like Fred Snyder, Truett Allison was a naturalist who hid out in the woods of psychiatry. Working at the West Haven VA hospital near Yale, he studied the sleep of animals and came up with what is still the best formulation of the relationship between a species' habitat, its survival behavior and its sleep.

Figure 11.3. The jaguar in the tree with his gazelle lunch.

Allison demonstrated that sleep was much more variable than we who focused on young human males had ever realized. Like Darwin himself, Allison made us more modest and more humble than our self-glorifying habits tended to inflate. We are animals, albeit brainy ones, and subject to the same laws of nature as other animals whom we tend to think of as beneath us. It could just be that psychiatry is a never-land of basic truth and the unsolved mystery of mental illness fosters surprising discovery.

Functional Hypotheses

Once the Freudian conviction that dreaming functioned to preserve sleep began to fade (after about 1970), there arose discussion of sleep's possible physiological, behavioral, and psychological benefits that were inspired by the discovery of and growing interest in REM. Integrating the phylogenetic perspective summarized in the previous section, Frederick Snyder advanced the "sentinel" hypothesis in response to the fact that the periodic NREM-REM cycle acted as an internal alarm clock, awakening animals at the end of each cycle. Snyder reasoned that such awakenings, occurring at different times in different members of a herd of prey animals (like gazelles in a savanna cohabited by tribes of hungry hyenas), served as a biological alarm clock. Wakened gazelles could scrutinize the field for predators and warn the sleepers to wake up, attend to threats, and move to safety, if necessary.

Such rhythmic arousal might also be useful to sleepers by allowing posture shifts to impede stasis and prevent the prolonged pressure attendant on the recumbent, immobile body of a sleeper. Alcoholics may suffer from skin lesions and pneumonia because of suppression of REM-induced posture shifts. Sleep apnea patients breathe normally only after they are awakened. These are all variants of Snyder's sentinel hypothesis. In brief, sleep interruptions, however unwelcome to light sleepers and insomniacs, may counteract the sleep risk incurred by external and internal threats to health, safety, and survival.

An attempt to link sleep and dream science to the solid base of ethology was made by the reticular activating system experimenter, Giuseppe Moruzzi. Ethology is the science of animal behavior founded by Niko Tinbergen, Erich von Holst, and Konrad Lorenz. Moruzzi proposed that pre-sleep rituals were appetitive phases which led to NREM sleep while REM sleep constituted consummatory behavior. The instinctual nature of dreams is not irrelevant to ethological theory. This theory was also the first to hint at the idea that REM was super sleep with critical implications for survival. The REM sleep-as-plasticity concept was advanced by Moruzzi long before the upsurge of experiments designed to test learning theory that continue to this day.

Allan Rechtschaffen (1928–2015) was a psychologist who worked with William Dement in Kleitman's lab in Chicago and ultimately became its chief. He claimed that dream consciousness differed from wake consciousness by being "single minded" as against dual or multiple minded. Later, he conducted a classical series of experiments showing that sleep deprived rats died within three or four weeks. En route to their demise, the rats developed defective energy regulation indicating that sleep engendered essential metabolic functions. Rechtschaffen's work thus unites the cognitive features of REM sleep dreams with the deep control of temperature and calorie regulation first proposed as brain functions by Helmholtz in 1866. Rechtschaffen joked that sleep was essential to life and not the greatest mistake that nature had ever made. The errors of design that are manifest as human disorders, especially narcolepsy, were spelled out as the first formulation of sleep pathophysiology.

In the 1990s, the last days of Nathaniel Kleitman's Chicago sleep and dream school, Allan Rechtschaffen conducted his heroic and definitive sleep-deprivation experiments. These experiments altered the course of sleep science and marked the evidence-based approach to sleep as an energy conservation function. Before doing so, Rechtschaffen had coined the term "single mindedness" of dreams without speculating about its functional significance. Single mindedness has become a pillar of the idea that REM sleep dreaming is a protoconscious precursor of waking awareness.

The Animal Mind

Are non-human animals conscious? Do they think? Do they have anything like human minds? Dog and cat lovers insist that they do but are hard pressed to suggest that they have enough language capacity to think and converse in the linguistic mode. Donald Griffin, a biologist at Rockefeller University in New York, argued for animal consciousness in the 1960s and '70s against his fellow scientists, who were skeptical about animal mind propositions. Following B.F. Skinner, but ignoring Charles Darwin, most scientists of the last half of the 20th century asserted that animal mind, like consciousness itself, could not be directly observed and was therefore outside the bounds of experimental inquiry.

Modern sleep and dream science provides a way out of this dilemma. If there are two kinds of consciousness and one of them (primary consciousness) precedes the other (secondary consciousness) in both individual and evolutionary history, then humans and other animals must share significant aspects of the first if non-human animals share with us only fragments of the second. Both humans and other mammals have overlapping kinds of consciousness and relate to each other in the shared domain. It is a mistake to assume that other mammals are exactly like us. Of course, they are not; but it is equally unlikely that they are not at all like us.

As the French say, "Vive la différence!" We are free to take advantage of the difference and celebrate relationships with pets that are always simpler and often more enduring than many of our human-to-human attachments. A more direct route to the same conclusion is the data indicating that man and other mammals share REM sleep. Whether animals dream or not is unknowable but they, too, have brain activation in REM. Their REM shares with us sensory integration and the anticipation of waking cognition and behavior.

Perhaps the most direct evidence for the sharing by man and other animals of consciousness components is the expression of emotion, recognized by Darwin as adaptive. It is the expression of positive emotion that bonds pets and their keepers. No one can doubt the recognition, greeting ceremony, and affection of dogs. No wonder that the dog is considered to be man's best friend. It seems preposterous to assert that bonding emotion is not subjectively perceived by non-human animals. This assertion is not proof but, as we will point out in the subsequent section, it is a good enough possibility for us to study animals in our quest for enriched and scientific self-understanding.

Animal Models for Consciousness Science

The technological advances which make human cognitive science so exciting today have still not eliminated the need for molecular and cellular-level neuroscience (because the resolving power of surface and depth recordings of electrophysiology and imaging analysis are still too weak to be sufficient). As the experimental work to be presented later will make clear, scientific progress would have been tragically curtailed had we not had animal models of human sleep.

None of our intellectual reservations about the theory of animal minds limit this biological power. Only humans can tell us about the subjective effects of brain activation but only animals can tell us about how deep aspects of physiological activation are instantiated. If dual-aspect monism is correct (and we think the scientific evidence for it is convincing), bijective mapping from the human mind to and from the brains of our animal friends is indispensable.

Foreshadowing this story, consider the memory differences between waking and dreaming. Are they due to dream repression (as Freud supposed) or are they due to state-dependent amnesia? Cellular and molecular-level studies using experimental animals decisively favors amnesia and helps us to forget about repression. The amnesia for dreams is physiologically mediated and we now know how. The why question remains to be answered more definitively but it is almost certainly not to preserve sleep as some diehard Freudians still maintain.

Dialogue 11. Are Non-Human Animals Conscious?

TH: What is it like to be a bat, a cat, or a rat?

AH: We can't answer that question any better than the one about the consciousness of human fetuses raised in the last chapter but we do know that all mammals, including flying animals like bats, have REM sleep brain activation.

TH: Why do mammals turn on their brains in sleep?

AH: As we have emphasized, they do this to ensure the temperature regulation that is essential to life. Here we consider the possibility that there may be cognitive functions that also depend on REM.

TH: Is this what the late Gerald Edelman called primary consciousness?

AH: Yes. Domestic animals are popular as pets because they share certain cognitive capabilities with us. These include sensation, attention, emotions, learning, and memory. According to Edelman, these are all elements of primary consciousness.

TH: Do animals need REM to guarantee those functions?

AH: We don't really know. Such functions, like memory, do not apparently suffer with sleep deprivation but many primary consciousness modules have not yet been carefully studied within the conscious states paradigm.

TH: As the REM sleep of the memory enhancement story already indicates, we may be in for some surprises here.

AH: Freud emphasized the primary process nature of dreaming by which he meant the primitive thinking and raw emotion that was often expressed in dreams. He also assumed that primary processes caused mental illness.

TH: In this sense, Freud anticipated protoconsciousness theory. However, his view that the dream expression of emotion functioned to safeguard waking consciousness contrasts with our current supposition that REM may function as a proving ground for emotion subserving the adaptive communication of emotion in the waking state that Darwin emphasized.

AH: Freud considered himself a Darwinian, but he was wide of the biological mark with his dream theory.

TH: Is there a corresponding lesson to be learned about what Edelman calls secondary consciousness and Freud's secondary process postulate?

AH: Yes, but the two viewpoints are diametrically opposed. For Freud, secondary process was considered to be defensive. For Edelman, secondary consciousness is creative.

TH: How can creativity depend upon a random process?

AH: Randomness is the soul of novelty. Many novel ideas are either false or useless, like Freud's dream theory. But others are true, like (I hope) thermoregulation and virtual reality cognition.

TH: How does one decide whether an idea is false or not?

AH: Science comes to the rescue here. The brain is a hypothesis generator but it is also an experiment organizer. Otto Loewi's dream, discussed in Chapter 9, is a case in point.

TH: If I understand you correctly, I should welcome disorder as an essential precursor of order.

AH: More than that. Artists have long celebrated the disorder of dreams. Surrealists, such as Salvador Dali, created image icons in honor of dreams as novelty machines.

TH: Can animals paint pictures?

AH: In Thailand, elephants can be trained to do so by timely reinforcement with peanuts but the human trainer really paints the picture.

TH: What about chimps? Don't they paint without training?

AH: Good question. Fred Snyder never told me if his pet monkey painted. The abstract expressionist, Willem de Kooning, painted pictures in his later years that were indistinguishable from monkey art.

TH: Wasn't de Kooning suffering from Alzheimer's disease at the end of his life?

AH: It could just be that a loss of cerebral cortex function releases the primitive forms that are the shared stuff of dreams and modern art.

Chapter 12. Deprivation

The discovery of REM sleep by Aserinsky and Kleitman in 1953 immediately triggered functional speculation: What was the purpose of dreaming? Because of the dominance of psychoanalytic theory at mid-20th century, the early studies by William Dement attempted to test the Freudian hypothesis that dreaming protected waking consciousness from invasion by repressed unconscious wishes. As we will point out in what follows, this idea was resistant to scientific proof or disproof and this lack of resolution has obscured evidence for the competing theory of dream consciousness as a preparation for wakefulness. Only empirical studies can decide this question.

In retrospect, an important reason for Dement's confusion was the identification of the physiological processes of sleep with the psychological experience of dreaming. There was, and still is, a regrettable tendency to equate rather than correlate the material and the immaterial levels of analysis. Because this tendency is so pernicious and so scientifically damaging, we describe here several examples of the error and emphasize alternative explanations of the dramatic findings.

Figure 12.1. An increasing number of awakenings is required to suppress REM sleep on successive nights of a deprivation program.

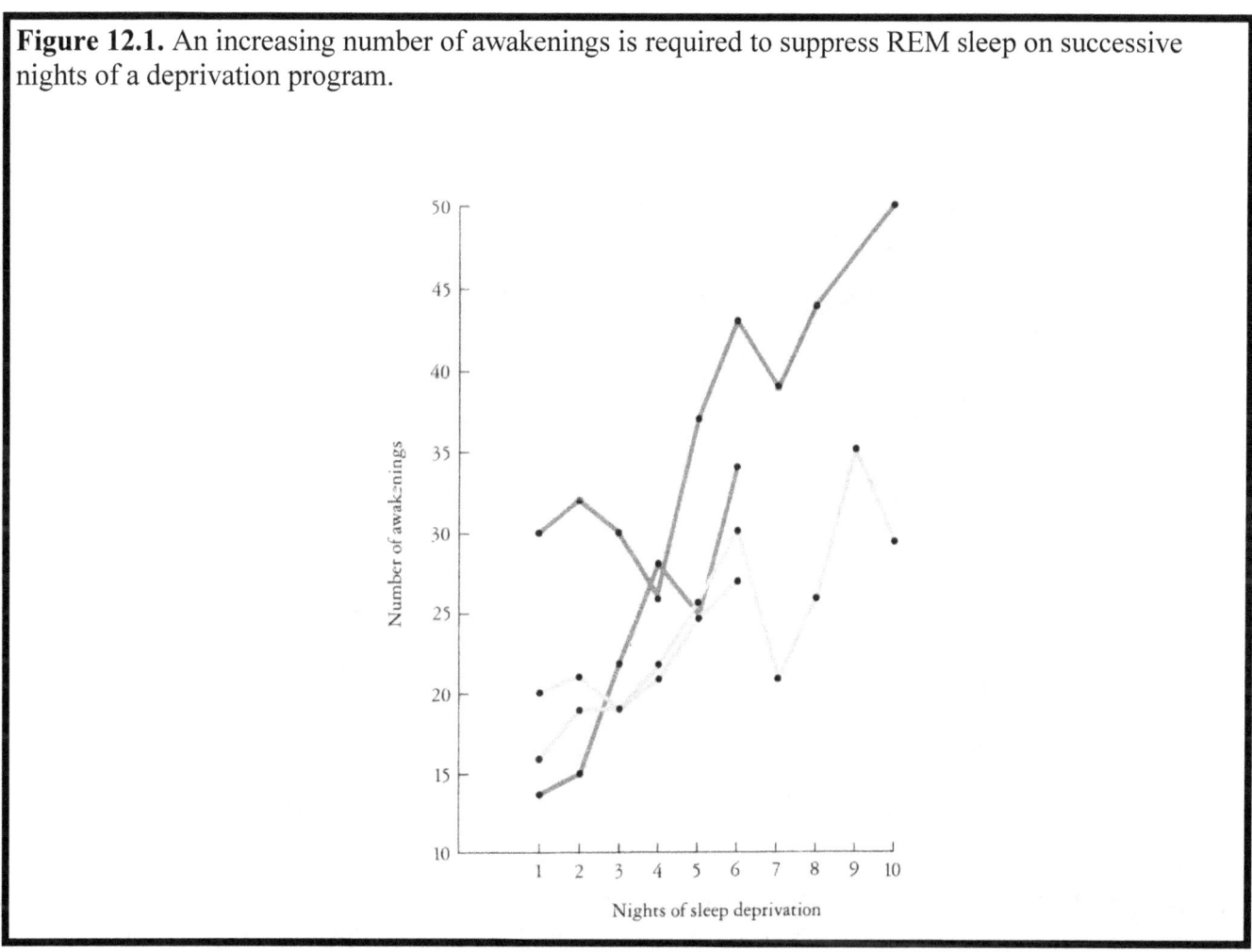

Dream Deprivation

After the discovery of REM sleep, psychoanalytic theory could finally be tested experimentally on the assumption that dreaming was essential to mental health. When William Dement awakened subjects from REM in the sleep lab, he observed that an increasing number of arousals were necessary on subsequent nights and that daytime waking consciousness suffered proportionately. Thus the cognitive benefits of REM were established but the psychoanalytic hypothesis was not confirmed. Anthony Kales later showed the same cognitive deficits were incurred by NREM sleep curtailment, indicating that it was sleep generally, not REM — and certainly not dreaming itself — that was essential to normal cognition.

What the Dement studies revealed was that REM was jealously conserved by the organism. This was a conclusion that was possibly justified in the cognitive domain, but it ignored other biologically significant adaptations revealed in the later sleep deprivation studies of Allan Rechtschaffen, Dement's colleague in the Kleitman lab. We devote a later chapter to these remarkable findings but wish to issue a warning that functional theories may be too narrowly construed. In other words, there is much more at stake here than mere dream theory. We confront a set of brain-mind phenomena that pertain to life itself as we reconstruct our analysis of REM sleep and dreaming.

The dramatic exigency of Dement's subjects to recover lost REM was a clue that REM was more than a mere vehicle for dreaming. So imperative was the "REM rebound" that it was literally impossible to deprive a human subject for more than about three nights. The power of REM rebound was equally evident in Michel Jouvet's cat experiments. The animal subjects typically evinced seizure-like motor behavior after three days of REM deprivation. All mammals strive to conserve REM as they strive to conserve life. This motive is well known to interrogators who exact fabricated confessions in exchange for permission to sleep.

William Dement (1928–) is a psychiatrist who understood the implications for dream theory of the Aserinsky and Kleitman discovery of REM. Immersed in the psychoanalytic hypotheses of Sigmund Freud, Dement pioneered the sleep lab awakening paradigm that occupied dream scientists for twenty years. When it became apparent that Freud's theory might be incorrect and that alternative models needed consideration, Dement had already become alert to the health implications of sleep and dream science. With Allan Rechtschaffen, he proposed that narcolepsy was a disorder of REM sleep control and, with Christian Guilleminault, he discovered that the cessation of breathing in normal males was exaggerated and complicated by airway obstruction in sleep apnea victims. His work thus spans the gamut of sleep physiology, from dream science to sleep medicine, of which he is the undisputed founder and world leader.

The Whole Body Beats to the Sleep-Dream Rhythm

Consciousness is an integral part of global organismic states. As we have already seen, many physiologic functions rise and fall together. With this principle, all scientists now agree, and this principle underlies the new science of sleep medicine. Cardiovascular and respiratory changes were emphasized in the beginning but many other systems, too numerous even to list, have since been added to the mix. Here we will select some that seem to us most relevant to the science of consciousness. The following discussion is meant not only to illustrate the multiple measures that can be made but to introduce functional theory: what survival purposes do sleep and dreaming serve?

At the top of our survey is the vital sign of body temperature. Body temperature peaks in the morning (when we are awake) and bottoms out at night (when we are asleep). Alert waking and obtunded torpor follow suit. In order to be awake, we need to be warm. When we are cool, we sleep and dream. It is thus a very surprising fact that when we REM and dream, body temperature control is surrendered by the brain to warm beds and nests. Yet our ability to control our wake-state temperature is dependent on REM. So we face a paradox: temperature control is abandoned in one state of consciousness in order to guarantee its efficacy in another. This shows that it is not mere rises and falls of a given function that we should be satisfied with. We need to look more deeply into the nature of things.

Motility can be real (in waking) or virtual (in dreaming). We run and walk in both states but one movement is actual, the other imaginary. Consciousness is usually unaware of this difference indicating both that they are very similar and implying that they may be related in as-yet poorly understood functional ways. One way may be procedural learning. Understood hyperbolically, we walk *virtually* in our dreams, the better to walk *actually* in our waking. Again, one state of consciousness serves another, this time positively rather than negatively as in the control of thermoregulation. The sleep scientists, Jana Speth and Clemens Frenzel, are investigating the effect of motor region brain stimulation on perceived dream movement.

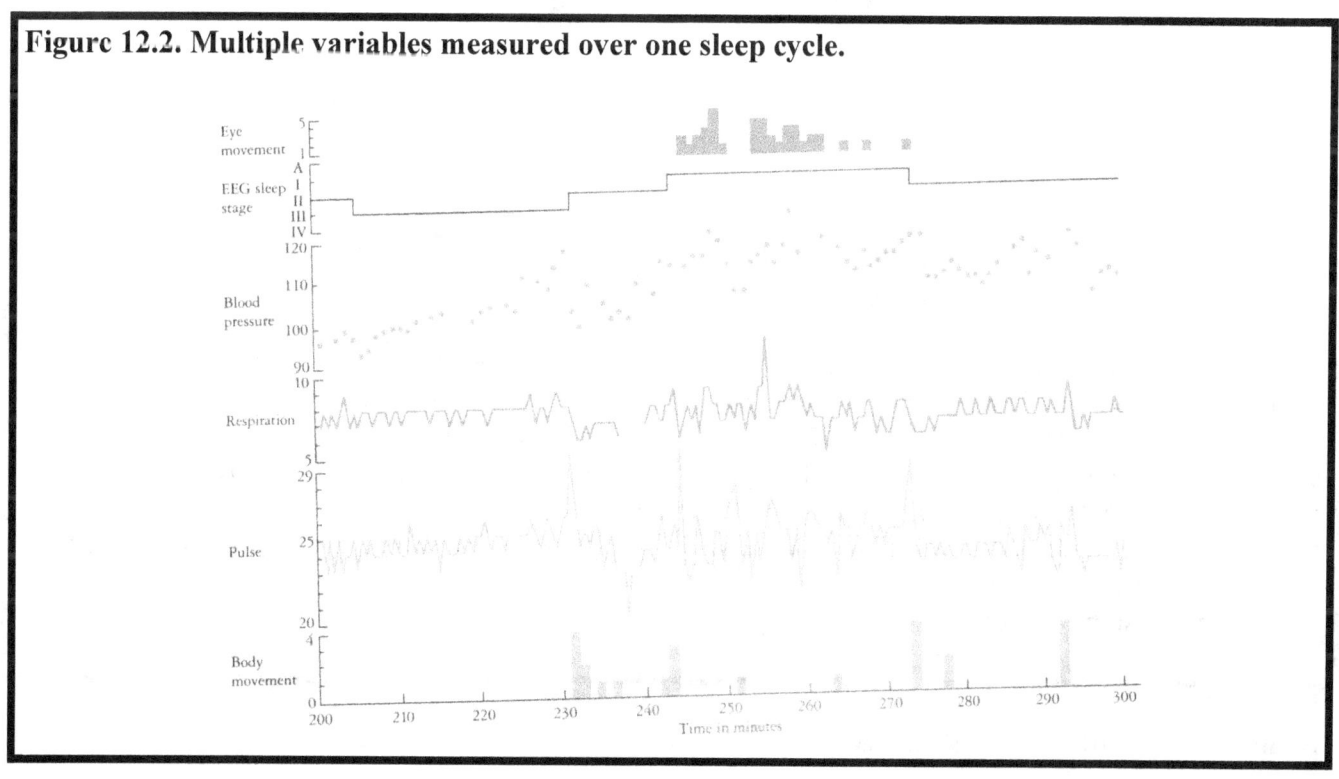

Figure 12.2. Multiple variables measured over one sleep cycle.

Why should men get an erection and women clitoral engorgement in REM whether or not their dreams are erotic (and, sad to say, they usually are not)? From a Darwinian perspective, we can theorize that sexual readiness is so crucial to procreation that we use the downtime of sleep to assure such readiness. But why then do we experience so few erotic dreams? Possible answers to this question are that the periphery of the body is cut off from the center in the brain and because a few erotic dreams is all it takes to sustain desire. Copulation and procreation are where it's at in evolutionary terms. Human men and women spend far more time courting than in the consummation of desire. Bonded pairs of all species are committed to offspring which guarantee propagation of the species.

Another great surprise is the almost exclusive early night release of growth and follicle-stimulating hormone (GH and FSH). When we are dead to the world, our bodies are primed to grow and our procreative capacity is enhanced. It has been estimated that at least 90% of the daily release of these hormones occurs in the first NREM (or SWS) period of the night. This hormone release is from Saper's flip-flop switch center in the hypothalamus. Obviously we need no awareness of this correlation and consciousness plays no part in the mechanism. On the contrary, it makes sense to have the most important biological functions occur automatically and reliably without mental interference.

Clifford Saper is a neurologist at Harvard Medical School, where he pursues his research on the neurobiology of sleep. Among Saper's important discoveries is the anatomical connection between the circadian biological clock in the hypothalamus and the sleep cycle oscillator in the pons. He has characterized the properties of these oscillators as "flip-flop" switches, an analogy borrowed from electronics to denote their tendency to rapidly change from one state to another. This quality is exactly what was earlier described for the pontine NREM-REM sleep oscillator on the basis of the single cell recordings. For the student of consciousness, this provides a physical basis for those rapid "changes of mind" that are so typical of our subjective experience.

Some Problems of Human Sleep Laboratory Science

Any one who has ever spent a night in a laboratory knows how difficult it is to sleep in an unfamiliar setting while a technician stranger attends your brain waves in an adjoining room. Female subjects are often made uncomfortable by a male technician, creating a gender bias in the sample. Males predominate. When a subject finally succumbs to sleep, he or she may be difficult to rouse, especially early in the night when NREM sleep prevails.

When awakened, sleep-deprived mathematics graduate students cannot subtract seven from one hundred, so powerful is their sleep inertia. Only in about half of the awakenings can a strongly motivated experimental subject give mentation reports of whose pre-awakening origin he is confident. Since most dream labs are in universities, the vast majority of subjects are undergraduates, greatly biasing the data set in the direction of youth.

David Dinges is the psychologist who conducted experiments on sleep cognition. Originally a University of Pennsylvania collaborator of the hypnosis expert Martin Orne, Dinges changed from the study of induced evanescent states in waking to the study of obdurately persistent states of sleep. He characterized his sleep-deprived subjects' difficulty in waking up as "sleep inertia." This moniker has stuck. It warns night workers and their employers to abandon their naïve assumptions about cognitive performance following awakening, especially in sleep-deprived persons (as many night workers are). Sleep is anything but evanescent; the brain is burrowing deep for its own good reasons.

The demand characteristics which mar most awakening studies are not completely eliminated by the deliberately falsified explanations of experimental intent often used to mislead subjects. If awakenings are performed, the subjects will know that dream reporting is expected and they will produce it, sometimes even by confabulation. Early in the night, the awakened subjects may be experiencing sleep inertia. They are temporarily disoriented and may be frankly demented; wishing only to go back to sleep, they then cough up what they rightly perceive to be the experimenters' target data, namely dream reports. The sleep deprivation that is induced may make it difficult for students to perform their day job, which requires them to stay awake in class, to concentrate upon their studies, and to learn consciousness science well. Hence the dropout rate is already high and will probably soon be higher.

Antidotes to these problems are considered in greater detail later but, for now, we focus on the desirability of the recruitment of younger and older persons sleeping at home. Consciousness science is of great import to everyone and more expert self-observers need to be recruited to normalize the database and make it more natural as well as broader-gauged. We suggest self-observation exercises at the conclusion of this chapter to make our point didactically clear and to include our students and colleagues in a cooperative research enterprise.

Sleep-deprived subjects complain that they can no longer focus on internally-directed cognitive programs in waking. We focus on disorders of attention in Chapter 15. Instead of concentrating, the attention of sleep-deprived persons often wanders and is seized by external distractors. It is as if consciousness were stimulus-bound rather than directed centrally by the subjects' own will. This tendency is particularly problematical for student subjects, whose school performance suffers critically. Wandering attention is a well-known handicap to student practitioners of late night cramming sessions. All-nighters may help a student pass tests, but they also guarantee the loss of learned material, since memory consolidation is dependent upon sleep.

Modern sleep science has demonstrated that even four hours of sleep curtailment incurs costs and that these costs are metabolic as well as cognitive. Eve Van Cauter has shown that sleep curtailed subjects develop the same glucose (sugar) intolerance as diabetic patients. Their blood sugar levels soar in the face of a dietary glucose load instead of retaining stable, low levels. This is variation on the theme of globally interactive co-variation of metabolic and cognitive functions discussed below. In order for consciousness to be effective, metabolic stability is required. The brain is not only exquisitely oxygen dependent but requires a regulated fuel supply as well.

Eve Van Cauter is a Belgian scientist who discovered that sleep functions to regulate energy metabolism. Working in the medicine department of the University of Chicago, Van Cauter curtailed the sleep length of her subjects, causing them to evince signs of diabetes. This experimental sleep curtailment diabetes was cured by recovery sleep. The nearly exclusive interest in REM as the substrate of dreaming is thus complemented by its evident role in energy regulation in the work of Van Cauter and her University of Chicago colleague, psychologist-turned physiologist, Allan Rechtschaffen. The wind of Lake Michigan fills the sails of modern sleep science and shows that parents are right to insist that their children precisely navigating the sea of slumber.

Why should internal medicine practitioners be interested in sleep? One answer is general: sleep is one-third of life. But more specific answers emerge from Eve Van Cauter's studies of partial sleep deprivation. Diabetic glucose metabolism may be induced by sleep loss. In other words, the energetic functions of sleep include proper management of the brain's most important fuel. The loss of cognitive ability with sleep deprivation is thus easily understood. The brain falls apart when it does not receive enough sugar. Eve Van Cauter has helped to make this point clear. Yet the persistent fantasy that sleep is a waste of time continues to delude the Thomas Edisons of our industrial world.

We can distinguish between the effects of chronic and acute sleep deprivation. Chronic sleep deprivation involves regularly not getting enough sleep — say five or six hours a night, when seven or eight are needed. This is the Van Cauter paradigm. Chronic sleep deprivation results in a range of physical and mental symptoms, including but not limited to muscle pain, tiredness, a general feeling of being unwell, headaches, obesity, and a clear decrease in mental functioning and alertness. Being sleep-deprived and tired can impair driving skill as much as being drunk. There is some evidence that chronic sleep deprivation can increase your chances of getting serious illnesses such as diabetes, heart disease, and psychiatric disorders.

In acute sleep deprivation, a person is deprived of all sleep for a short period. We have tried, in the interests of science, depriving ourselves of one night's sleep, and even that is not easy or pleasant. Particularly around the time of normal waking, we experienced nausea, extreme tiredness, and an inability to concentrate. Intrusive visual hallucinations are not welcome even if we recognize them as deprivation-induced.

The documented record period of sleep deprivation without any kind of stimulant is 264 hours (11 days) is held by the then student Randy Gardner of San Diego, CA, and established in 1964. Towards the end of this record attempt, studied by William Dement, there were clear behavioral and cognitive changes; although Gardner was able to engage in normal conversation, he was unable to maintain simple tasks, such as repeatedly subtracting seven from a number. He also became very paranoid. After completing his marathon, Gardner slept for nearly 15 hours the first night, and 10 ½ hours the second night. He very quickly recovered and returned to normal, without any apparent long-term effects, but human sleep deprivation is not popular for moral reasons. If sleep really does benefit cognition, how can we conscionably prevent it?

Whether or not it is dreams that are important (as Antti Revonsuo maintains in his threat-avoidance scheme), it is clear that the brain needs sleep and that waking consciousness deteriorates when it doesn't get enough of it. REM may well be super sleep, an idea that we will entertain in Chapter 11 when we consider sleep's necessity for life itself. Energy is every bit as important to consciousness as information as we will make clear when we flesh out the theory of predictive inference in later chapters. In summary, we have come a long scientific way since the early days of dream deprivation.

REM Sleep Eye Movement Direction and Dream Gaze

The association of REM with dreaming almost certainly goes further than mere state-to-state correlation. The visual intensity of dreaming suggests a specific contribution of REM sleep eye movement to the direction of hallucinated gaze. Again, it was William Dement who, with Howard Roffwarg, masterminded the first experimental test of this theory. Fitting dream reports with laboratory physiological data led to the *Scanning Hypothesis*, which asserted that the dreamer was looking at what he saw: the dream image was the stimulus and the eye movement was the response. Unfortunately, replication was not achieved by others who could not find even a relationship of eye movement to any aspect of dream vision. A functional implication of these studies was that REM sleep actively benefitted visual system development.

To enjoy success with this and any other psychophysiological study of dream consciousness, a scientist must work backward from the time zero of the experimental awakening. The first person experience and the third person behavioral event must be perfectly synchronized. This is a difficult, if not an impossible task. The time of a given shift of hallucinated gaze must be estimated from the verbal report. With which recorded eye movement is it to be correlated?

There are two related alternatives to this stimulus-response dilemma: one is to abandon the search for perfection and settle for establishing the correlation between eye movement frequency and dream vision intensity (this has been done); the other is to establish the correlation between eye movement direction and its central physiological correlates, the famous PGO waves (this kind of study is now possible thanks to the work of Charles Hong). An advantage of the latter approach is the reinstitution of the stimulus-response paradigm, but at a central rather than a peripheral level: one part of the brain (in this case the oculomotor system) creates the stimulus (an eye movement command) and sends it to the thalamocortical brain (the substrate of subjective awareness). Both alternative approaches have suggested that the brain-dream hypothesis is not just state to state but involves a more precise connection between subjective vision and its physiological substrate. We will return to this admittedly complex but important argument later.

To understand this key concept, it may now be helpful to imagine that such correlations must exist. In order for the brain-mind to evince consciousness, there must — repeat *must* — be a solid foundation in precise sensorimotor integration. In order for us to see anything, be it external or internal, the brain must determine the effect on visual image formation of visual search. To do so requires, as Helmholtz long ago surmised, a feed-forward system from the motor to the sensory side, an "efferent copy" as Helmholtz put it. We cannot see anything unless we move our eyes, and if we move our eyes we need to keep the image continuous. Sensorimotor integration is all about keeping the subjective experience of consciousness in order.

Hermann von Helmholtz (1821–1894) was a German physiologist with a strong interest in psychology. He was violently opposed to vitalism, the then-popular idea that a mysterious "life force" or "soul" animated matter. Instead, Helmholtz insisted, the mind is a physical force which could be understood without recourse to mysticism. His original insights included the theories of free energy and the brain's prediction of the consequences of sensory data collection. Helmholtz argued the body needed to keep free energy to a minimum in order to reduce the probability of surprise, and that the brain predicted perceptions and movements by internal computations. He was famous in his day for his rigorous scientific approach and was forerunner of today's effort to understand waking and dreaming consciousness without invoking non-material energies.

Post-Psychoanalytic Dream Theories

Many psychologists never took Freudianism as seriously as physician psychiatrists did. But as behaviorists and even as cognitive scientists, they often went so far as to deny that dreaming was significantly different from waking. This surprising position was also taken by some physiologists. We will state our reasons for disagreeing respectfully with these colleagues in subsequent sections but report here some of their views.

Continuity Theory (championed by such psychologists as William Domhoff, Michael Schredl, and Marino Bosinelli) holds that waking and dreaming are concerned with the same content themes. Moreover, the states are substantially continuous with each other. This view is supported by the common experience of the representation of yesterday's experience in tonight's dreaming. These informational leftovers are what Freud called "day residues" which he thought triggered the revealingly symbolic transformation of unconscious infantile wishes. For the non-Freudians, even the assumption that the immediately previous day's experience was predominant in triggering dream content was challenged by evidence that the peak incorporation of prior experience was 6 days before the night of the dream. Admittedly, continuity theory still holds whatever the time constant of incorporation.

It was also observed that many dream items, some highly charged with emotion, had no identifiable source in prior experience. Acceptance of continuity theory is linked to the need to explain dreaming as exclusively backward-looking, a manifestation of the usual direction of stimulus and response: any conscious event must be caused by some antecedent experience. This adaptive assumption seems reasonable: what possible use can waking or dreaming consciousness be if it is not shaped by the past? But is this the whole story?

A brain-mind doomed to repeating what it already knows is denied originality and creativity, undeniably useful aspects of consciousness. Some psychologists have shown that problem solving depends upon REM sleep dreaming, especially when thinking "out of the box" is required.

A motive of continuity theorists is to cast doubt upon physiologically random explanations of dream bizarreness and to emphasize the meaningful aspects of conscious experience, no matter how nonsensical some dream content may appear. Many psychological theorists are reluctant to give up the clinically appealing practice of dream interpretation. They have a vested interest in their theory. It sells. Again, there is no possible quarrel with continuity, as far as the theory goes. Many suppose it does not go far enough.

Randomness is anathema to rationalists but the recombination of disparate, loosely associated bits of cognition may depend upon discontinuity, not continuity, and upon randomness, not orderly predictability. Both continuity and discontinuity must co-exist. As with many scientific controversies, the truth is probably *both-and* rather than *either-or*. Other reasons for the expansion of our thinking about consciousness as invention as well as faithful reproduction will inform more versatile dream theories that integrate physiology, psychology, and philosophy. We need to move ahead beyond Freudianism, beyond behaviorism, beyond physiological nihilism, and beyond philosophical rationalism.

Dialogue 12. The Human Sleep Lab May Be a Cloudy Lens

AH: When you sleep in the lab, are you a valid scientific instrument?

TH: Scientists love laboratories but laboratories are not always useful or even reliable. I am afraid I could not fall asleep knowing that someone was watching me.

AH: This common fear frustrates dream science. It especially limits the experimental study of consciousness.

TH: Sleep is, by definition, an altered state of consciousness. But scientists rely on prompt awakenings from sleep to describe its vicissitudes retrospectively.

AH: Sleep inertia is a term coined by David Dinges to describe the persistence of sleep, especially following early night awakenings from the NREM phase.

TH: I don't know about you, but I sometimes take at least half an hour to wake up in the morning, especially if I have slept well. Is this sleep inertia, too?

AH: Very probably. David Dinges reported that graduate students in mathematics could not perform subtractions of seven from one hundred if their NREM sleep inertia was intensified by 40 hours of sleep deprivation.

TH: They were still asleep when they were supposed to be awake? That's bad news for students and patients who consult physicians at night as well as for sleep and dream scientists.

AH: The assumption that sleeping and waking have sharp boundaries is an illusion.

TH: This must be why many students become glassy-eyed even in my most stimulating lectures.

AH: Don't take it personally. They may just be sleep-deprived. Tell a better joke and you can get their attention back.

TH: This must be what you mean when you say that consciousness is dynamic: continuously graded, constantly changing in depth and focus.

AH: Consciousness is always dynamic: it is always a mixture of external and internal awareness. How else could it be as useful as it appears to be?

TH: How can science proceed in the face of these alarming facts?

AH: Flexibly, resourcefully, and deliberately. Sleep and dream science must use a panoply of methods to corner the central truth.

TH: Once again I feel daunted, and even a bit defeated, by what you say. How can students confront this complexity and not only learn but also play a scientific role?

AH: Students don't want to hear this because we sound like their parents, but the answer is simple: get a good night's sleep; say no to drugs; and study your own consciousness. Consciousness is your most powerful talent.

TH: Consciousness science can be used or abused, studied or ignored, taken apart and fitted to everyday experience. We can all begin by studying ourselves.

AH: The pinnacle of consciousness is awareness of awareness. We humans are uniquely qualified and morally obliged to try to understand ourselves. This is the essence of education.

Chapter 13. Field Studies

As indispensable as is the sleep and dream laboratory, there are advantages to the complementary approach of direct observation in the field. Among the advantages are: (1) the naturalness of observing phenomena in context (as, for example, seeing tigers in the savannah as against in a zoo); (2) more plentiful data (studies with sample sizes of thousands instead of tens); (3) greater efficiency (cost reductions of ten to a hundred fold); (4) a wider range of subject types (including people under the age of five and over the age of thirty); and, most important, (5) study of the brain via study of the mind. This chapter explores some of these advantages.

We begin with the last strategy on our list, study of the brain via study of the mind and take up the story of Otto Loewi's dream discovery of acetylcholine, alluded to earlier. Before we accept his claim to have dreamt of the crossed frog perfusion experiment, we would like to ask the following questions: What kind of dream was he talking about? The visualization of his dream-self doing the experiment (suggesting a REM sleep origin) or a thought conception (suggesting NREM mentation)? Other cues would be the time of night of the awakening, the clarity of recall, and other phenomenological details about the experience. Scientists can reliably determine brain state from subjective reports. Word length alone is a good predictor, with 50 words reliably separating REM (50 and above) from NREM (below 50). I strongly suspect that Loewi's Nobel Prize-winning dream was NREM mentation.

An important reason for raising these questions is to avoid the mystique of folk psychology. If we follow popular assumptions, we are tempted to carry over the idea that dreaming has prophetic and magical attributes that denote an otherworldly origin and a benevolent architect. This is not to say that Loewi's sleep mentation was not meaningful: it certainly was. But would he not have conceived of his experiment in waking within a day or two? The problem was on his mind and the solution just "popped out" as forgotten names often do when we put them in automatic memory search.

The same questions apply to the celebrated conception, by August Kekulé, of the ring structure of benzene (via his daydream or reverie of a snake biting his own tail, forming a circle). Like Otto Loewi, Kekulé was a scientist specializing in structural formulae who was puzzling over a question and he answered it when dozing on a streetcar. Was he not about to have this vision while wide awake? We all want to believe that dreaming is creative and aids incubation in problem solving, but, as scientists, we must always seek rational answers to our hypotheses about consciousness. In any case, both stories are anecdotal evidence of the value of field studies. They give impetus to research. Does sleep abet problem solving? How often do dreams suggest erroneous solutions to problems? Only systematic research has a chance of answering these questions.

Sleeping deeply and comfortably in a sleep lab is not easy, especially, as is increasingly often the case, the lab is in a hospital in the combat zone of a large city. Even in a peaceful setting, people are wary of being watched. As we have stressed before, female subjects are often anxious when monitored by male technicians and everyone is more comfortable when sleeping at home rather than abroad. It used to be that the first two nights of laboratory sleep were thrown away as unreliable; at today's prices, that comes to thousands of dollars and it assumes that subsequent nights are normal when they are more likely recovery from the deprivation and poor sleep caused by the laboratory situation.

The value of objective, third-person data is correctly stressed by sleep lab adherents. But most of the physical measurements can now be made by portable, miniaturized, or algorithmic proxy tools (see Nightcap section below). EEG, EMG, and EOG can be recorded in analog or digital form and replayed visually or printed out for subsequent quantitative analysis. Telephonic connections allow physiological and psychological data collection and automatic arousals from preordained times and brain states can be computer controlled, eliminating the need for technician and experimenter alike. These persons, too, can sleep restfully at home.

Increased breadth and size of sampling have been little attempted or achieved because dream laboratories have tended to be university-based. Professors study students, which greatly limits the observational window since both are unusual people. Home-based studies can now extend to important questions regarding consciousness, along the lines of: What is dreaming really like? Prediction: Different from what we now suppose. Is it visually the same or different from waking? What about other sense perceptions? Prediction: Different. Is it possible to describe in words? Prediction: No. Are men's and women's dreams more the same than different? Prediction: Yes.

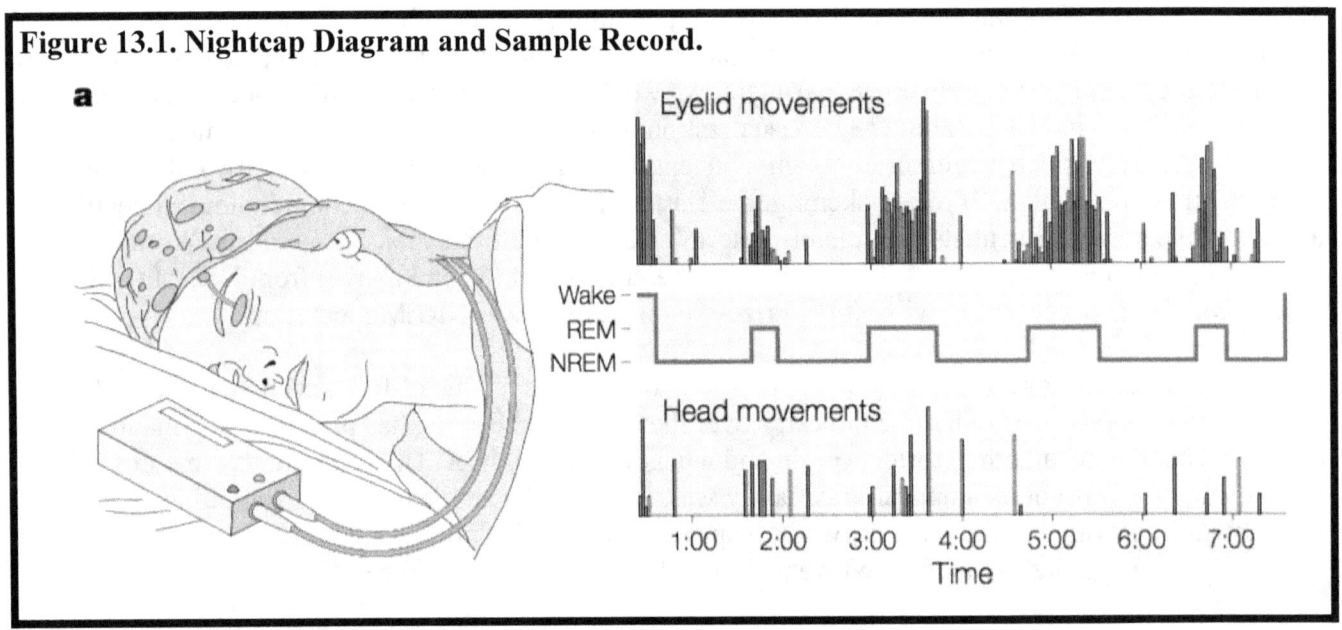

Figure 13.1. Nightcap Diagram and Sample Record.

The future of dream research in relation to consciousness study depends upon the strategic shape of conception and methodology. Only in this way can consciousness begin to understand itself.

Direct Observation

So sure were we that sleep was a loss of consciousness that we thought there would be nothing to see by simply watching it but, as Lucretius realized 2,000 years ago, there is a lot to see (and relate to the inside) from the outside. REM sleep is directly observable. We have had to rediscover it and should not be blind to our senses simply because we have high-tech devices to help us look within. The implications of this fact put consciousness research in the hands of everyone as this chapter, and the exercises suggested in it, are designed to make clear.

Begin by convincing yourself that the movement of the eyes is easily seen in your bathroom mirror and in sleep because the corneal bulge deforms the closed lid. Muscle twitches of the face, fingers, and limbs also announce the internal activation of the brain, suggesting that dream consciousness is present. This example shows behaviorism at its best because it ties spontaneous acts to their subjective concomitants in a way that you can comprehend holistically, that is, by integrating three levels of analysis: psychology, physiology, and philosophy.

REM sleep is particularly revealing to direct observation in the young because it so often occurs at sleep onset, is longer lasting, and is more vigorous (because the skeletal muscle inhibition is less effective at suppressing motor output than in mature animals). This is strikingly true of newborns. Nursing mothers may incorrectly assume that the decline in sucking at the breast and its replacement by smiling is an outward sign of an infant's gratitude. Mothers ignore the grimacing because they are looking only for positive reinforcement, not punishment, for their mothering. The recognition that adaptive emotional programs are being run never occurs to them.

For those of you who are lucky enough to have bed partners as accomplices, the early morning is an ideal time to experiment. There is enough light to see the abundant and vigorous REMs, feel the muscle twitches, hear the throttled vocalizations and, with permission, awaken your partner and elicit a report. It is almost as intimate as other bed sports. The discussion of dreams makes breakfast conversation lively and dreamers can censor unwanted candor. Cats and dogs make good, if nonverbal, subjects in which the eyelids can be opened without derailing the REM engine. The spasmodic display is activated and uninhibited motor output.

You can thus see REM behavior in bed partners, babies, pets, and other mammals in zoos and in the wild. In his charming book, *Le Château des Songes* (*The Castle of Dreams*), Michel Jouvet chides all of us for not seeing the visible REMs for, lo, these 20 centuries of neglect and mystification. Jouvet is particularly concerned that, since the dawn of science in the Renaissance 500 years ago, REM sleep went undiscovered, and he wonders what other phenomena of relevance to the science of consciousness are we now ignoring. Do subjects really close their eyes when wishing to concentrate their attention and thoughts? Does their gaze shift in one direction or another when they close their eyes to better reflect on a question? Are they more thoughtful and imaginative in the dark than in broad daylight? Do an experiment yourself and publish your findings.

Time-lapse Photography and Video

Complementing direct observation are the more tireless technical visual aids. Available since the time of Sigmund Freud and Eadweard Muybridge in the late 19th century, time-lapse photography (TLP) was not used to record sleep systematically before 1970, when Theodore Spagna began to photograph himself, his friends, and other fellow animals. His work is both aesthetically appealing and scientifically illuminating. The flash color images reveal both the peaceful repose and postural dynamism of human sleep as well as the decorated bedrooms in which these socially meaningful behaviors unroll.

Figure 13.2. Image from one of Ted Spagna's Sleep Photography Studies.

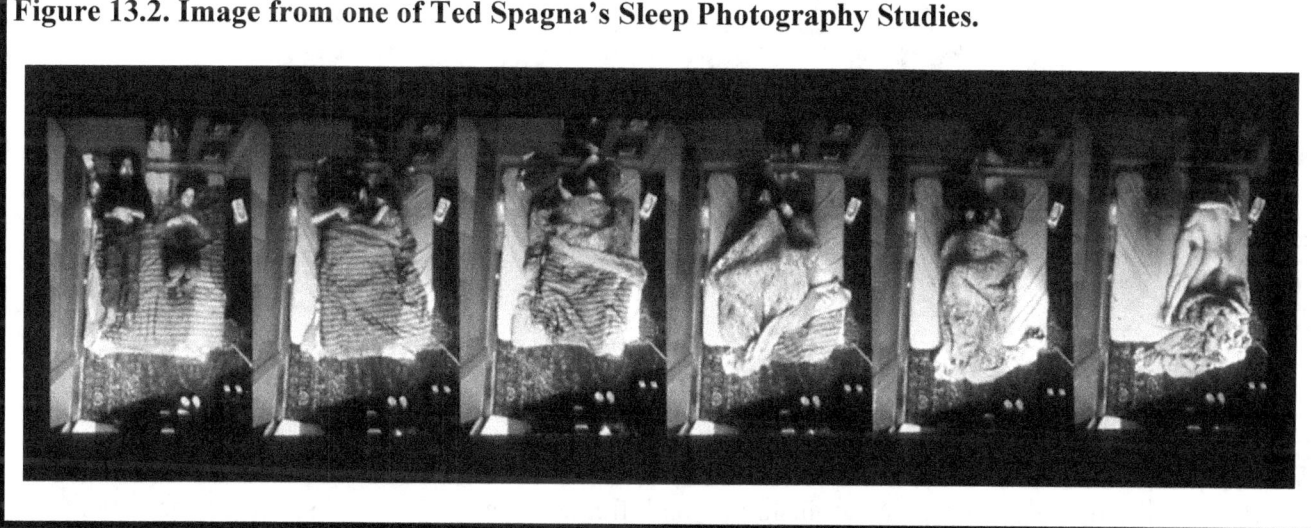

Even at the relatively long fifteen-minute intervals of Spagna's early photographic studies, it is possible to discern consistent and meaningful patterns: the first half of the night is more immobile than the second half (in keeping with the predominance of NREM sleep stages III and IV); posture shifts are periodic throughout the night and correlate strongly with the beginning and end of NREM and REM periods (a fact established by combining TLP with sleep lab recording). Posture shifts vary in number from 15 (2 per NREM-REM cycle) in good sleepers to 30 or more in insomniacs. Posture shifts can be reliably quantified with bedspring monitors, a cheap and easy way of measuring goodness of sleep.

Increasing the frequency of images in photographic, video, and film studies revealed more and more detail and greater temporal accuracy. The transition from photography to time-lapse video (TLV) was not only scientifically cogent (in studies of adults, children, and pets) but also economical (because TLV technology was underwritten by the huge security surveillance market). One particularly informative study showed that the sleep and waking behavior of mother-infant cats was highly organized, to produce a precise transition from REM sleep to waking motor behavior. This presumably automatic sequence suggested the protoconscious programming of the brain, which is the basis of the virtual reality theory of consciousness.

Figure 13.3. Cat REMs. Giovanni Berlucchi filmed the eye movements allowing successive ocular positions to be precisely measured.

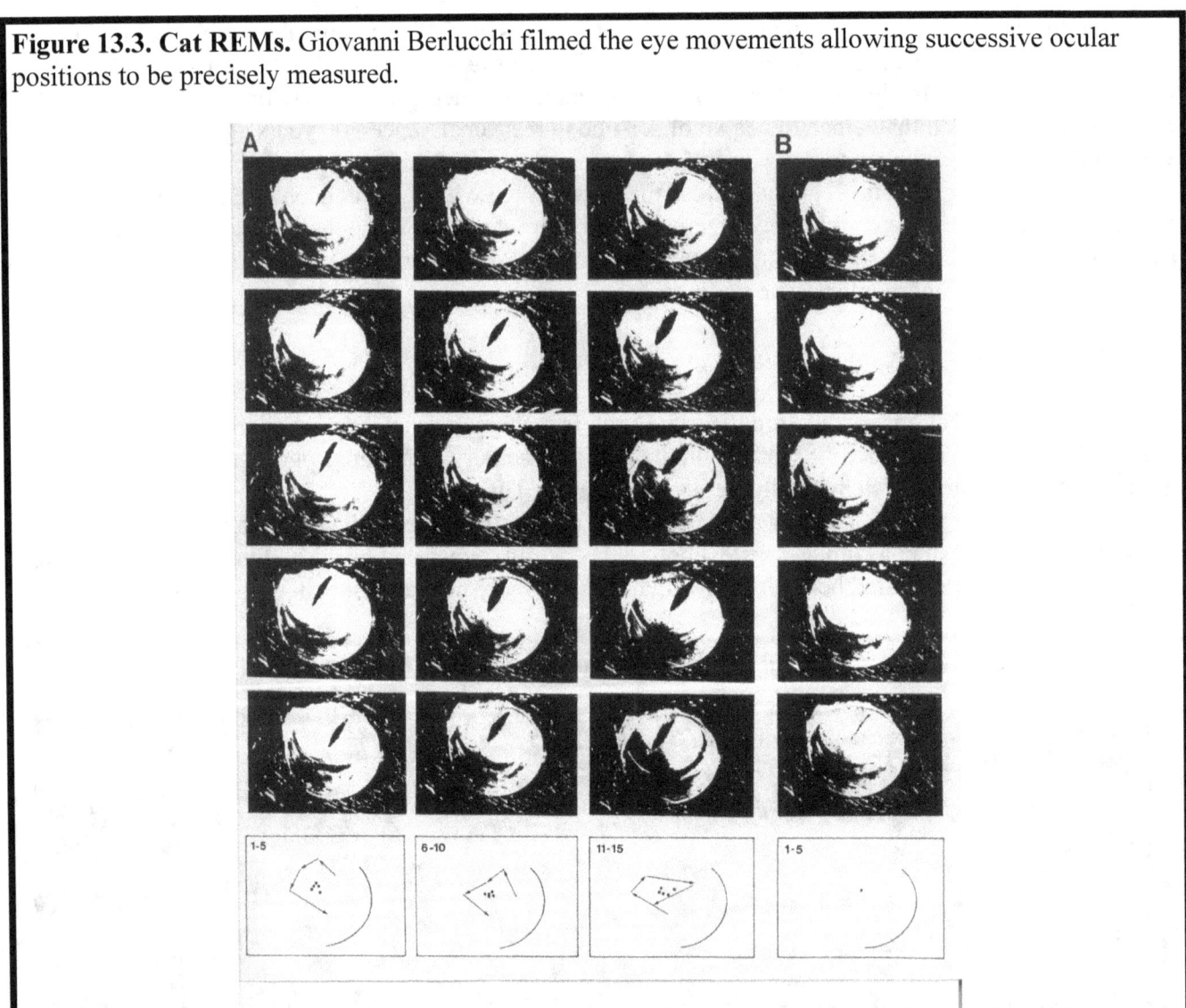

While a postgraduate student in Giuseppe Moruzzi's lab in Pisa, Italy, Giovanni Berlucchi filmed cat REMs. His film directly reveals the astonishingly high frequency and variety of sleep eye movements. An irresistible implication of this film (which could easily be replicated and extended to other mammals including humans using digital technology) is that REM constitutes a proving ground for the sensorimotor foundation of waking consciousness. This does not prove that the scanning hypothesis of dreaming is incorrect but introduces a broader functional context for that and other theories of dream consciousness.

The Nightcap, a Simple Sleep Recording System

Sleep can be objectified by taking advantage of the movement of the eyes (as filmed by Berlucchi) and the posture shifts (as photographed by Spagna in Figure 13.2). See Figure 13.1 for a schematic representation of the technique. These two movements create high-voltage outputs which are usually considered to be unwanted artefacts (or noise) but, in the Nightcap home sleep recording device, they are instead treated as desirable signals. Because they are of such high voltage, they swamp out the low-level EEG, EOG, and EMG measures of sleep collected in the sleep lab. But their high voltage makes them easy to record with little amplification and their fidelity to brain state reduces the necessary sampling rate many hundreds of times.

The posture shifts punctuate sleep and the eye movements positively identify REM. The movement-induced voltage changes can be digitized at the data intake stage. The net result of these advantages allows conscious state to be algorithmically diagnosed. Signals in both channels are associated with waking, and signals in neither with NREM; meanwhile, REM — like NREM — has no posture shifts, but abundant eye movements. The algorithm requires one estimate per minute and thirty nights of sleep data can be saved in a cigarette box-sized recorder which is battery operated. The sensors are easily self-applied so home sleep can be recorded without the need for a technician. Currently available iPhones could be adapted to home sleep studies.

Despite an appreciable loss of sensitivity and reliability, the Nightcap assessment of conscious state has 90% agreement with sleep lab recording. Ambiguous or uninterpretable data can be discarded, a small price to pay for the numerous advantages of this simple technique. As we will discuss later, report sample sizes of 3000 have been obtained in only two weeks using the Nightcap as physiological data source. A similar study in the sleep lab would take at least two years, an efficiency ratio of 365:1 in favor of home-based recording.

A distinct disadvantage of the Nightcap, in its current industrial configuration, is the unacceptably high cost of disposable eye movement sensors. Improvement in the device can and should be made by non-commercial bioengineers. Meanwhile, the proof of concept is undeniably promising and worthy of application in a wide variety of settings that have an interest in consciousness, including clinical and experimental psychology, psychophysiology, and sleep medicine laboratories.

Self-Observation

One's own consciousness has long been an object of theory and experiment. With the decline of organized religion and the increasing ease of access to psychoactive drugs, this tradition has assumed a new form. Meditation is a potent means of altering consciousness by withdrawing from the domination of the world to the relative peace of inner contemplation. It may be a provocation to suggest that meditative states are more often touted as panaceas than described dispassionately. Insofar as this is true, transcendental meditation is a secular religion. Its proximity to Buddhism proves this point. We could do better by adopting a scientific stance and studying meditation more objectively without destroying it with skepticism.

The study of dream consciousness provides an inspiring example. Describing dreaming without hasty interpretation is as honorable as it is entertaining and this chapter provides guidelines that can lead to the scientific contemplation of one's own mind. The conscious mind is a theater of endless fascination and a bottomless font of insight. Admission to this nightly theater is free and open to all. Every student of consciousness should consider self-observation as an integral part of education in scientific humanism. How can I overcome normal dream amnesia, you might ask? Otto Loewi has already answered this question: pre-sleep autosuggestion and diligent transcription. Before going to sleep, a dream collector tells himself to become aware of dreaming and to awaken when that awareness occurs. The recognition of dreaming while dreaming is the crucial step toward lucidity, a special substate of consciousness to which we will return. If the dream catcher does not want to turn on the light, to write on his bedside journal, or tape record his report, he has only to lie in the dark, as awake as possible, and repeat his narrative to himself for transcription in the morning. It is also now possible to buy digital voice-activated recorders cheaply or download inexpensive or free apps providing such functionality on smartphones.

Acolytes of scientific humanism record their recollections in a dream journal and very soon have a large collection of self-observations for their own reading pleasure or scientific analysis. Surprisingly, even well-remembered and detailed reports are soon forgotten, as if dream recall were not a very high priority for conscious memory. What may be important about dream consciousness and the sleep that underlies it is the process and not the product of that process. This same technique can be applied to waking consciousness, its process, and its product.

Very little scientific attention has been paid to fantasy. Fantasy lies in the borderland between waking and sleep. This borderland is occupied by mindfulness meditation but differs from it in being a more continuous background stream of review and rehearsal of the foreground content of waking awareness. Fantasy is as evanescent and difficult to capture as dreaming itself. Are fantasy and dreaming continuous processes, differing only in degree? This question is difficult to answer but scientists should try harder since high-tech approaches must only complement low-tech self-observation.

Figure 13.4. Bicycle built for two drawing from the Engine Man's dream journal.

Scientific Humanism

The philosophical position of *scientific humanism* undergirds the goal of self-observation. This philosophy asserts that consciousness can and must learn to study itself. It is self-contained but avoids solipsism by resort to third party and to the objective data of brain science. Scientific humanism is open to any and all self descriptions including diverse political, religious, and moral accounts. Echoing the American psychiatrist Harry Stack Sullivan, science states that "nothing human is alien to me."

Humanism is the spirit of the Renaissance, which began to shift emphasis from God, spiritual immortality, and other worlds back to mortal man and this world. It takes up a tradition that goes back to classical Greece, a tradition which was usurped by the imperial Romans and forced underground for the millennium of the Dark Ages by the mystical Christians. Humanism does not dispute the cosmological yearnings of man but ascribes them to social and evolutionary adaptations arising from the living brain and its conscious states.

John Antrobus (1933–) and his collaborator Jerome L. Singer pioneered the study of fantasy. Antrobus went on to a distinguished career in sleep and dream research. Skeptical of the claims of activation-synthesis dream theory, Antrobus claimed that waking mentation was every bit as fragmented (and therefore just as bizarre) as dreaming. His earlier work on fantasy is cited by continuity theorists who see a flow of unconscious mentation across all states of consciousness. John Antrobus was not only a distinguished scientist but also a whimsical artist whose sense of humor was evident in his collage assemblages.

Scientific humanism began to emerge from Darwinism and Herbert Spencer's social construal of evolution in the second half of the 19th century. Scientific humanism continues to grow. While still traditionally spiritual, it has become more earthly and experimental through the transcendental idealism of Ralph Waldo Emerson, Henry David Thoreau, and William James. At the same time, humanism is rendered neurological by John Hughlings Jackson. These lines of protoscience culminate in the turn-of-the-twentieth century research of Ramón y Cajal and Sherrington. After the modern Dark Age of behaviorism and psychoanalysis, scientific humanism, as we now define it, was born again in the mid-20th century.

Today, some seventy years later, brain physiology, experimental psychology, and philosophy unify and march together under the banner of scientific humanism. Superficially atheistic, it does not deny mysticism but supposes, as Sigmund Freud asserted in 1928 (the same year that Hans Berger discovered the EEG), that religious mysticism may well be a genuine and deep human trait but it is an illusionary scientific tool. Contentious criticism aside, there is no obstacle to adding scientific humanism to whatever creed you choose. The Japanese are admirably non-sectarian; they respect and practice Buddhism as well as Confucianism. We should all be so inclusive.

Scientists are all prone to mysticism. That is a fact. But should we seek scientific truth about the origin and functional meaning of consciousness through mysticism? The question is rhetorical, but our answer is: do not exclude scientific humanism, whose growth is palpable and whose answer is clear. The proper study of humanity is being human.

Dialogue 13. Mirror, mirror on the wall...

TH: …who is the fairest of them all? So asked the wicked queen, Snow White's nemesis, in the Walt Disney animated film classic.

AH: Snow White was produced in the 1940s, before REM sleep was discovered, but the mirror can be used today to recognize REMs outside the laboratory.

TH: I talk to my face image when I shave every morning. I won't tell you what I say but it is not, "You, oh queen," the queen's reassuring answer to her own question about beauty.

AH: REM recognition is less engaging but more instructive than a beauty contest: close one eye, hold its lid down, and see the eye move from side to side as the corneal bulge travels to and fro under the closed lid.

TH: I have seen this phenomenon in REM sleeping bed partners in the early morning light.

AH: You have no doubt also perceived the body twitches that accompany the REMs of your bed partner. You may have even heard the muffled cries when your sweetie dreams that she is shouting.

TH: Sometimes, her muffled vocalizations do tip me off to REM so I can elicit REM sleep dream reports, even in the dark.

AH: When I wake up spontaneously with vivid dream recall, I can be fairly sure it came from REM by both its bizarreness and by my persistent erection.

TH: *Pace* Yogi Berra, you mean that its not over when it's over. Some REM features can characterize the waking that follows dreaming.

AH: This is a variation on the theme of brain state inertia we talked about in the previous chapter. In this case, the dream lens is crystal clear, not cloudy, and the imagery is helpful to cognition, not a hindrance.

TH: Taken a step further, dreaming may not only prepare us for waking but contribute to its imaginative potential. Most people certainly believe that dreaming is linked to creativity. Otto Loewi is a famous example.

AH: That's because there are plenty of stories about it, but as you know, science cannot proceed by anecdote alone. We need to track creativity around the clock in a large, prospective experimental study.

TH: Good luck in getting that one funded.

AH: It would cost much less than an intercontinental ballistic missile that is designed to destroy both a few evil and many innocent people.

Chapter 14. Reciprocal Interaction

We now consider what we know about the dual-aspect monism account of consciousness. We begin by summarizing the cellular- and molecular-level analysis of conscious states and, in the following chapter, show how that science fits with a quantitative analysis of the subjective experience of those states. In other words, we tie as best we now can the deepest level of the bottom-up physiological observations to the top-down data of psychology. An obvious limitation of this endeavor is the necessity of relying on animal data for cellular and molecular details while it is the human mind that we most want to understand.

This integration focuses mainly on dreaming but points the way forward toward waking, which is now accessible to investigation because of recent and revolutionary increases in neuropsychological power. Textbooks often restrict themselves to what is securely known or to widely shared opinion. Because this field is so new, we present an approach that is novel and, insofar as it is noticed at all, controversial. In fact, many scientists might say that ours is mission impossible.

It may seem ingenuous to invite students to learn material that is clearly the product of work in progress but we are convinced that the prospects for future work are bright and that the subject is certainly central to philosophy, psychology, and physiology, three fields that have been separated at the expense of each of them. We thus hope that this book will inspire as well as educate, and aim to indicate career opportunities as well as to certify intellectual competence.

Single-Cell Recording

The neuron has been seen as the fundamental structural building block of the brain since Cajal's enunciation of the neurone doctrine around 1890. Yet the ability to record from *single units* (individual neurons) in unanaesthetized animals was only realized in the 1960s. After Ralph Gerard showed that this was possible, scientists began to link their study of neuronal activity to sleep, in part because they thought that easy answers would be obtained and that the answers were of relevance to neurology and psychiatry. They were proved right about the point of relevance but wrong about how easy it would be to clarify the point.

A half-century later, it is now difficult to realize how widespread was the assumption that the loss of consciousness in sleep was caused by a cessation of brain activity. Dream activity was thought to be incidental to waking up. Sigmund Freud believed these myths and so did Ivan Pavlov, B. F. Skinner, and Frédéric Bremer. This colossal mistake is a product of inappropriate reasoning from first-person subjectivity to third-person objectivity and constitutes strong evidence in favor of the practice of, not just the belief in, dual-aspect monism. Today, we suspect that many people succumb to equally erroneous superstitions and we freely admit that we might be among them.

David Hubel, like so many other great scientists, thought that sleep was caused by the brain turning off. Working with Herbert Jasper at the Neurological Institute of McGill University in Montréal, Canada, Hubel developed a micromanipulator that enabled him to probe the cat cortex while his animal subjects slept. To Hubel's astonishment and consternation, as many cat cortical cells increased as decreased their firing rate at sleep onset and none turned off altogether. It was obvious that Hubel's working hypothesis was wrong. Although the mean firing rate of neurons did decrease significantly, brain activity was never quiescent, indicating that it was continuous and that sleep involved the reorganization, not the cessation, of brain activity.

Moreover, Hubel found that in REM sleep brain cells resumed and sometimes exceeded their level of firing in immobile waking. That clinched the reorganization hypothesis. Edward Evarts improved the micromanipulator in preparation for the study of neuronal activity in the motor cortex of monkeys. At the National Institute of Mental Health in Bethesda, Maryland, Evarts also overcame the formidable quantitative and statistical obstacles that were not required by the erroneous brain activity cessation idea. Using these techniques, Evarts was able to suggest the first specific mechanistic hypothesis of the neuronal activity of REM sleep vs. waking.

The Evarts proposal that recurrent inhibitory interneurons were silenced in REM was invalidated by Mircea Steriade at Laval University in Québec, Canada. Steriade, who by recording inside brain cells, (not just outside them as Hubel and Evarts had done), took the science of consciousness to a new level of sophistication when he demonstrated that many of the cellular changes in waking vs. NREM and NREM vs. REM could be explained in terms of thalamic gating and aminergic modulation. As we have already hinted in Chapter 10, the stage was set for a testable model of the brain basis of conscious state determination. Brain activity was continuous in all states, but with what aspects of subjective experience were the changes in neuronal activity linked?

Edward Evarts (1926–1985) was a psychiatrist-turned-neurobiologist who championed sleep and dream research as scientific avenues to a deeper understanding of the mind. Best known for his technically and conceptually elegant studies of the cortical neurons involved in movement, Evarts used sleep dependent changes in the organization of motor control to elaborate the first specific cellular level theory of conscious state dependency. Motor control is an essential component of the brain basis of consciousness. We move with unconscious sensorimotor integration in waking to achieve behavioral goals. These waking movements depend upon a brain model of movement that is established in sleep. Our dream movements are the subjective manifestations of sleep practice and rehearsal of this otherwise unconscious sensorimotor integration.

Brain Stem Neuronal Recordings

As early as 1962, the work of Michel Jouvet had indicated that whether a brain was awake, or in NREM or REM sleep, was a function of the brain stem and especially its pontine subdivision. Jouvet's brainstem localization invited cellular investigation. This invitation was answered by research groups in Lyon, France and in the US at UCLA, the Salk Institute, and Harvard Medical School. At Harvard, the quantitative analysis of dreams was matched by a quantitative analysis of neuronal activity in relation to REM sleep.

Having overcome formidable technical obstacles, the Harvard team succeeded in recording neurons throughout the pons and classified and measured their properties as follows:

REM-on cells: The highest selectivity ratios for discharge in REM sleep compared to waking were found in the medio-pontine reticular formation. They were therefore accorded tentative REM generator status.

In keeping with their hypothesized generator status, the REM-on cells fired in intense bursts prior to the eye movements of REM sleep. They projected locally to oculomotor neurons, as would be expected if they were part of an attentional orientation system.

The REM-on cells were large, rapidly conducting neurons which projected directly to the spinal cord, where they could mediate the muscle twitches of REM by synapsing directly on spinal motor neurons. They might thus serve to integrate posture with eye movement.

The REM-on cells were found to alter their discharge rate continuously throughout the sleep-wake cycle suggesting that their excitability was modulated by some intrinsic or extrinsic source.

Figure 14.1. REM-off and REM-on cells in a descent through the locus coeruleus to the giant cell region of the PRF.

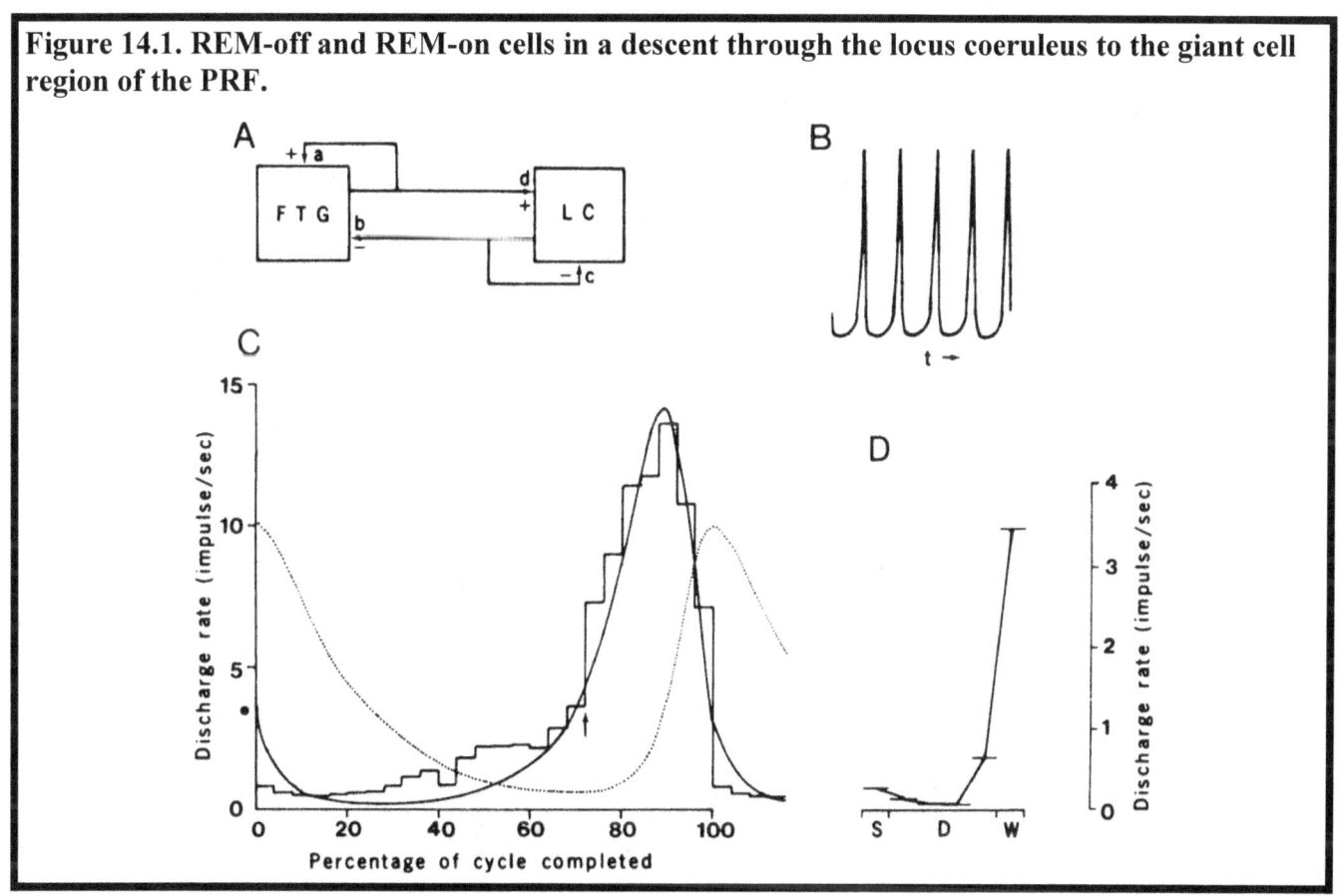

REM-off cells: The lowest selectivity ratio for discharge in REM compared to waking was exhibited by neurons of the noradrenergic locus coeruleus and serotonergic raphé nuclei. Norepinephrine and serotonin were thus candidate sleep cycle modulators.

REM-off cells fired continuously at low, regular rates during waking, when REM-on cells were often silent in the absence of body movement.

While REM-off cells occasionally fired in bursts in association with the REMS, they tended to be silent during REM.

The decrease of firing by the REM-off cells began at sleep onset, reached a low point in REM, and began to increase to peak waking levels before the end of REM.

In summary, REM-on cells are the best candidates for a generator role while REM-off cells are the best candidates for a modulatory function. Their respective roles suggested a model for sleep-cycle control with strong implications for dream theory.

The Reciprocal-Interaction Model of Sleep-Cycle Control

The observations presented in the preceding section prompted McCarley to consider a formal model of reciprocal interaction between the two cell groups that distinguished themselves as REM-on and REM-off neuronal populations in the pontine brain stem. As the name implies, the basic idea of reciprocal interaction was that two neuronal populations were in continuous and balanced opposition. When one group was on, the other was off and vice versa. They were like the paired half-centers that Thomas Graham Brown had postulated in 1912.

In addition to the reciprocal rate curves, there was evidence of reciprocal excitation and inhibition, mediated by acetylcholine and the aminergic neuromodulators, respectively. In waking, the brain was under the dominant influence of serotonin and norepinephrine, which held acetylcholine's influence in check. When the aminergic brake was removed in sleep, acetylcholine excitation ran wild and took over the brain. Dreaming might well be the result.

One attractive aspect of the reciprocal interaction model was that it specified not only the physiology of the neurons that were involved but the chemistry by which their interaction was effected. That added modulation (M) to the activation (A) and input-output gating (I) factors that had already been known to be operative in mediating the shift from waking to REM. The three factors activation (A), input-output gating (I), and modulation (M) lent themselves to the elaboration of reciprocal interaction as in the AIM model that was described in Chapter 9.

Figure 14.2. Volterra-Lotka equations.

$$\frac{dx}{dt} = \alpha x - \beta xy$$

$$\frac{dy}{dt} = \delta xy - \gamma y$$

Meanwhile, back at his computer and mathematics formula library, McCarley recognized that any dynamic system with properties of reciprocal interaction would obey the rules of the *Volterra-Lotka* formulae for the cyclical fluctuations of prey and predator animal populations known to field biology. Volterra and Lotka were two mathematicians hired by the French government to explain why populations of lynx, the provider of valuable fur pelts, fluctuated so dramatically from year to year over a timespan of years. It turned out that the number of predator lynx was a function of the number of rabbit prey. When one species flourished, the other diminished and vice versa. In a boom year for rabbits, the lynx can flourish; if the rabbit populations crash as the lynx over feed, then there isn't enough food for the lynx and their population falls accordingly.

Figure 14.3. Schematic representation of the process of REM sleep generation. The network is represented as comprising three neuronal systems (aminergic, reticular, and sensorimotor) that mediate rapid eye movement (REM) sleep electroencephalographic (EEG) phenomena (right). The actual synaptic signs of many of the aminergic and reticular pathways remain to be shown and, in many cases (such as thalamus and cortex), the neuronal architecture is far more complex than is indicated here. III, oculomotor; IV, trochlear; VI, abducens; AHC, anterior horn cell; BIRF, bulbospinal inhibitory reticular formation (for example, gigantocellular tegmental field, parvocellular tegmental field, magnocellular tegmental field); CT, cortical; LC, locus coeruleus; LDT, laterodorsal tegmental nucleus; mPRF, meso- and mediopontine tegmentum (for example, gigantocellular tegmental field, parvocellular tegmental field); P, peribrachial region; PGO, ponto-geniculo-occipital; PPT, pedunculopontine tegmental nucleus; PT cell, pyramidal cell; RAS, midbrain reticular activating system; RN, raphé nuclei; TC, thalamocortical.

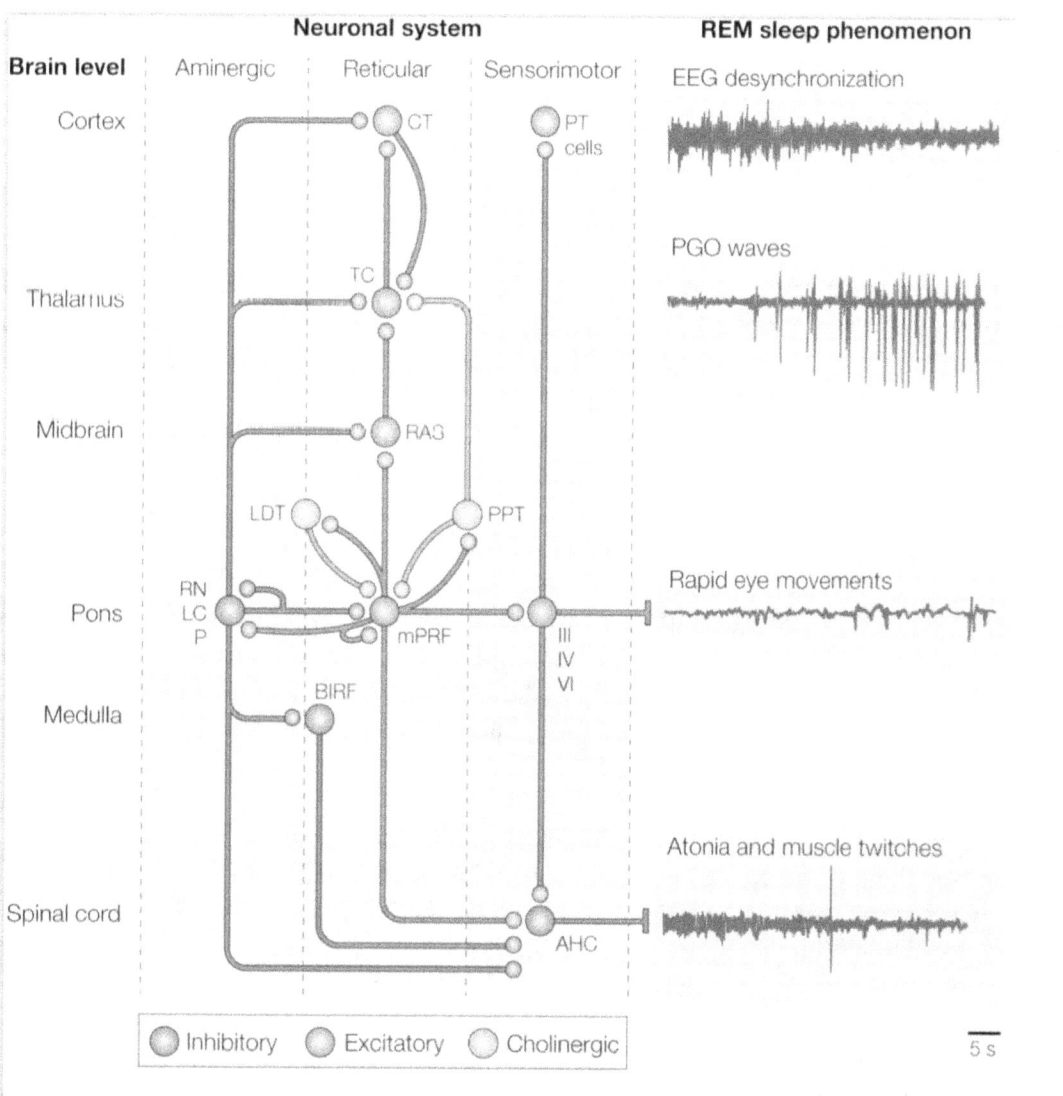

In this model, the probability of REM sleep might be seen as the peak number of prey rabbits, while waking is most probable at the peak number of predator lynx (and vice versa). Whatever the most understandable analogy might be, the math was expressed as two simple differential equations which McCarley used to fit to the experimental data, the rate curves of the REM-on and REM-off cells. The fit was good and subsequent research has made good use of the model to explain the variable time constant of the interaction.

In addition to providing a simple test of the plausibility of the reciprocal-interaction concept, the structural and physiological model created a platform for ten years of hypothesis testing using primarily pharmacological experimentation. The most important result of this work was affirmation of the cholinergic theory of REM sleep generation and the role of REM-on cells in mediating the effects. While not themselves cholinergic neurons as the model originally assumed, the REM-on cells were the recipients of cholinergic excitation which turned them on as part of the REM generation process. REM sleep (and human dreaming) were thus brought under experimental control.

Resistance to Reciprocal Interaction

Physiologists Jerome Siegel and Dennis McGinty of UCLA at first resisted the claims of REM-on cell selectivity because they found abundant cell discharge in waking (in free moving cats, which was not seen in the Harvard group's head-restrained subjects). They claimed that REM-on cell selectivity was an experimental artefact. This criticism did not obviate the cogency of the reciprocal-interaction model to which McGinty had earlier contributed through microwire demonstration of raphé REM-off cell discharge. Rodolfo Llinás, a neurobiologist of the New York University Medical School, opines that REM sleep dreaming consciousness is waking minus sensory input. In other words, like Antrobus, Llinás subscribes to activation (A) but adds input blockade (I) to the mix.

The Centrality of Reciprocal Interaction

Just as the brain stem is central to other parts of the brain, so is reciprocal interaction central to an appreciation of how the whole brain-mind system might conspire in the creation of conscious states such as waking and dreaming. The development of the activation-synthesis model, culminating in AIM, proceeded by showing how the brain stem could influence the forebrain (above) at the same time that it influenced the spinal cord (below). By means of its connections and chemical signals it was ideally suited to command all conscious states.

A schematic representation of these influences is visible in Figure 14.3.

The whole brain is orchestrated by reciprocal interaction such that waking and dreaming consciousness may be generated and differentiated. Both are characterized by activation of the forebrain; this is effected, in both states, by reticular-formation excitation of cortical neurons (among others). Consciousness is thus turned on. This is activation, Factor A of AIM. In the figure it is seen as the anterior portion of the central column.

In both states, eye movements are executed, again as motoric responses to brain stem command. In waking we can see the outside world because information is admitted to the brain. In dreaming we can see only our own internal stimuli. These are produced by the PGO waves visible in the figure. Muscle tone is simultaneously inhibited, suppressing movement. These variables can be measured and expressed as Factor I of AIM. They are visible in the central and rightmost columns of the figure.

The psychologies of waking and dreaming consciousness are different not only due to the changes in Factor I. They are also determined by modulation, Factor M of AIM, shown in the leftmost column of the figure as the unifying effect of aminergic neurons, again of brain stem origin. These are the REM-off calls of reciprocal interaction. When they are on, we wake. When they are off, we dream.

Revisions of Reciprocal Interaction

The simplicity of the original model has been modified in the light of subsequent research. These alterations are schematized in Figure 13.4. Of course the physiology of the brain stem is far more complex that we knew in 1975 but the basic concept of reciprocal interaction is unchallenged by complexification. The person responsible for the schema seen in Figure 13.4 is our colleague Edward Pace-Schott. Ed has developed a critical review of the neuropsychological findings and published it with Cambridge University Press. Among the important insights are the roles of glutaminergic neurons on the excitatory side of the model and gabaergic neurons on the inhibitory side. These modifications help us understand how the brain creates the mind and how the widespread parts of the brain work together. In other words, consciousness is the product of the entire brain, not just one part of it. Consciousness varies in intensity, focus, and content by virtue of the interconnection of those parts.

Dialogue 14. Modeling is an Exciting Game

TH: When I was a boy, I made model airplanes. Now I try to model sleep and dreaming with a view to understanding consciousness.

AH: Slim women model fashion designs on runways, architects model building designs in drawings and scientists model everything from economics through neurophysiology to space/time.

TH: Modeling is a good example of secondary consciousness at work.

AH: Infrahuman animals are not impressive model builders.

TH: Do mathematicians dream of writing their waking formulae on imaginary blackboards while they sleep?

AH: I don't really know because I never asked but I doubt it because I never even pay money or make change in my own dreams. I have a linguist colleague, Andrea Moro, who calls this dream trait "acalculia." He wonders if we dreamspeak or if we are aphasic, as Emil Kraepelin, the father of descriptive psychiatry, asserted. With collaboration and funding, we can find the answers to these interesting questions.

TH: In waking, models are made by scientists to test the logic and validity of their concepts about conscious states.

AH: Robert McCarley was the first to do this for the NREM-REM sleep cycle but he used mathematical formulae that were a century old and created to explain variations in fur pelts from prey and predator animal populations in Canada. The circadian clock had previously been mathematized and its model still attracts scientific tinkerers.

TH: Alex Borbély integrates the circadian and NREM-REM rhythms in his influential super-model which ties the circadian clock to the NREM-REM cycle.

AH: For scientists like Karl Friston, consciousness itself is a model. The brain-mind, for Friston, is a hypothesis-testing device.

TH: He is right when he points out that the world which we perceive and the image we have of our selves are virtual realities. The German philosopher, Thomas Metzinger, refers to a "self-model". Are we then only a set of instantiated mathematical principles?

AH: Most people would recoil at his idea but Friston thinks that, just like scientists, we constantly test hypotheses and live by the theories that our brain-minds elaborate.

TH: Many people would fear that this is all a house of cards that could come tumbling down at any time.

AH: I remember my adolescent chagrin when my kid brother trampled a model airplane that I had labored over for weeks. I was hurt but I got over it. It was just a model after all and I overvalued it because it was my creation.

TH: Thomas Henry Huxley, Darwin's bulldog, defined a tragedy as a theory murdered by a fact.

AH: Scientists are enjoined by Karl Popper to welcome disconfirmation of their theories. That's the way that scientific progress is made.

TH: But still, like Sigmund Freud, we all hope we die before someone comes up with a better model than ours.

AH: It looks like I will make it out with AIM intact. This book is my legacy.

Chapter 15. Activation-Synthesis

At the same time that the reciprocal-interaction model of sleep cycle control was being developed, the new brain physiology and dream psychology data were combined with a view to creating a brain-based dream theory. It was named "activation-synthesis" to connect two ideas about what was happening in the forebrain: activation meant that the brain and the mind were simultaneously turned on; synthesis meant that the cerebral cortex fabricated the dream by integrating internal signals from the brain stem with other internal data, including memory information. Dreams were thus composed of both historically meaningful and biologically random mental products.

Since the most coherent and widely known dream theory of the day was the disguise-censorship hypothesis of Sigmund Freud, activation-synthesis was promulgated as an alternative to the psychoanalytic viewpoint. The activation was considered to be a motivationally neutral dream consciousness instigator in contrast to Freud's theory of oneirogenesis by the release of repressed infantile wishes. The synthesis of random neuronal signals was an alternative to Freud's idea that the dream was only apparently nonsensical and could be interpreted as symbolic disguise of the unconscious infantile wishes. Other aspects of the theory included the novel views of perception, emotion, and memory that we will consider in more detail later.

Quantitative Analysis of Dreaming

Inspired by the additional neurophysiological findings to be described in the next section, Robert McCarley, a psychiatrist with a strong mathematical bent to whom this book is dedicated, began to analyze dreams from a formal rather than a content perspective. It is once more important to understand this difference. A formalist, like McCarley, asks whether dreams are, say, visual rather than whether the dreamer sees his mother; do they have an auditory component, rather than does the dreamer hear his mother speak. McCarley defended his choice on scientific grounds. In tune with dual-aspect monism, he wanted to match the formal aspects of dreams with the formal aspects of brain physiology. Until he had done this, he thought he would be prone to interpretive error. The interpretation of a dream of one's mother has the psychological appeal of content analysis but the formal approach offered a golden opportunity for genuine scientific progress rather than mere speculation. In the view of activation-synthesis dream theory, psychoanalytic dream interpretation is speculation. Figure 15.1 highlights these differences in approach.

McCarley began by asking what sense modalities were represented in dreams. The observed preponderance of vision correlated well with the observed activation of the brain's visual system in REM sleep. "What else is new?" you might well ask. "Why are dreams often described as strange?" McCarley might ask you and add, "What does dream strangeness boil down to?" and, "To what aspects of brain physiology might the strangeness of dreams correspond?" These questions proved more difficult to answer but tentative responses were forthcoming and led to unexpected similarities between dream formalism and the delirium of organic psychosis. It is now plausible to suggest that dreaming *is* an organic psychosis. This topic is further explored in Chapter 21.

Since this was a judgment call, disagreement about the plot would be expected to be high, and it was. To get around this reliability issue, sensitivity was sacrificed to reliability by including for further analysis only those items that all three judges identified as strange. The identified items were then assigned to the dream plot (A) or the dreamer's thought about the plot (B). An example is shown in Figure 14.2. The inter-judge agreement was high and any disagreement caused rejection of the item. You might suppose that little would be left after such stringent cutting but there was still plenty of strangeness left over to analyze. Dream consciousness is dramatically stranger than waking consciousness.

Figure 15.1. The mind and brain conspire to create dream imagery.

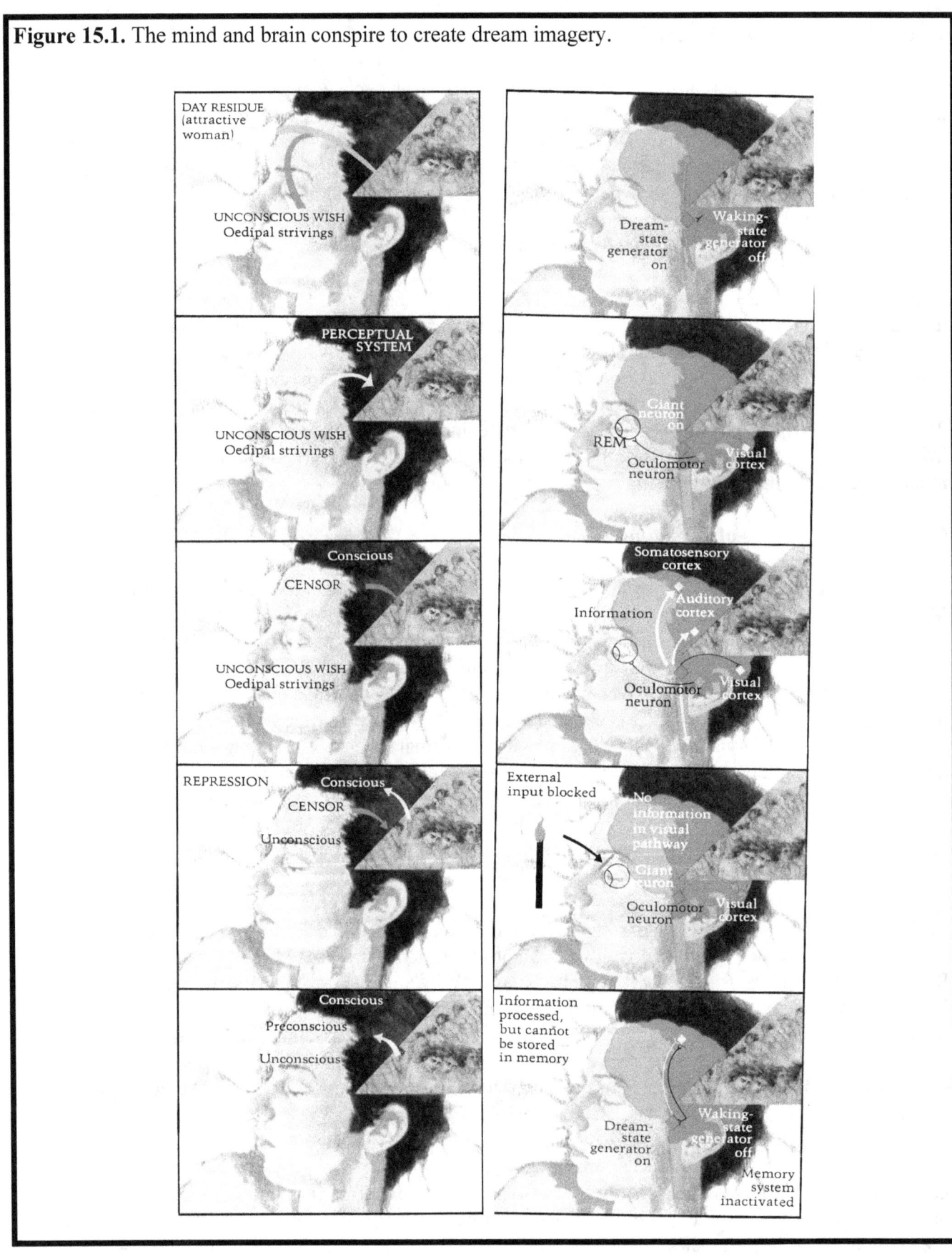

The first step was to ask three judges to identify strange aspects of dream reports. "Strange" was defined as a plot item or thought about that item that would have a 5% or less chance of occurring in waking.

Strangeness was renamed *bizarreness* and the three judges were asked to decide whether the bizarre item was: 1) discontinuous, 2) incongruous, or 3) ambiguous. Discontinuity might be noted when a character explicitly disappears or when entire the scene suddenly changes. Incongruity was defined as plot items that didn't belong together, like a church that was in a skyscraper. Ambiguity was defined as explicit vagueness (e.g. "it seemed like" or "I wasn't sure if."). The two judgments were then combined as, for example, A-2 (for plot incongruity), which turned out to be the most frequent form of dream bizarreness.

We dwell on these details for two reasons: to show the difficulty of any scientific rendition of conscious experience; and to demonstrate that perseverance does pay off if one sticks to the task. What this study means is that: 1) dream bizarreness can be defined as cognitive discontinuity and incongruity, 2) these formal features have no apparent connection with unconscious wishes and, 3) dream discontinuity and incongruity are therefore more likely to be caused by the distinctive brain physiology to be reviewed in the following section. This is admittedly a small step forward but the results can be confirmed (or not) and extended to waking consciousness to answer the question inspired by William James: Is there really an unbroken stream of consciousness or is the stream as choppy as it appears to be in dreaming? The answer can be quantified and the interpretation grounded in biology.

The activation-synthesis theory ascribed dream bizarreness to the neurophysiology of the REM state. Since orientation, memory, thought, and other aspects of cognition were so conspicuously deficient in dream consciousness it seemed reasonable to suppose that the root cause was organic, not psychological. This was Freud's original supposition too. Of course activation-synthesis never suggested that dreams were entirely meaningless, only that some aspects were nonsensical while others certainly revealed the dreamer's preoccupations and were thus worthy of discussion. But a line was drawn in the sand: activation-synthesis was scientific; psychoanalysis was speculation.

A Storm of Controversy

To say that activation-synthesis unleashed a storm of discussion is an understatement. Psychoanalysts were (and still are) vociferously negative. They see the new theory as denying dream meaning instead of recognizing that the essence of the new theory was criticism of the Freudian concept of repression as the determinant of dream nonsense. Many psychologists thought Freud's theory was obsolete anyway and others said that they had never taken it seriously. They often had theories of their own which were at odds with both psychoanalysis and activation-synthesis. Alternative theories will be discussed in the subsequent section of this chapter.

Meanwhile, hypothesis-testing psychological research on dreaming was conducted over the following 35 years. The preceding section of this chapter has dealt with dream bizarreness. Dream logic, dream emotion, dream perception, dream memory, and dream interpretation have also been investigated and the results are surprising. The interested reader can find an annotated bibliography in Hobson's 2009 *Nature* article listed as recommended reading.

Activation-synthesis disappoints many people because it explains so little of what they really want to know: what do dreams really mean? The answer of activation-synthesis is that they may not mean as much as we would hope when we ask the question that way. But their meaning may be greater when we regard dreaming as instructive with respect to the broader question: how and why are we conscious? For the first time in recorded history, this question has preliminary answers.

Alternative Dream Theories

Many psychologists are more impressed with the similarities between waking and dreaming consciousness than the difference that activation-synthesis uses to explore the brain basis of consciousness. Some even assert that no differences exist that are important. Many of these critics suppose that the brain is of no help to psychology and many physiologists share their disdain for integration by belittling psychology as of no relevance to brain science. This book is dedicated to the positive conviction that studying dreaming in comparison to waking must focus on the differences and that the differences are actually informatively large.

Psychologist John Antrobus is a sophisticated behaviorist working at CCNY in New York who holds that dreaming and waking do not differ from each other in any measurable way. He also asserts that both waking and dreaming are subjectively more or less intense as a function of activation (A) only. He thus denies the relevance of REM sleep input-output blockade (I) and modulation (M). His skeptical position is as useful as that of the modern Freudians in providing an antithesis to the thesis of activation-synthesis and AIM. His early studies of fantasy (with Jerome Singer) link his scientific stance to the continuity arm of dream psychology that emphasizes similarities and minimizes differences between dreaming and waking. Fantasy is undeniably continuous with the waking consciousness that it parallels but fantasy never contains formal features such as the character change (A-1, plot discontinuity) of dreaming. So waking, fantasy and dreaming are all continuous but they are discontinuous too.

Figure 15.2. Custom House dream from the Engine Man Journal.

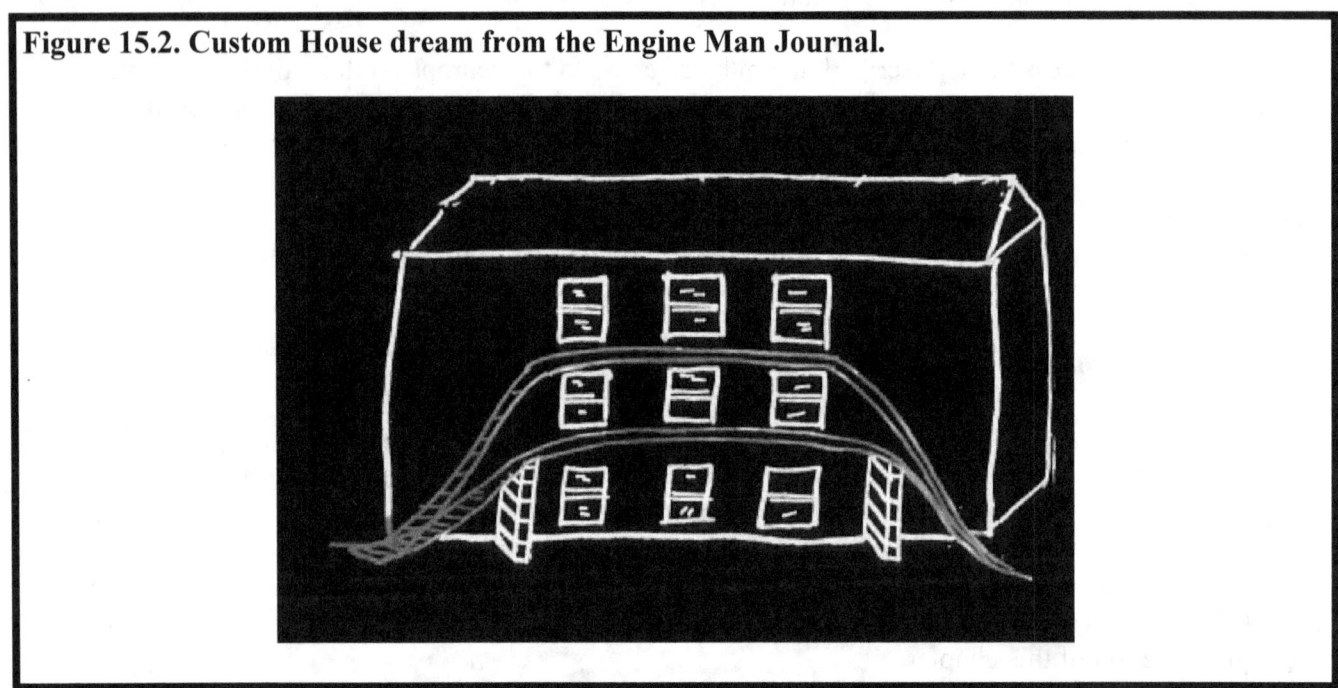

William Domhoff of the University of California, Santa Cruz and Michael Schredl of the University of Mannheim in Germany are continuity theorists who measure thematic content and find statistically significant similarities in reports of dreaming and waking consciousness. Activation-synthesis theorists accept both these findings and the concept of continuity but ask, is that all there is to it? What about discontinuity? The truth may be both/and rather than either/or. Dreaming evinces both continuity and discontinuity, but it just could be that the discontinuity is of greater scientific import, at least as far as the integration of physiology and psychology is concerned.

Extension of Activation-Synthesis Dream Theory to Other Conscious States

In retrospect, the activation-synthesis theory of dreaming was evolved to account for brain-based features of any conscious state. It is thus clear that waking must involve activation and synthesis too. Indeed the AIM model treats the two conscious states as identical with respect to activation. When we are consciously awake, our brains are turned on and information is processed. Following Tononi, we might suggest that large amounts of information are processed and that phi measures that amount. But what phi doesn't do that activation synthesis does is to specify two important differences between dreaming and waking. One is the source of the information, almost exclusively internal in dreaming and significantly external in waking. This is Factor I of AIM, which attributes the difference to input-output gating. The other is what is done by the brain with the information that is processed, whatever its source. In waking, significant amounts of the information are stored in memory, while in dreaming almost none of it can be recovered. This is Factor M of AIM which attributes the difference to aminergic demodulation.

In both waking and dreaming information is synthesized. Waking consciousness feels to us like a continuous, realistic subjective experience. So does dreaming. We think that we are awake when we dream. Our activated brains in both states synthesize information and create narrative continua that are quite similar in the retelling. Part of the reason for this narrative similarity is that dreams are reported in waking, as we will later emphasize and explain. The brain activation and narrative synthesis common to waking and dreaming should have tipped Freud off to the likelihood that what he called the unconscious was really a part of consciousness, a part that increased its power rather than threatening its integrity.

In both waking and dreaming consciousness we integrate information from both internal and external sources. That is, we synthesize information actively in both states. It is the brain's job to process as much information as possible and, *pace* Giulio Tononi, it processes a lot of information. A functional implication (that we will make much of later) is that waking and dreaming are mutually beneficial rather than competitive, as Sigmund Freud would have us believe. Thus the debate with the Freudians — and any who hold that dreams are prophetic rather than predictive — is a welcome nuisance. It helps us to make our point over and over again: Consciousness is due to brain activation and the activated brain synthesizes information.

In the early, intentionally provocative days, a particularly needling claim was that "in dreaming, the brain made the best of a bad job." How could the brain be expected to synthesize information as seamlessly as it appeared to do in waking if it was only equipped with its own internal signals? A galling assumption was that some of these internal signals were random. The idea that anything mental was due to chance really irritated the Freudians, who believed, almost religiously, that every aspect of the mind was reducible to psychoanalytic interpretation. This arrogant and absurd assumption is still unfortunately shared by diehard Freudians who often sound more like religious fundamentalists than dispassionate scientists.

The word "random" should now be replaced with the word "chaotic," where chaos is defined as orderly disorder. No one can predict a series of REM sleep eye movements any more than he can predict a night's dream content. Both are very variable and we should be grateful for the freedom that our chaotic brains grant us. Retrospective dream interpretation may be amusing — and even helpful — but scientific probity is restricted to prediction. If you are a mammal you will evince REM sleep. If you are a human mammal you will dream in REM sleep.

Reciprocal interaction and activation-synthesis have been revised and extended in the 39 years of their scientific life but remain unchallenged as conceptual and empirical advances in sleep and dream science. Many scientists agree with us that activation-synthesis is the only brain-based theory of dreaming. It must thus be recognized that the theory as a solid cornerstone of consciousness science.

Dialogue 15. Dream Sense and Nonsense

TH: How did you come up with the activation-synthesis dream model?

AH: I wanted to emphasize two aspects of brain physiology in REM sleep, activation of the cortex by the brain stem and synthetic integration of internally-generated data by the forebrain.

TH: I think I understand the activation but I need your help in unpacking the synthesis side of the model. What do you mean by integration?

AH: The brain turns itself on (activation) and sends itself messages which must be integrated to produce the dream experience.

TH: You got into hot water when you suggested that the dream experience was both sensible and nonsensical.

AH: My critics derided my theory as a return to the pre-Freudian nonsense era.

TH: I can certainly understand why. My students love Freud because he insisted that all dreams were meaningful.

AH: Most people feel that way. I myself love to interpret dreams. I never said that they were meaningless.

TH: But how can you reconcile the implication of randomness with the concept of meaningfulness?

AH: Simply by considering randomness as helpful rather than a nuisance to be avoided.

TH: That's a tall order. No wonder you had such a hard time selling your theory. Say more about the utility of randomness.

AH: Randomness provides a liberating escape from the Newtonian determinism which Freud accepted and incorporated in his model.

TH: But how can randomness convey meaning in dreaming or any other state of consciousness?

AH: Randomness is orderly disorder. The function of the brain-mind is to create meaning out of chaotic noise.

TH: How does that idea express itself in your new model of virtual reality?

AH: Virtual reality is always the informational substance of consciousness.

TH: Virtual reality sounds like science fiction. Do you really think that consciousness is synthetic?

AH: Yes, I believe that virtual reality is science fiction. We live by the stories we synthesize about ourselves. Dreaming is meaningful and directly revelatory of this synthetic process.

PART III. PSYCHOLOGY

Introduction and Summary, Part III.

In this section, I review the scientific and clinical evidence for AIM. I begin with chapters on memory and attention, two cardinal faculties of the mind that have long been of central interest to psychologists. I demonstrate how the brain-based approach of AIM is enlightened by and enlightening of these faculties of the mind. How brain lesions affect consciousness is discussed in the chapter on neuropsychology. I present evidence that the restoration to dream consciousness of memory and attention in "lucid dreaming" may be occasioned by frontal lobe activation. In the subsequent chapters, I review sleep disorders and psychosis in the light of AIM. In these chapters, I go beyond what the various subspecialties of medicine have been able to discern. To me, this exercise constitutes proof of concept and warrants further exploration. I conclude this section with a like-minded consideration of epilepsy and migraine and then say what I mean by "altered states" of consciousness with special reference to Freud's bugaboo, hypnosis. I consider hypnosis to be the most powerful instrument of medical psychology. The scientific study and legitimization of hypnosis is a goal now within reach.

Chapter 16. Attention*

Consciousness unifies the psychological and physical components listed and characterized with respect to their state dependency in Figure 16.1. Instead of trying to explain each component in turn, the conscious state paradigm tries to understand how all of these faculties might be united by the brain. The analysis of consciousness components separately is nonetheless useful, as I will demonstrate by focusing in this chapter on one of them, the faculty of attention. Attention is an inviting subject for three reasons: it is considered by many scientists to be the essence of consciousness; a great deal of scientific attention (sic) has been paid to it; and attention is dramatically state dependent.

Having first discussed some of the research on attention in waking, we will then introduce the data that show the state dependency of this psychological faculty.

Attention and Consciousness

When we are awake, we are conscious of that which is the focus of attention. In this book, we follow psychological convention, and use attention in the strict sense of selective attention: the top-down selection of a particular stimulus array for processing. A frequently given analogy is that attention is like a spot light. We present a summary of the spotlight hypothesis of Francis Crick later in this chapter. The terms are so closely related that many writers use consciousness and attention interchangeably, but others argue that it is possible to distinguish between them. We are among the latter group in that we distinguish between psychological faculties in order to facilitate understanding of their physiological underpinnings. We are supporters of the former group in that we aim to explain how the state-dependent aspects of conscious experience are engineered.

Attention is notoriously dissociative. For example, we can have attention (to stimuli beneath some visual threshold) without conscious awareness, and we are probably quite often consciously aware of a stimulus without paying attention to it, an obvious example being the cocktail party phenomenon, whereby we can be paying intense attention listening to someone but then become aware that someone else has said our name elsewhere in the room. We do need to be careful about the words "consciousness" and "attention."

We will emphasize the reasons for this caution as we proceed. The box at the end of the chapter makes plain the debate between psychologists like Trevor Harley, who fear that subjective experience may never be understood, and brain scientists, like Allan Hobson, who are certain that progress can be made via the strategic use of neurobiology as detailed in the preceding chapter.

The Psychology of Waking Attention

Psychologists have distinguished between two types of information processing for some time. Different names have been used, and there has been controversy as to whether there are two distinctly different types, but most agree that there is a broad difference between skilled and unskilled behavior. To understand this argument, it is important to define working memory as the conscious recall of a specific procedure or fact. On the one hand, we have information processing that is automatic, occurs relatively quickly, and doesn't require any working memory resources; on the other hand, we have processing that is slow and that requires concentration and memory, called controlled (or attentional) processing.

To understand this distinction, consider automobile driving. When you began driving, many aspects of the task required your full attention: you had to think about how to change gears, how to steer, and when you should start braking, and you might well have got many of these decisions wrong early on. You seemed to have too much to remember and might well have had to follow verbal instructions. You probably were fully occupied by driving; you might have had difficulty even talking to your instructor when fully engaged with the car and the road. All of these are characteristics of controlled processing. As you improved with practice, these things became automatic. Skilled drivers will carry out much of their driving without realizing it, and will also be able to talk, listen, and sing. They might even have difficulty explaining how they're doing what they're doing to another person.

Automatic and controlled information processing differ also in their involvement of awareness and consciousness. We're aware of what we're doing when learning. Once a complex procedure is learned, automatic processing requires no attention. How often have you done something like driven around for some time and then realized that you haven't paid any conscious attention to what you've been doing or the road that you have travelled? And yet you haven't crashed, or jumped a red light, or anything like that. This phenomenon is called "highway hypnosis."

Michael Posner (1936–) is a psychologist who immediately understood the utility of neuroscience (especially clinical neuroscience) to experimentation on humans, especially on attention. Working first with Marcus Raichle at Washington University in Saint Louis, Posner was able to show that the brain mediated shifts in attention via the thalamocortical system and specific areas of the cerebral cortex (such as the dorsomedial frontal regions of the forebrain which are selectively reactivated in lucid dreaming). Posner was quick to appreciate the state dependency of attentional shifting capacity and has written about this phenomenon as a discussant in my recent book *Dream Consciousness*.

Highway hypnosis indicates the normality of a dissociation, a psychological process that is often considered abnormal or at least unusual. Instead of taking this view, we must recognize that dissociation is both normal and universal. There is no getting away from hypnosis and many good reasons for accepting, understanding, and exploiting it. We return to this subject in the later Chapter 22 on altered states of consciousness.

Things to which we're paying attention are in focus: they take up space in working memory. The definition of controlled processing implies that what is being processed is stored in working memory, while automatic processes are not. On the other hand we need not — indeed cannot — be conscious of all the contents of our working memories. Hence awareness, attention, and working memory are all closely related, but ultimately distinguishable, concepts. Researchers have also found that they involve different parts of the brain.

Attention is dramatically state dependent. Hence it is important to point out that while we often refer in this chapter primarily to attention in the waking state, we later discuss the degree to which attention is lost in REM sleep dreaming. This loss of attention may be due to the sensory gating and/or chemical modulation of the brain in REM. These are Factors I and M in the AIM model described in Chapter 9. The loss of attention cannot be ascribed to differences in brain activation (Factor A of AIM) because activation levels are high in both REM and waking. Scientists often make the mistake of identifying attention with consciousness. We will come back to this caveat later, but for now let us consider some of the attentional processes which have been studied in waking subjects.

Semantic Activation Without Awareness

How much processing can we carry out without conscious awareness? This topic has been one of considerable controversy. This is another way of saying that even in waking, attention is state-dependent. Early experiments suggested that in some circumstances we can carry out a considerable amount of processing without being aware of the stimuli.

Typically these experiments made use of visual masking, wherein a pattern similar to a word is shown on a computer screen very soon after the target word. If the duration of exposure of the word is brief enough, then participants cannot consciously identify the word. Nevertheless, if a word related in meaning is shown immediately afterwards, then participants are faster to identify that second word if it is related in meaning to the first than if it is unrelated — a well-known phenomenon called semantic priming. So participants can more quickly identify the word NURSE if it is paired with doctor as shown in Figure 16.1.

Figure 16.1. Semantic Priming.

Associated Pair of Stimuli: DOCTOR — MASK — NURSE
Dissociated Pair of Stimuli: BREAD — MASK — NURSE

This advantage could only be gained if the first, masked word (DOCTOR) is processed long enough to gain access to its meaning. Hence, these experiments suggest that it is possible to gain access to relatively abstract aspects of a target, such as its meaning, without being consciously aware of it.

These experiments are difficult to carry out and the effects are relatively small. In particular, they are sensitive to the amount of masking: too much masking obliterates recognition completely, so there is a small and variable window between no processing at all and fully conscious processing. To demonstrate semantic activation without awareness, we need to get the timing exactly right by getting into that window for every participant every time. An influential critique by Hollander questioned whether anyone had managed to achieve this goal.

Experiments have also shown that we can be unaware of surprisingly prominent stimuli. The most famous example of this phenomenon of inattention blindness is the invisible gorilla studied by Simons and Chabris. In this experiment, participants are asked to watch a short video and carry out some fairly repetitive task: for example, watching a game of basketball and counting the number of bounces or passes of the ball. At some point in the video, something bizarre happens: for example, a man in a gorilla suit slowly walks across the scene, pausing half way, and waving at the camera. Usually about half the participants fail to notice the gorilla at all. It is amusing to watch their reactions when they are shown the video a second time knowing that they have to look for something. People are amazed that they could not have been aware of something so obvious. Conscious awareness is capricious: we cannot rely on seeing everything or getting an honest appraisal of what we have seen. I now lead the reader to the consideration of a set of topics related to attention and the state dependency of consciousness.

Shifting Attention

Waking attention is dynamic. The survival value of this trait depends on our ability suddenly to shift our attention from one focus to another. A scientist who has pioneered the study of shifting attention is Michael Posner of the University of Oregon.

At the same time that Posner was pursuing the neuropsychology of attention in Oregon, Raichle reported that subjects who were off task in the PET or MRI scanner underwent subjective mind-wandering. Other parts of their cortical brains became selectively activated. Raichle characterized this tendency to fantasize negatively with respect to the sustained attention of waking but, for the student of conscious states, the localization of the cortical seat of fantasy was in itself positive and capital. Working first in Munich, Germany and more recently in Nimwegen, Holland, Martin Dresler has shown that specific cortical areas are activated in REM sleep. It might be said that the brain and its mind had shifted their attention from the outside world of sensory perception to the inner world of fantasy (in waking) and dreaming (in sleep).

Figure 16.2. Dresler et al. image of the regional cortical activation in lucid dreaming.

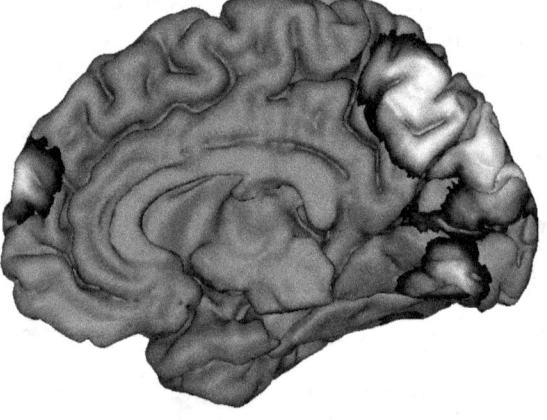

Working with Manfred Spitzer, a psychiatrist from Heidelberg, Germany, we investigated the hypothesis of state-dependent attention by asking subjects to perform the semantic priming task on awakening from REM and NREM sleep. Subjects more rapidly and accurately performed the task on awakening from REM than NREM but both sleep conditions were less efficient than waking. These results were interpreted as evidence for the hypothesis of state-dependent alterations of attention. More specifically, they suggested that the state dependency of attention is related to our ability to focus our minds in waking more easily than in sleep. The increases in semantic priming speed in REM compared to NREM help to explain the hyperassociative character of dreaming as well as the increased activation level of that state compared to NREM sleep.

While waking attention is top-down (i.e. it can be controlled by the upper brain), dream attention is driven from the bottom up (i.e. it is preferentially controlled by the brain stem). The fixation of attention thereby suffers and we find that we cannot focus on the bumpy stream of conscious events in dream consciousness. Our minds flit from one scene to another and we are helpless to arrest this process. It may well be that this dream vacation of the associative attentive mind benefits waking consciousness.

This idea will be developed when we later discuss the functional implications of cognitive state dependency.

Francis Crick's Spotlight Hypothesis

Thalamocortical interaction has long been known to be anatomically sectorial. That is to say, thalamic neurons in a given nucleus project to and receive feedback from regionally specific cortical neurons. This is the anatomical basis of external sensory segregation. It also provides a structural substrate for internal attentional focus. One of the scientists who articulated the "spotlight" hypothesis of attention was Francis Crick. After successfully modeling the molecular structure of DNA, Crick turned his attention to neuroscience, hoping to solve the riddle of the brain basis of consciousness.

Francis Crick (1916–2004) was a physicist who, with James Watson, successfully modeled the DNA molecule, the mediator of genetic inheritance. This achievement was foundational to molecular biology and may be regarded as according Darwinism its most significant scientific status. With Christof Koch, Crick later turned his attention to the brain basis of consciousness, focusing on the visual system because so much was known about how the brain processes luminous stimuli. Along the way, he immersed himself in sleep science, which held that sleep was positively reinforcing of learning and memory. Together with Graeme Mitchison, Crick advanced the countercurrent idea, first put forth by David Hartley, that we dream in order to forget. Outspokenly critical of the work of others, Crick was as much feared as revered by his peers. He embodied a skepticism that could be withering, but he was also a person of great charm and loyalty to colleagues even if he often disagreed with them. When plans were made to celebrate the 50th anniversary of his DNA discovery, he urged that the date be moved up 5 years in case he might die before being feted. As it turned out he did survive his prophecy.

According to Crick, the thalamocortical system regulates attention by facilitating one or another of its reciprocal systems, much as a spotlight illuminates one or another part of the external visual world. We voluntarily and reflexively shift attention as the thalamocortical spotlight changes its target. When we examine the possible consequences of our behavior, we can use the spotlight to focus our thoughts and perceptual imagination on imaginary outcomes and choose between them. This is a possible substrate of the free will which we suppose, with William James, that we have. When an unexpected stimulus arises in the outside world, our brain-minds can rapidly evaluate its threat or irrelevance to our survival and our equanimity.

One of the reasons that this searchlight hypothesis is so appealing is that it conforms well with the theory of activation that I will develop shortly and because it fits so well with the ideas of the late Gerald Edelman and his student, Giulio Tononi, both of whom recognized the importance of the thalamocortical activation underlying consciousness. Tononi has advanced the phi theory of consciousness and is testing it by studying changes in thalamocortical excitability in sleep compared to waking. According to Tononi, consciousness can be defined and measured as the total sum of information processing by the brain. This is an admirably concise and quantitative proposition but, even if phi is correct, Tononi realizes that the differentiation of conscious states remains to be investigated.

Another reason for enthusiasm for the spotlight idea is the scientific research of the late Mircea Steriade, considered in more detail when we took up the topic of slow wave sleep in Chapter 6. The marked reduction of conscious awareness is possibly related to the loss of spotlight capacity owing to blockade of throughput in the thalamus in slow wave sleep. If the thalamic throughput is blocked, the spotlight is put out of commission. Although vision is internally illuminated, attentive consciousness is in the dark, as it were.

A variation on the theme of spotlight theory has been developed by Dr. György Buzsàki and Rodolfo Llinás, who postulate that the cerebral cortex is not only illuminated but also functionally scanned by electrical signals arising in the thalamus. This internal code system allows the thalamus to read and hence "know" what is going on in the entire cortex, not simply in the sectorial regions illuminated by the spotlight. This theory is beyond my comprehension but its plausibility equals that of the hierarchical model of my collaborator, Karl Friston. The upshot of all this speculation is that we are well over the threshold of gaining a scientific fix on consciousness and that attention to conscious states will play a key role in future advances.

Adrian Morrison's Startle Theory of REM Sleep Generation

The collaborator of Ottavio Pompeiano, the principal architect of AIM Factor I, was the veterinarian Adrian Morrison. After showing that the mechanism of REM sleep motor inhibition was the post-synaptic inhibition of motor neurons, Morrison studied the internal stimulus generation of the PGO waves originally described by Michel Jouvet. In other words, the brain stimulates itself at the same time that it shuts off external sensation and movement. Quiet genius at work.

Adrian Morrison regarded PGO waves as the expression of a startle reflex network. It was known that unexpected external stimuli could commandeer attention by triggering the brain stem-mediated startle response (as in, "oh, you scared me"). But it was Morrison's insight to recognize that animals were scaring themselves in REM sleep, as if they needed to practice survival techniques while they slept. As if this were not enough, Adrian Morrison also pioneered the humane practice of animal experimentation. As a veterinarian. he was well-positioned to be instrumental and effective in codifying regulations governing scientific research using live infra-human species.

In dreaming, our attention is seized by the onward rush of internally-generated percepts. We cannot "stop the music" and critically examine our conscious experience. We are up shit's creek without a paddle. It is all we can do to stay afloat and hope that the solid rock of waking may save us from drowning in the turbulent stream of our inattentive consciousness. We have lost top-down control to bottom-up forces. This is the power of the unconscious mind which seized Freud's attention but it is more sensorimotor in its expediency than he ever imagined. To be generous and Darwinian, we might cede Freud a wish to survive as well as a wish to procreate. Survival must come first.

Adrian Morrison (1935–) is a veterinarian who collaborated with Ottavio Pompeiano on the scientific basis of AIM Factor I. At his University of Pennsylvania laboratory, Morrison showed that anatomical connections between the locus coeruleus and underlying pontine tegmentum were essential to the normal function of Factor I. He later developed the idea that PGO waves were produced when the decreased inhibition of Factor I rendered the startle response more excitable.

Dialogue 16. The Modularity of Consciousness

TH: Psychologists like to carve the mind up by its joints. According to the cognitive module chosen for study, consciousness is deduced from the modular data. At any one time, multiple psychological processes are active.

AH: But consciousness is the integrated product of all those modules, so there must be a unifying system in the brain for pulling them all together.

TH: Agreed, and we will talk about this unifying system in more detail later. In a way, models such as the Global Workspace Theory try to explain how multiple cognitive data are brought together to form one item of consciousness. Isn't this unification the process of EEG synchronization proposed by Wolf Singer?

AH: Yes, but what is it that orchestrates the synchronization and what differentiates waking and dreaming, both of which are associated with the same sort of EEG rhythmicity?

TH: Now I understand why you regard dreaming as a state of consciousness. You want to explain its differences from waking. EEG synchronization alone won't do that. But I'm also struggling to see how EEG synchronization is sufficient to explain all types of cognitive unification.

AH: We have known since Moruzzi and Magoun's discovery of the reticular activating system that the brain stem controls both EEG synchronization and desynchronization.

TH: Jouvet's report that REM sleep was a brain stem function fits right into the brain stem hypothesis of consciousness, doesn't it?

AH: Yes, but the theory needed elaboration to propose that a single process might unify all of the cognitive modules at once.

TH: That's an ambitious but important task. I'm still struggling to see how EEG synchronization applies to combining the sound of my blackbird singing with the image of a little blackbird on top of a tree moving its beak.

AH: A first step is to give up for now on the blackbird as well as on other aspects of conscious content. Instead, we can list formal cognitive modules and define the changes in each of them when dreaming is compared to waking. These are shown in the following table.

TH: In this and the following chapter, we emphasize attention and memory. These are only two of the ten functions listed in the Table of Figure 4.1.

AH: Attention and memory are two important psychological skills that suffer in dreaming compared to waking. We think we know why: aminergic demodulation deprives the brain of the capacity to attend and to remember.

TH: You say that other psychological functions are enhanced in dreaming. How can that be if all the systems are unified?

AH: The unification is physiological but the psychological effects are differentiated by the very same physiological processes. Perception and imagination are enhanced at the same time that attention and memory are diminished.

TH: No wonder that artists have celebrated dreaming while scientists have ignored it.

AH: Even scientists need the enhancement of imagination to conceive new theoretical models and new experiments. A cogent example, Otto Loewi's Nobel Prize winning dream, is recounted in Chapter 8.

TH: Acetylcholine not only regulates the heart, it also turns on REM sleep and dreaming. Still, it is hard to believe that a single molecule enhances psychological perception and imagination in dreaming.

AH: This is reductionism at its best. A single hypothesis accounts for a multitude of mental state features. Psychology is scientifically explained, not eliminated, by these considerations.

TH: If you are right, psychologists may soon be out of business.

AH: You don't have to be a weatherman to see which way the wind is blowing.

Chapter 17. Memory*

No topic is more central to psychology than memory. For a century and a half, scientists have sought to define and explain it. After our necessarily brief introduction, we will endeavor to show how sleep and dream science can help us to understand memory via its brain state dependence. Our general position is that the vicissitudes of memory across conscious states are an experiment of nature which constitutes an indirect but revealing set of insights about what memory is, how it is engineered, and what its functional significance might be. Just as memory is central to psychology, so is memory central to sleep and dream science.

The distinction between procedural and semantic memory has its place in the science of consciousness as seen from the vantage point of sleep and dream science. We know how to walk without thinking about it. Such innate procedures arise as part of our sensorimotor capacity and are independent of the semantic knowledge that we acquire in school.

But exactly how is procedural memory instantiated? Certainly learning has something to do with it. We are taught to walk but automatic motor programs, clearly evinced in REM sleep, are innate and may constitute the *a priori* blueprint upon which ambulation is constructed. Our dreams are constantly animated, indicating that procedures are reliably rehearsed even as we lie in bed paralyzed as we dream of moving.

An analogous paradigm may pertain to semantic memory. We learn to speak a given language and acquire a culture-specific lexicon but a universal grammar may be prescribed by the genome and spelled out in sleep. The integration of grammar and language may thus be fostered by sleep. In sleep, we talk to ourselves in the twilight zone between waking and NREM and these self-to-self conversations serve to structure the day to come. This is not so much language or memory per se but our nocturnal musings are semantic memories in that they build upon yesterday and predict tomorrow. This is a semantic memory, a record of the past as it informs the future.

Figure 17.1. Original reciprocal action model and its subsequent elaboration.

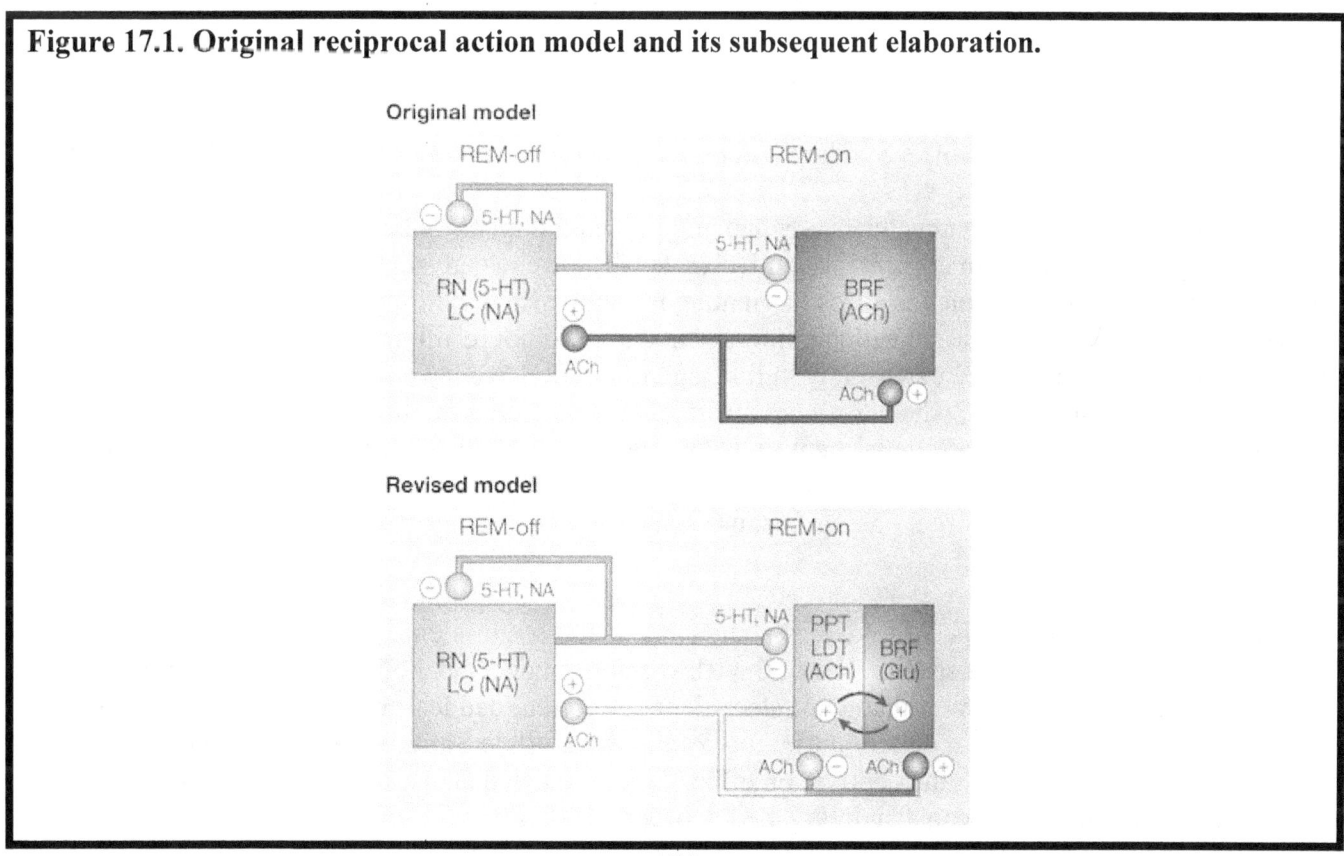

In the integrative sprit, memory may be defined as the codification of our genomic past in the service of our performance. Memory is essential to our ability to tell ourselves stories about our movement in and knowledge of the world we live in. This definition must not be construed to mean that procedural and semantic memory are created only in sleep. Rather we aim to suggest how sleep might play a part in their creation. In our opinion, such a role has, until recently, been entirely neglected. As we will now show, the scientific evidence for sleep as a memory enhancer is overwhelming and can only grow stronger as investigators take advantage of the paradigm of state dependent alterations of conscious state.

Dream Amnesia

In 1900, Sigmund Freud attributed dream amnesia to the repression of unconscious infantile wishes which, if not repressed, would invade and awaken consciousness. Both implications of this hypothesis have been overturned: repression has little or nothing to do with dream amnesia, and unconscious infantile wishes have little or nothing to do with dream instigation. In this section, we will focus our attention on the alternative theory of dream amnesia suggested by modern sleep and dream science: dreams are forgotten because memory for them is physiologically defective.

The bolder and more intriguing claim is that the amnesia for dreams is part of a physiological enhancement of procedural memory.

Here is an example of long-term amnesia from Oliver Sacks' patient Jimmy G.:

"What year is this, Mr. G.?" (It was 1975).
"Forty-five, man. What do you mean? We've won the war, FDR's dead, Truman's at the helm. There are great times ahead."
"And you Jimmie, how old would you be?"
He hesitated for a moment … "Why, I guess I'm 19, Doc. I'll be 20 next birthday." (He was 49.) (I showed him his face in a mirror.) "Jesus Christ, Doc! What's going on? What's happened to me? Is this a nightmare? Am I crazy? Is this a joke?" He became frightened, panicked.
I stole away … Two minutes later, I re-entered the room. Jimmie was gazing out with pleasure …
"Hiya Doc! Nice morning! … I can't say we've met before... I'd remember that beard if we had!"
Jimmy G suffers from Korsakoff's Syndrome due to alcoholic destruction of parts of the limbic system.

How might the paradox of dream amnesia-enhancing procedural memory be explained? Most simply, by shifting emphasis from dream content (which fascinates the psychoanalyst in all of us) to dream form — in this case movement — (which is a less fascinating but universal dream feature of critical survival function). In our dreams, we are practicing movement and need not recall the movements that we practice as long as we practice extensively in time and repertoire. We have a built-in, automatic motor program routine which is run, whether we are aware of it or not, every night of our lives. It is run because the physiological substrate of memory is disenabled. This is the mechanism of dream amnesia. From a functional point of view, we do not have to remember the dream exercise to profit from its existence. We can afford to forget dreams because something more exigent than the recall of semantic memory is guaranteed.

The evidence for this theory is twofold: first is the clinical and basic science story covered elsewhere in this book; second is the more specific test of the REM sleep enhancement of learning. Beginning in 1970s, a series of studies showed that REM sleep deprivation retarded learning and that the restoration of REM enhanced it. Among the more intriguing results was Carlyle Smith's identification of a "REM sleep window" through which the deprivation effects were mediated. This set of experiments showed that a surprisingly long time constant intervened between the deprivation stimulus and the learning response. This factor must be metabolic, in keeping with other basic science studies of learning and memory implicating protein synthesis.

Carlyle Smith is a Canadian psychologist who conceptualized the "REM Sleep Window." According to this theory, external information stored in the brain during waking is transferred to long-term memory by an active brain process occurring later in sleep. He performed extensive experiments to test his hypothesis and demonstrated a significant delay between memory acquisition and incorporation. This work inspired other scientists such as Robert Stickgold and Matthew Walker who have advanced and embellished Smith's ideas. One of the functions of sleep is to facilitate learning and memory.

The Brain and State Dependence of Anterograde and Retrograde Amnesia

Clive Wearing was a musician until an attack of the herpes simplex encephalitis virus almost completely destroyed his hippocampus (and some of the surrounding temporal lobes). What was preserved? Although he can remember nothing new, Wearing still recognizes his wife (although each time he sees her he thinks that it's for the first time). He can still play the piano, and he can still write. Nevertheless, as his wife observes, he has only a "moment to moment consciousness."

Amnesia appears to cause a loss of the continuity of self. And yet while we experience amnesia both within and after our dreams, we never lose the sense of self-as-agent while dreaming. This dissociation calls into question the assumption that the sense of self is only mnemonic. Wearing's persistent cognitive and motor skills are further evidence of this point. We need to know more about the brain basis of self and sleep/dream science can be expected to help us in this endeavor.

Sleep and Memory Studies

Since 1990, Robert Stickgold and his co-workers have delved into more specific links between REM sleep and procedural learning. This line of work is easily understandable, given the obvious and strong relationship of REM sleep dreaming and motor program activation mentioned above. Semantic priming has been used to quantify these effects: a verbal stimulus is flashed on a computer screen and the time taken to recognize proximal and remote word associations is recorded. REM sleep is correlated with enhanced procedural learning as if the brain was running and updating its motor programs in sleep.

Semantic memory has also been shown to be improved by both REM and NREM sleep phases by Jan Born. Born's work has been honored by the prestigious Leibniz Prize. Sleep is clearly not simply rest. The brain is active throughout the night and, while rest and energy savings are effected, adaptive cognitive processes are simultaneously achieved. Sleep learning may not consist of information transfer from a textbook placed under one's pillow but a previous day's study is consolidated via incorporation into brain memory.

Memory and Self-Modeling

According to the German philosopher Thomas Metzinger, the self is a functional state of the brain-mind. As such, the self is not a fixed structure but rather a dynamic construction that is compounded of past experience (as recorded in remote memory), recent experience (as recorded in recent memory), and instinctual motivation (a constantly changing, internally-generated program). This radically novel theory is consonant with sleep and dream science in its emphasis on continuous dynamic flux (vs. a static structure), change (rather than permanence), and forward-looking molding (rather than backward fixity).

Thomas Metzinger (1958–) is a German philosopher who argues that the self is a functional state that is constantly subject to reinvention. Instead of an ego structure, we are all passing through what Metzinger calls an Ego Tunnel, a passage from one second to the next. The reason for our erroneous view of our minds is the transparency of the psyche, whose connection to the brain is not perceptible. Thus we tend to see our selves as independent entities, spirits or souls. One antidote to our self-illusion is the critical analysis of phenomena such as out-of-body experience and lucid dreaming. With his colleague, Jennifer Windt, normal sleep and dreaming have also been scrutinized and made an integral part of the philosophy of mind. William James' dream of a triad of physiology, psychology, and philosophy is now being realized.

Some aspects of Metzinger's dynamic self-model are foreshadowed by Freudian formulations but these are superseded by a less rigid view of individual history and a greater acknowledgement of spontaneity. Of particular significance is Metzinger's incorporation of phenomena such as lucid dreaming and out-of-body experience, whose physiological basis is now partially understood. Most important is his experimental approach to hypothesis testing in the domain of body image construction. His studies clearly reveal the context sensitivity of dynamic self-modeling.

Coupled with the unreliability of memory discussed in the following section, our view of ourselves as reliably stable structures is undermined by brain dysfunction. Mental illness is as unpredictable as it is disruptive. At the same time, the relative freedom implied by iterative self-construction demands an optimistic therapeutic philosophy. We are free to be determined and determined to be free. Genetics may shape our direction and our "destiny" but our philosophy can select and implement voluntarily any one of a vast menu of developmental options.

Learning, be it Pavlovian (classical) or Skinnerian (operant), plays an important part in shaping our behavior but we are free to choose the training program that best suits our personality and our goals. Students of psychology should be in the vanguard of self-determination using what they learn to tailor satisfying, constructive, and responsible lives. There is no more powerful force than the critically educated human mind.

False Memory

It has long been held that memory was perpetual, stored forever in the brain. The most recent adherence to this indestructibility of memory hypothesis is psychoanalysis. Freud's theory of the unconscious maintained that infantile sexual wishes were banished for consciousness, disguised and censored for eternal banishment to the unconscious. These unconscious memory elements pressed for release from their unconscious prison, causing dreams in sleep and neurosis in waking.

Not only is this theory unsubstantiated but it now clear that in addition to the existence of motivationally neutral amnesia, it is possible for the mind to fabricate entirely false memories. Thus, we may not only forget many true memories but hold on to others which are entirely untrue. This scientific fact strongly impacts psychology and jurisprudence since false memories can be suggested by clinical interviewers, criminal investigators, and even disinterested parties.

As an example of a disinterested false memory, we have been informed by a responsible and credible colleague of his firmly maintained memory of a gift stored in the cellar of a friend's home. When he expressed hope that the long-lost gift might be recovered, he was informed by the owner that the house in question had no cellar. The house was, in fact, built upon a concrete slab and stood entirely above ground. This anecdote is telling because our distinguished professorial colleague could actually visualize and describe in detail the imaginary cellar and specify the location of the shelf on which his gift was stored. Suppose our colleague was a psychoanalyst or a witness in a trial. His outstanding scientific credentials might have helped his false memory to be believed.

More pernicious examples of false memory abound in the case reports of misguided therapists in the domain of trauma psychology. Young persons who are conflicted about their own sexual behavior are easily induced to "remember" an abusive parent, teacher or priest in order to confirm the untrue theory of a well-meaning but deluded clinician. Thus have innocent people been deprived of rights, made to pay fines, and even sent to prison for "crimes" they never committed.

False memory is comfortably consistent with Metzinger's model of dynamic and continuous memory construction. It is also compatible with the well-established data of post-hypnotic suggestion. Best of all, false memory elaboration conforms to virtual reality dream theory, which has a solid foundation in sleep physiology. We may be deluded, in dreams, to believe that we are actually doing things that we have never done and are often incapable of ever doing. Our minds are normally prone to imaginary scenarios in waking fantasy as we conjure up adaptive but impractical behavioral strategies. The normal human brain has a built-in, hard-wired hypothesis generator that is clearly manifest in our imaginative propensities.

Déjà Vu

In déjà vu, we find time-stamping errors such that the present is coded as if it were the past. Input is coded as memory. This normal variant of false memory is enhanced in patients with dementia and diffuse temporal lobe pathology. Confabulation is often associated with déjà vu in these patients. Normal dreaming shares so many of these formal features that it is reasonably considered to be a delirium caused by normal alterations of temporal lobe function. This theme will be echoed time and again as we proceed in this section. That this may not be sheer speculation is guaranteed by the solid evidence of temporal lobe activation in normal human REM sleep and the extensive cellular and molecular studies of REM sleep in animals.

Sue Llewellyn is a Manchester University sociologist with a strong interest in psychology. She theorizes that dreaming and REM sleep serve to reorganize memory in a useful and meaningful way. In developing her theory, she resorts to the ancient art of memory (AAOM) by which adepts store information (such as a random sequence of playing cards) by associating it with already well remembered material (such as the architecture of a house). According to Llewellyn, the brain-mind refreshes its memory in REM dreaming by tying new experience to old, well-established records. Her theory should be tested experimentally.

Memory and Imagination

Perception is compounded of memory of the past and prediction of the future. As we perceive the world, it feels like we were only passively recording outside events, but we are also actively anticipating that which we perceive. Our brain-mind is a camera but it is also a scriptwriter and a director. Our perceptions may not be worthy of praise as art but they are nonetheless artistic in a deep formal sense. Our consciousness is experienced as a story-film-painting of ourselves in our private subjective world. James Thurber's *Secret Life of Walter Mitty* can be read as attesting intermittent lapses into fantasy but it seems likely that fantasy is not intermittent but continuous, as continuity dream theorists claim.

Figure 17.2. A Smithsonian insect specialist dream designed this desk to help him work. The lettered flags allowed him to open alphabetically ordered files representing species of interest. This device is a forerunner of the desktop computer.

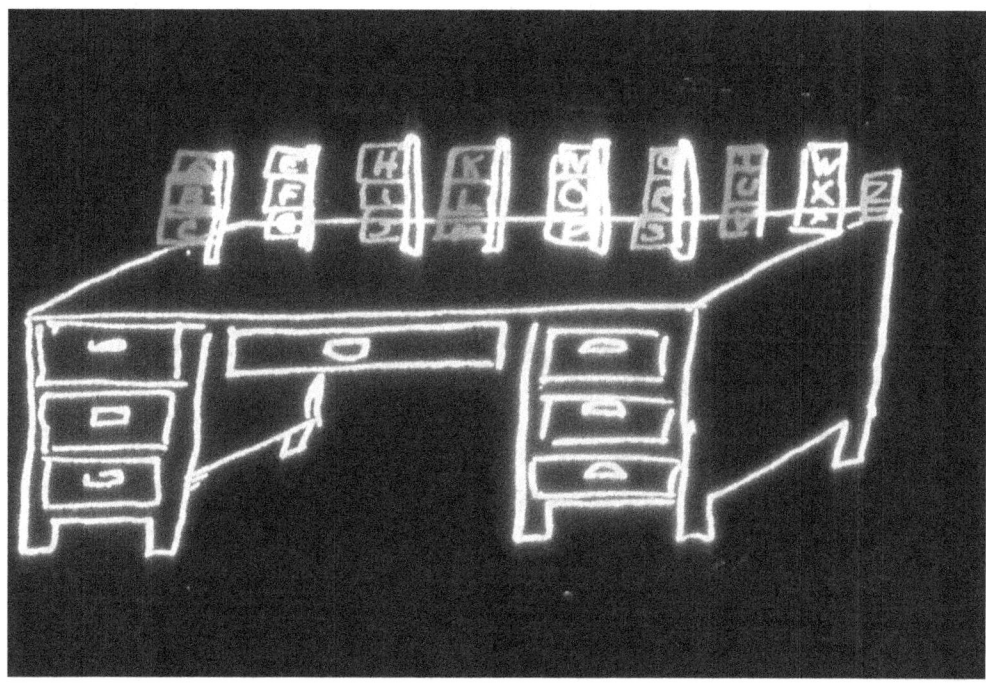

Day Dreaming

We are forced to admit that "day dreaming" is an integral part of consciousness. When we daydream, we create fictitious plots, many of which are impossible and most of which are never realized.

What do we do with our imaginative rushes? Remember them? Or leave them as trash on the cutting room floor? We could not possibly remember all of them and the few that we may remember are more likely to be delusional than veridical. Dream consciousness makes this creative delusional point about memory even more clearly. When our brain-minds are turned on in sleep, their creative products are as formidable as they are forgettable. It seems possible, née probable, that consciousness is warming up for another day's work. In Metzinger's, it is the self remaking itself by incorporating recent memory fragments into a novel synthesis with remote experiential and ancient biological forces.

When we read a gripping novel or see an engaging movie, we lose our selves in the fictitious world of another. At first glance, this might pass for escapism and be derided as mere dreaming. But as much as escaping from the humdrum of our unimaginative and boring waking, we are seeking, and sometimes finding, fictitious worlds that contrast but complement our own. Thus our selves are remade in the image of the other, just as in dreaming we remake our minds as we sleep.

An artist is one who helps us to lose and find ourselves as we do in our dreams. No wonder artists are such connoisseurs of dream consciousness. They help us to forget in order to remember.

Dialogue 17. Dream Amnesia

TH: If I am conscious in my dreams, why can't I remember them?

AH: Because your brain memory system is disenabled.

TH: How so?

AH: While your upper brain is buzzing away and transporting you into dreamland, the lower brain chemical systems which allow you to remember during waking are turned off.

TH: From a Darwinian perspective, it would seem that survival is enhanced by dream amnesia.

AH: Which may mean that you do not have to know what happens during your nocturnal warm-up sessions, you just have to have them.

TH: How do you know that dreaming is so important?

AH: It may be that dreaming itself is not so important but the demodulated brain activation of REM upon which dreaming is built has been shown to be essential to temperature control and to life itself. It may also be crucial to survival by allowing belief and imagination.

TH: This brings psychology on to center stage and promises a scientific status it has long sought.

AH: The key to success for psychology today is its liaison (or marriage if you will) to physiology.

TH: All marriages are difficult and this one is no exception.

AH: The physiological husbands think that they don't need their psychological wives, and vice versa, but neither can reproduce without the other.

TH: And they have to learn to live together to nourish their offspring.

AH: I like the marriage analogy and would like to further explore what each member of the metaphorical couple must do to assure a happy conjunction.

TH: Isn't dreaming just a window on the brain?

AH: It is much more than that. Dreaming informs us of otherwise inaccessible laws of the mind.

TH: Isn't that what Sigmund Freud meant when he proposed the unconscious?

AH: Exactly, but the memory story nicely illustrates the strength of the new formulation: it replaces Freud's dubious psychological hypothesis of repression with the more solid and demonstrable fact of aminergic demodulation.

TH: How do you know people dream a lot if they have no memory of it?

AH: There are two ways: one is to teach them to awaken from REM at home and the other is to awaken them from REM in a sleep lab.

TH: It sounds like everyone can participate in the new science of consciousness.

AH: And analyze themselves at the same time!

TH: Are you saying that dreams have meaning after all? I thought you opposed this view.

AH: I never opposed the meaningfulness of dreams. I opposed the Freudian hypothesis that dreams arose from a need to conceal their meaning and that only a psychoanalyst could interpret them correctly.

TH: If you prevail in this controversy, you will put a lot of psychiatrists out of business.

AH: Oh, don't worry about that, most psychoanalysts, including Freud himself, never took his dream theory seriously. They all felt, as I do, that dreams are intrinsically meaningful.

Chapter 18. Neuropsychology*

You appear to have free will and believe that you are in charge of your own actions. Your mind makes decisions and your body acts upon them. You can choose what to do at this very moment. You can choose freely between the alternatives in front of you.

At least, that's how it seems to us. But we have already seen from the work of Libet that the research on the timing of decisions shows that the brain appears to have made its decision before that decision reaches a level of conscious awareness. That is, our brains have already decided before that moment in which we think we decide. Our will may be less free than we thought.

Free will, our ability to choose, has been one of the issues that has most troubled philosophy. We certainly think we are free to choose at any point. If we wished, we could stand up right now in the coffee shop and shout "God save the Queen!" The problem is that we live in a *deterministic* universe. By deterministic, we mean that everything is determined by the laws of physics.

Imagine a snooker table and hitting the white ball with your cue. In principle, if you could measure accurately enough, given the initial state of the table, and the way you hit the ball — the exact force and direction — you could, given enough time, work out exactly what is going to happen; you could work out the precise timing and location of each subsequent collision, and then repeat the calculations until you had computed the final outcome. Damn, we've potted the white. Given the initial state and the laws of physics, there is no possibility of any other outcome. Let us put aside the objection that we cannot (currently) measure the initial position and force of billiard balls and neurons accurately enough, and that it is not feasible to carry out such complex calculations in any reasonable time; the outcome, the final position of the balls, is determined by the initial state.

On this account, we are effectively billiard balls. The outcome is determined by the laws of physics and our initial state. We may think we are free to stand up and shout "God save the Queen!", but that choice has already been determined by our past. Our free will is illusory. The future is already determined. The hypothesis was first clearly articulated by the French scientist Pierre-Simon Laplace in 1814. If some super being (a demon, say, hence leading to the name *Laplace's demon* for this proposal) knew the exact position of every atom in the universe, they could use the laws of classical physics to compute any past or future position. There is no scope for variation.

Some may claim that quantum mechanics provides a way out of this conundrum. It's true that at the quantum level we are not able to measure with sufficient accuracy (if we don't wish to give up other sources of knowledge), and we will be forced to deal with probabilities rather than certainties. But at the macroscopic level, which is the one that concerns us, the state of the world, including our brains, will still be determined at the microscopic level. We have no choice but to accept this uncertainty.

Philosophers have struggled for centuries to reconcile this picture of a deterministic world with our sense that at any moment we are free to choose what we do. Freedom seems such an integral part of our being and consciousness that it's troubling to think that we are in fact prisoners of our past. We caution the reader that reservations about free will do not impugn the assertion, developed throughout in this book, that the mind is causal.

Working Memory Revisited

In order to read this book and retain some of what you have read, you need to pay attention and to mobilize what has been called "working memory." As the name implies, working memory involves the selection from your vast memory stores, specific items that become conscious (e.g. I am reading a book hoping to learn about the brain basis of consciousness). I am about to read a chapter on dreaming in which it will be alleged that I can become aware of my dreams while they are occurring.

Frontal lobe activation is associated with dream lucidity, as discussed in the subsequent chapter and with the activation, unusual in sleep, of working memory. The concept of working memory has a long history that is independent of consciousness science but it dovetails nicely with the story that we are developing. We hope that this story will take its place in your consciousness even after you have demonstrated your mastery of this material for your exam at the conclusion of this phase of your study. In brief, we hope to become a part of your working memory.

Cognitive scientists have studied working memory via experiments which demonstrate the importance to normal waking behavior of mobilizing knowledge structures in the service of ongoing tasks, like reading, paying attention to and understanding this text. One leading investigator is Alan Baddeley, who conceived the working memory process and localized it to the frontal lobe of humans. A synonymous term is executive ego function, the "conflict-free ego" of the neo-Freudian psychoanalyst, Heinz Hartmann. Informative work on its precise neuronal underpinnings in the frontal lobes of monkeys was done by the late Patricia Goldman-Rakic. Relevant terms related to the paradigm of working memory include: self-reflective awareness, orientation to time, place, and person, the concept of self-as-agent, and a variety of other consciousness attributes that enable us to function cognitively in waking.

Patricia Goldman-Rakic (1937–2003), shown here with her recording instruments, was a neuroscientist who pioneered the experimental study of working memory. She trained her monkey subjects to perform motor tasks while she recorded from their brain cells. The brain region most closely related to working memory is in the frontal cortex, later shown by Ursula Voss to be deactivated in sleep and reactivated in lucid dreaming (see Chapter 19). In order for us to enjoy voluntary control of our thoughts, we may need to activate our frontal cortices.

Of particular significance to the scientific story that we are telling is the impairment of all of these working memory functions in sleep, most prominently in REM sleep. Dreams are typically characterized by disorientation, recent memory loss, imprecise identification of dream characters, and the substitution of confabulatory information for the precise references of working memory. As detailed in a subsequent chapter, the compromise of working memory constitutes the degradation of waking consciousness in the quasi-delirium of dreaming sleep. These cognitive deficits may be a function of the persistent deactivation of the frontal cortex in human REM sleep. A person without a working memory is a dreamer. We hope this wakes you up if you are dozing or day dreaming now.

Brain Damage and Belief

There is a set of neurological syndromes resulting from brain damage that affect our beliefs about our selves and the world. Some of these syndromes seem as bizarre as dreams, but they show how modular our mental states can be in that we can still function very well in spite of having these almost unbelievable beliefs. In this chapter, we consider some of these disorders in terms of the light they shed on the importance of the brain to orientation, especially our orientation to our selves as entities and agencies, as well as illustrating how such vital psychological functions can be disrupted by even small and contained brain damage.

People with *Capgras syndrome* believe that people around them, particularly their loved ones, have been replaced by impostors. Capgras syndrome is an example of a wider class of delusions called *delusional misidentification syndrome*, wherein affected individuals question the identity of those around them. In the related *Fregoli's syndrome*, people think that strangers are their loved ones. These delusions arise in paranoid schizophrenia, but also in some cases of widespread brain damage and dementia. Such misidentification syndromes arise because faces trigger an incomplete representation of the person — for example, you might see your husband or wife, but not get the appropriate emotional reaction. In dreaming, characters identified as family members are often perceived as strangers.

In some delusions, people doubt the nature of those around them. In *derealization* states, people think that reality itself isn't real: the external world seems dreamlike. A related condition that often co-occurs with derealization is depersonalization, the feeling that you yourself are not real. Significantly, we think, this is rarely true of dreams. In dream consciousness, the first-person self is relatively constant while the other dream characters may change identity and even gender. A mild version of derealization is not uncommon, being experienced by many people at some time in their lives as a result of drug use, migraine, epilepsy, psychosis, sleep deprivation, or trauma, among other conditions. In these disorders our representation of the world is incomplete or unstable in some way. An extreme disorientation of this sort is the *Cotard delusion* (also called Cotard's syndrome). The sufferer thinks he doesn't exist, or even believes that he is dead.

It's difficult to imagine the conscious experience of someone with the more extreme of these delusional states. But the people suffering from them are apparently conscious when they are awake, and most of their consciousness is functioning as usual. It seems that we can still be conscious while thinking that the world isn't real, or that we aren't real. In dreaming, consciousness may be intensely vivid even though we are deluded in supposing ourselves to be awake. We theorize that this paradox is due to a widespread neuropsychological process. So far, the dreams of people suffering from disorders of cognition in waking have not been studied. We predict that their cognitive problems will persist because the same disabled brain is activated in REM as in waking.

This bestiary of disorders is not something to be gawked at, but its fascinating range shows how consciousness can continue fairly normally in the most bizarre circumstances. Again, we can separate core consciousness, awareness, and self-awareness from peripheral consciousness — the rest.

The Inner Voice

Many of us equate waking 'thinking' with that inner voice we all report hearing in our heads. For many years, the inner voice was considered beyond the reach of psychology. As usual, all it takes is sufficiently ingenious experiments to unmask the phenomenon. Among the first experiments were Dell's studies showing that people reported "slips of the tongue" in internal speech, and that inner speech is as prone to tongue twisters as overt speech.

More recent work on internal speech has been able to make use of techniques such as the fMRI explained in Chapter 19. While awake, we clearly do routinely make use of a form of internal speech that shares many of the early stages of overt speech production. One question is at what stage do overt and covert speech part company?

Emil Kraepelin (1856–1926) was a neuropsychiatrist before Freud created the new and independent field of psychiatry. Kraepelin founded the Munich, Germany hospital, around which the modern Max Planck Institute for Psychiatry formed and which, today, again generates evidence for the brain basis of mind. For Kraepelin, psychiatry was a branch of medicine whose strength lay in careful diagnosis and scientific clinical study. Consciousness and especially dreaming were subjects that Kraepelin thought were brain functions. The Freudian psychoanalytic revolt turned away from Kraepelin as neurology became divorced from psychiatry. One of the victims of this revolt was the formal mental status exam, honored in the breach by psychoanalytic practitioners. The formal analysis of dreams restores this canon of neuropsychiatric medicine to its rightful place in consciousness science.

Thinking is not the same as inner speech that doesn't quite make it to overt articulation, as shown in one psychology experiment. In this study, the experimenter allowed himself to be poisoned by curare. Curare completely paralyzes the body — including the muscles of respiration, so the person has to be maintained on breathing support until the effects wear off. It therefore naturally paralyzes the throat muscles, and those involved in working our larynx, or voice box. If thinking is nothing more than very quiet sub-vocalization, the experimenter should not have been able to think while paralyzed. However, he could still report later on the contents of consciousness while unable to articulate speech. He is the hero of this section in showing that there is more to thought than speaking.

Inner speech is much reduced in REM sleep dreaming. This datum is referred to as diminished self-reflective awareness and the finding suggests that the inner speech of waking may be lost together with other cognitive capacities (such as attention and memory). These losses of cognitive capacity have been ascribed to diminished activation of such forebrain regions as the basal frontal lobes and the parieto-occipital junction of the cortex. Waking consciousness is functionally useful in part because it permits us to recognize our selves, to be aware of other selves, and to recognize that other selves have minds like ours. We all have a theory of mind even if we are entirely unaware of it.

The scientific study of linguistic capacities in sleep and dreaming are now possible to contemplate. The famous German psychiatrist, Emil Kraepelin, was convinced that dreaming was aphasic and kept a dream journal of his own to test this idea. The rise of neuroscience outlined on page 30 has triggered a resurgence of organicist theories. We will say here only that we wish linguists like Andrea Moro, our colleague in Pavia, Italy success with their efforts in this field. Our own dream experience convinces us that language, including inner speech, is dramatically state dependent. Proving that this is so awaits experimental attention.

The Embodied Self

You seem to exist in your body. You almost certainly locate your consciousness in your head, and your head is attached to your body. For most of us, "I" am somewhere in the middle of my head behind my eyes and between my ears. We also know that our body belongs to us and that there is a very special boundary between our body and the rest of the world. This idea that our self is located in our body, and that we are located in our body and particularly head is called *embodiment*.

Like everything else, embodiment can go wrong. An example of a transient difficulty with embodiment is the *out-of-the-body experience* (OBE), in which a person sees himself from outside his body (a phenomenon known as *autoscopy*) — typically above his own body as if he was floating just above it. Reports of OBEs are common in dreaming; it is estimated that up to 10% of the population reports an OBE at some point. However, every year we ask in class if anyone has had an OBE, and no one puts his hand up. Neither of us recalls having had one either but the German philosopher, Thomas Metzinger, assures us that OBE's are as real as dreams themselves. Our students must be embarrassed as Metzinger himself to admit their flight from their bodies.

OBEs can be prompted by brain injury and mechanisms that alter the state of consciousness, such as psychedelic drugs and sensory deprivation, but they can also occur spontaneously. OBEs are not paranormal experiences and do not involve the soul temporarily leaving the body; instead, they occur because of a transient malfunction, rather similar to dreaming itself and, particularly, lucid dreaming. Our models of where we are and where our bodies are no longer coincide.

Our sense of embodiment can be permanently affected by brain damage. One consequence of damage to the area around the junction between the temporal and parietal lobes on the right side (called the *right TPJ — for temporoparietal junction or the parietal operculum*) is that patients see an image of themselves outside their physical body. An example of such a case is a patient of Zamboni, known as BF, who suffered damage to her occipital cortex during childbirth. She saw a translucent mirror image of her self one meter in front of her; the image behaved in the same way as she did, and she even saw actions happening to it.

Another example of permanent disorders of embodiment are the *visual-spatial disorders*. Vision and our representation of the space around us are intimately connected. The best example of a visual-spatial disorder is *neglect*. Patients suffering from neglect appear to be unaware of one half of their body and visual field. They ignore stimuli in one half of the visual field. The painter, Anton Raderscheidt, who painted himself as he recovered from a stroke, can see an example of the experience of neglect in a series of self-portraits.

Neglect usually results from right-hemisphere damage, causing left-field neglect. Neglect isn't just a perceptual disorder; the neglect is also spatial and attentional. Patients are reluctant even to look into neglected space. They may dress only half their body, putting on just one trouser leg, or eat only half a plate of food. Neglect is not a permanent loss of information, as shown neatly in an experiment by Bisiach. Bisiach asked neglect patients to imagine the cathedral square in Milan, which they all knew very well. At first, they could describe only what was on the right-hand side of the square. But then they were asked to imagine the scene from the opposite end of the square. They then were able to describe the missing part (but not the part they first described). It is not the knowledge that is missing, but the ability to access that knowledge. Neglect, as such, does not occur in dreaming, possibly because the brain is intact even though its state has been altered.

How do these patients cope with such a strange experience? Neglect is usually accompanied by agnosagnosia, which is defined as the inability to recognize that one is impaired. Essentially the patient is in denial. Bisiach gives an example of a patient who sits at a table with his hands on the table in front of him while the doctor sits beside him and does the same. The patient concludes that a man must have three hands. Denial is found in other conditions as well, and often involves a considerable amount of confabulation (making things up) in order for the patient to be able to explain away what is happening. For example, a paralyzed woman asked for her knitting needles to give her "something to do." All of these examples tell us that the brain is the organ of consciousness and that the state of one determines the state of the other and vice versa.

Figure 18.1. Rubber hand illusion experiment.

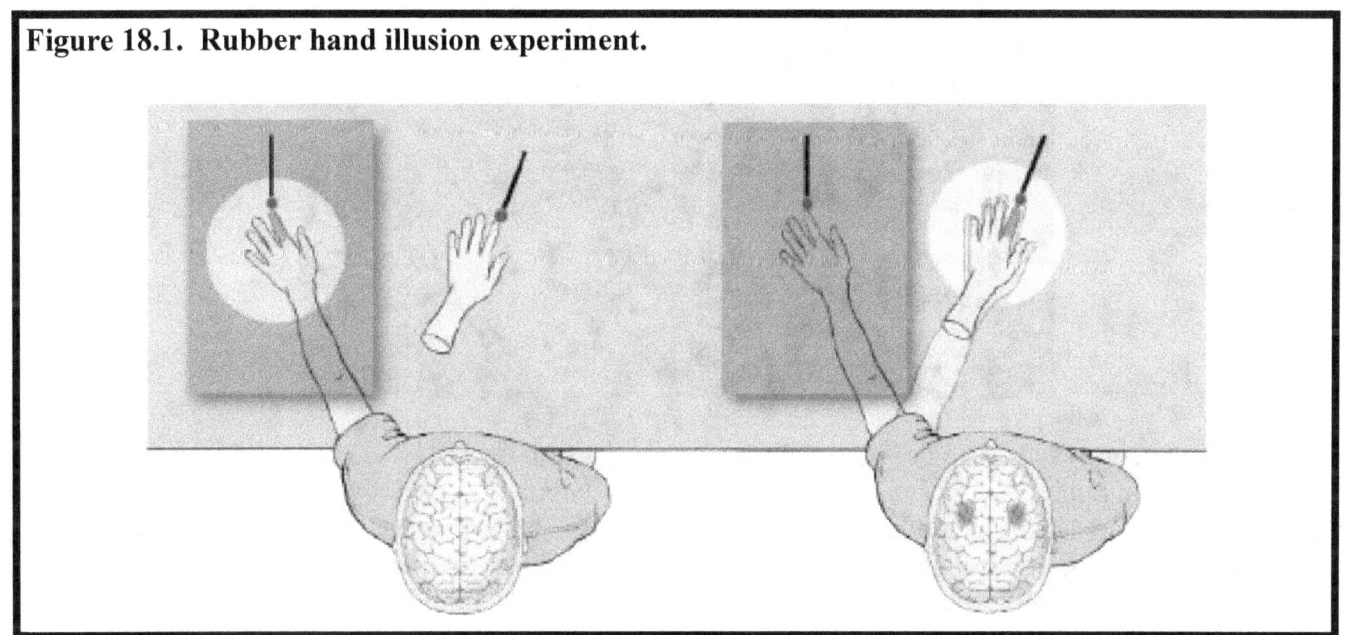

The Fragility of the Sense of Embodiment

It is possible to get some idea of how vulnerable our sense of embodiment is to disruption by a simple demonstration known as the *rubber hand illusion* (RHI). In this illusion, people see a dummy rubber hand being stroked with a brush, while their own hand, hidden from their view, is stroked at the same time in the same way. If the tactile and visual information are simultaneous, then people may think that the feelings on their hand are coming from the rubber hand, and even that the dummy hand is part of their own body. So our mind can be tricked into changing its body image. This result means that we must be able to construct and adjust our body image on the basis of new incoming information. The illusion also suggests that the visual modality is usually the dominant one — if vision and the other senses conflict, vision wins out (at a preconscious level). The same domination of consciousness by vision occurs in dreaming.

Along the same lines, but with far more serious consequences, is the *phantom limb* sensation, where people who have had limbs amputated continue to feel sensation — particularly, unfortunately, mostly unpleasant ones such as severe itching and even chronic pain — in the absent limb. The phantom sensation must arise through the mental representation of the body rather than directly from the body itself. In some cases, the itching can be alleviated by stroking another part of the body; one of Ramachandran's patients, who had lost his left arm, could relieve phantom itching in his left thumb by scratching his face. Phantom limbs show that the brain starts to rewire itself immediately after bodily damage. Ramachandran has also worked extensively with mirrors to generate the illusion that the missing limb is replaced.

Hence although our consciousness is embodied, this embodiment can be disrupted. Although extraordinary things can go wrong, our inner core seems to be preserved. The Portuguese-American neuroscientist, António Damásio, has conducted research in this area, and has studied how brain structures relate to conscious experience. He argues that in these cases, the "core consciousness" of a person (equating to self-awareness) is intact, but his "extended consciousness" (equating to the concept of himself in space) is damaged.

António Damásio (1944–) is a Portuguese-born neurologist who trained with Norman Geschwind at the Harvard Medical School before establishing an influential school of cognitive neuropsychology at the University of Iowa. Damásio and his wife, Hanna, have pioneered the application of quantitative testing of defects in consciousness evinced by patients with brain lesions (work that they continue today at the Salk Institute). An example of the Damásio's originality is their anatomical analysis of the damage to the brain of Phineas Gage, a railroad worker who suffered a personality change after he was accidentally lobotomized when a dynamite explosion forced a crowbar through his brain. Unusual is Damásio's sensitivity to subjective experience, especially emotion (e.g. *The Feeling of What Happens*) and his historical reading of philosophy (e.g. *Descartes's Error* and *In Search of Spinoza*). His pellucid writing style has appealed to a wide circle of readers wishing to educate themselves about the new science of consciousness.

Consciousness and Perception

For most of us, awareness of the auditory and visual worlds plays a very important role in our mental life, dominating the contents of our consciousness. Indeed, many models of consciousness give a central role to visual awareness in consciousness. We don't emulate these authors, because there is more to life than vision, but we acknowledge the prominence of vision in perception. Our position is that vision contributes powerfully to perception, whether it be real or virtual.

There are several situations that demonstrate that we cannot always trust our visual and auditory senses. The existence of visual illusions shows how easy it is for our mind to be misled: certain configurations of lines can induce all sorts of non-existent movement and color. Hallucinogenic drugs cause visual and other sorts of hallucination that have no or very little basis in a sensory input. People with schizophrenia hear voices; these are auditory hallucinations which must be internally generated. When between sleeping and waking, in the hypnagogic state, we occasionally hear clear voices that aren't of external origin. Our capacity for perceptual illusion is strong, giving weight to our activation-synthesis hypothesis.

The visual system can be disrupted by brain damage in many ways. Damage to the core visual processing pathways results in a deficit known as *agnosia*, which causes people to have difficulty in recognizing everyday objects. In the condition known as *prosopagnosia,* they have particular difficulty with faces. Face recognition is impaired in normal dreaming.

Steven Laureys (1968–) is a neuroscientist who studies the brain basis of normal and abnormal states of consciousness at the University of Liège in Belgium. He collaborated with Pierre Maquet on the first PET studies of human REM sleep which revealed a similar brain stem activation pattern to that seen in animals (See also Allen Braun cameo on p. 230). His model of brain damage effects on consciousness is similar and compatible with AIM. Most notably, coma is the result of brain stem damage. Together with Giulio Tononi he has published the first full-length book on the *Neurology of Consciousness.*

In the rare condition known as *Anton's syndrome* (also called the Anton-Babinski syndrome), patients are blind as a consequence of damage to the occipital lobe, yet they deny their cortical blindness and insist that they can see. Obviously this denial (anosognosia again) is accompanied by an extreme amount of confabulation. For example, the neurologist, Oliver Sacks, described a patient who said he enjoyed watching television although "my eyes aren't that good." Sacks proposes that "the patient has lost the idea of seeing." The reasons for the denial of the blindness are not currently understood. So Anton patients think they can see but cannot.

In contrast, *blindsight* patients think they cannot see, but can. They report being blind, at least in part of the visual field (they have a blind spot, or *scotoma*), yet they can retrieve some information from the visual field. When asked to guess (for example, if a light is flashing or not in their scotoma), they perform well above chance. Furthermore, they make appropriate saccades to objects. The most famous and best researched case of blindsight is patient DB studied by Lawrence Weiskrantz. Blindsight arises because visual information is processed by at least two routes: the midbrain-orienting visual pathway is still intact in blindsight patients. According to Weiskrantz, stimuli are seen but do not reach the "commentary stage."

Note that people with small blind spots (called scotomata) don't go around reporting big gaps in their awareness. The visual system fills in for them: for example, they will complete straight lines (even if there is a gap in their scotoma line). In fact, we all fill in the visual world as our experience of the ubiquitous *blind spot* (where the optic nerve passes through the retina) shows. We need to fiddle about quite a bit before we can become aware of the blind spot — and even then we don't see a black hole, but instead the brain fills in material from the immediate environment.

We fill in the auditory world, too. In the famous phoneme restoration effect, people hear the sentence "The state governors met with their respective legislatures convening in the capital city." The /s/ sound in "legislature" was replaced with a cough. Nevertheless, people report hearing the /s/ sound, and the cough sound is moved to between the words.

So the brain fills in gaps in the perceptual world for us, and we are unaware of things that are missing. Furthermore, we can distinguish between perceptual processing and awareness of processing, and our awareness is not always reliable. Dream consciousness reveals the synthetic nature of consciousness; in dreams, the images arise entirely within the activated brain.

Filling In and Dream Perception

Dramatic examples of filling-in are the dream reports of virtual visual perception in the congenitally blind, virtual auditory perception in the congenitally deaf, and virtual movement perception in the congenitally paralyzed. These reported data need further evaluation but point to an impressive ability to create conscious experience from internal sources. To discuss these data critically requires more than the Sherringtonian reflex model and the Freudian model of dreams as wish fulfillment.

As intriguing as are the over-inclusive claims of disabled dreamers is the curious absence of disability from their reported dream content; in their dreams, the blind do not use white canes or seeing-eye dogs, the deaf do not use hearing aids, and the disabled do not use wheelchairs although all three aids are used by day. The brain must be endowed with a confabulatory skill of momentous power.

The Executive Self and Free Will

Consciousness research at the clinical level addresses several issues related to free will. Any sense of free will seems to be lost in the neurological disorder known as *akinetic mutism*, which is characterized by a total inability to initiate action. The executive commands to move or think or feel emanate from the frontal cortex.

Frontal lobe, or executive, disorders change the nature of the executive self and result in what is known as the *dysexecutive syndrome*. For example, consider the famous case of Phineas Gage, who was a railway worker when, in 1848, an accidental explosion drove a large iron rod through his head, almost completely destroying his frontal lobes. Among the many changes the accident induced were a "personality change," from a pleasant mild-mannered man to an irascible, coarse, rude fellow, who had difficulty in planning and following plans (for at least some period of time). The frontal lobes host our ability to plan and choose effectively. Further evidence for this claim comes from the study of the effects of *lobotomy*, a form of psychosurgery once widely used as a medical treatment. Formerly placid, pleasant people become irascible, unreliable, and aggressive — and vice versa. Furthermore, the ability to plan effectively is disrupted.

These findings show that our freedom to choose is at least in part illusory. We do what our brains tell us to do.

The Personal Self

We have already talked about free will — or perhaps we should say — if you are convinced by the story of Laplace's demon, the *illusion* of free will. Our personality is our inclination to respond in a certain way in a particular circumstance, with that inclination determined by our genes and environment. So an introvert has a good excuse: I had no choice but not to go to that party I was dreading — my genes and environment made me refuse the invitation.

There are interesting ramifications of this argument for society. If free will is an illusion, can criminals be held responsible for their crimes? If I couldn't have done anything other than rob that bank just now, if I had no choice, how can I be held responsible and punished? Here we enter the field of *neuroethics*, considering issues of right and wrong in the light of what we know about the brain.

If "normal" people cannot be held responsible for their behavior, what about people with *personality disorders*? The best known personality disorder is psychopathy, where affected individuals appear to have no conscience; they act without guilt or remorse, manipulate others, are antisocial, and are strongly inclined to get whatever they want; however, sometimes they can appear to be superficially very charming.

There is a genetic element to psychopathy, and brain imaging suggests that the brains of psychopaths are different from "normal"; when seeing or imagining others in pain, they fail to show activation in the parts of the brain associated with empathy (particularly the amygdala and ventromedial prefrontal cortex); in contrast, when asked to imagine themselves in pain, their brains light up as normal. If psychopaths have abnormal brains, to what extent can they be held responsible for their abnormal behavior?

A major problem of consciousness science is to link the microscopic neurons and molecules to thought and emotion, as we attempt to do in this book. Several examples will now be presented to indicate how this might occur. A few parasites can change the behavior of their insect hosts. One of the best examples is the Lancet liver fluke *Dicrocoelium dendriticum*. The adult lives in the liver of a cow, where it mates and lays eggs, which are excreted by the cow. A snail eats the infected feces. The eggs hatch in the snail, and the parasites then provoke a "coughing response" in the snail. An ant then eventually eats the infected mucous. The flukes then make their way to the ant's head, and in the dark they make the ant climb up a blade of grass and hold tight, until eventually the ant is eaten by cows. And thus the cycle begins again.

Ants and snails have little brains, so perhaps it's not surprising that they can be controlled by parasites. Surely humans are too big, too complex, and too clever for their behavior to be influenced by tiny creatures. But this may be a mistaken assumption. Toxoplasmosis is a disease caused by the protozoan (a single-celled animal that can move), *Toxoplasma gondii*; although it mainly affects cats, humans can also catch the parasite, usually through ground or grass contaminated with cat feces. It's been estimated that half the world's population is infected by Toxoplasma, usually without symptoms. However, a recent study has shown that the suicide rate is higher in people with toxoplasmosis, and also that sufferers are more likely to be involved in traffic accidents. In general, it seems that toxoplasmosis can elevate distractibility, making people more accident prone. The idea is that the parasites act by making distractible animals more prone to predation by the primary host (e.g. a cat is more likely to be able to catch a mouse that is jumping up in the middle of road saying "eat me", or "run me over"). The precise mechanisms that convert parasite infestation in the brain into behavioral change are currently not understood, although presumably attention is involved.

Our behavior can be affected by the smallest things. Neurons are small and several billions of them conspire to make us conscious.

Dialogue 18. *Mind Blowing Brain Injury*

AH: The brain and its mind are so subject to error that it is a wonder that we do as well as we do.

TH: It helps to ignore our faults and to focus on our successes.

AH: Isn't that the power of positive thinking, the triumph of optimism over pessimism?

TH: The same cup of water that is seen by optimists as half full is also half empty when a pessimist looks at it.

AH: I like the metaphor even if we are more than half ignorant. I am optimistic about progress in consciousness science while you see even brain research as falling short.

TH: The effects of brain damage nonetheless emphasize how vulnerable consciousness is to disruption.

AH: And the normal vicissitudes of our states of consciousness caused by brain state change underlines this point.

TH: We have enough trouble with normality to beware of making things worse intentionally. Some people seem determined to destroy their minds by altering brain chemistry.

AH: I was always afraid of drugs but I poisoned myself with alcohol on weekends from early adolescence through late middle age.

TH: Alcohol and drugs are now an integral part of Western culture. Can we compete with teetotalling religious fundamentalists and more abstemious Eastern peoples?

AH: We seem to be at war over these very issues. Functional brain damage can weaken our technological edge.

TH: Let's strengthen our edge and give heed to the self-destructive effects of mind-blowing.

AH: That's never been a popular prescription.

TH: No wonder Sigmund Freud postulated a death instinct to account for our suicidal tendencies. No wonder he himself abused cocaine.

AH: I hate to give religion credit for anything but I have to admit that moral and physical purity are both valuable to us.

TH: Scientific humanism can adopt the same values as orthodox sects.

AH: Consciousness science provides a useful basis for secular humanism and it may help to keep people off the streets, out of bars, and out of brothels.

Chapter 19. Lucid Dreaming

This chapter examines what little we know about the content of dreams. Our emphasis is on the psychology of lucid dreaming, wherein dreamers have some awareness that they are dreaming. Do non-lucid dreams have any meaning, or is their content randomly generated? We will suggest that the answer to both questions is yes. We favor a both/and rather than either/or position.

For much of history, it was thought that dreams contained premonitions of the future, and many people still believe this idea today. We must say that there is no scientific evidence to support this idea that dreams can reliably predict the future when prediction concerns floods or famines. The Bible contains accounts of dreams, of which our favorite is that of Nebuchadnezzar who had a dream of great significance, but when he woke the next morning he was unable to remember it, following which "he was greatly troubled." We have described the antidote to prophetic dream forgetting as practiced by Otto Loewi, the discoverer of acetylcholine.

Our virtual reality theory (detailed in Chapter 24) asserts that all states of consciousness are prophetic in a narrower, sensorimotor sense. They tell people, without words, what to expect next but this is not what most people mean when they talk of dream prophecy. Of course when most contemporary people think of dream meaning, they think of Sigmund Freud and his interpretation of dreams theory. Freud believed that dreams disguised hidden content, the real meaning, because it was unpalatable to the ego, and might wake or distress the dreamer. So, for example, a Freudian penis might become the Eiffel Tower. For Freud, dreams were *symbolic*. The meaning of these symbols could only be uncovered through dream interpretation with a knowledgeable therapist. However, there is scant evidence that dreams are symbolic or that they disguise our fears, wishes, or repressed sexuality.

Along similar lines, the many dream dictionaries that you can buy telling you that seeing a wasp in a dream signifies the expression of angry thoughts, or a spider stands for female creativity, have little basis in scientific research. We belabor these points because the temptation to interpret dreams is so strong an impediment to science. Dream interpretation is fatally linked to the hope that dreams foretell the future. Freud claimed that he could free a person from the prophecy of his past.

Antti Revonsuo (2000) argues that dreams help us deal with threats. In particular, they help us simulate threats and design healthier ways of responding to them. However, analysis of dreams provides only mixed support for this idea. While a good proportion of dreams contain threats or elements of threats, it is difficult to estimate what a baseline rate of threat would be, and only a small proportion of recurrent dreams contain life-threatening situations. Furthermore, the dreamer rarely succeeds in simulating escaping from the threat, casting doubt on the idea that finding a way of escape is the primary purpose of dreams. Finally, some dreams are so positively delightful that only a Freudian could regard them as concealing threat.

Dreams are often permeated by an atmosphere of anxiety, sometimes involving the replaying of stressful events in life. When we were younger, we had dreams about exams: we would turn up at the examination hall without having prepared for the exam. Now we have dreams about giving talks and presentations: we turn up without the slides. In as much as dreams are involved in learning, it is not surprising that some of their content involves everyday events, or fragments of the previous day's events. All of these theories pale before scientific import of the main theme of this chapter, lucid dreaming.

Definition and Historical Details

Given its now apparently universal prevalence, it is surprising how little attention was paid to lucid dreaming before the second half of the 19th century. Almost everyone experiences, at one time or another, the recognition that one is dreaming while dreaming. The term "lucid" means insightful. Instead of erroneously thinking oneself awake, one correctly identifies his conscious state as sleep. It is thus clear that at rare times there may coexist two states of consciousness that are usually discrete. This fact might have attracted attention repeatedly throughout history because it is so philosophically, psychologically, and physiologically intriguing and informative. But, until recently, lucid dreaming was ignored or denied.

Stephen LaBerge (1947–) is the leading popularizer of modern dream science. He is convinced that lucid dreaming increases the power of the mind and is determined to convince others that this is so. After pioneering the laboratory study of dream lucidity, he has created a private institute for its promulgation and a variety of techniques for its enhancement. His most ambitious invention is the "Dreamlight" which converts the EOG signals of REM into flashes of light behind the sleeper's mask to facilitate awareness that he is dreaming. Unfortunately, LaBerge has not published the results of his private enterprise and many scientists are leery of his hucksterism but there can be no doubt of his passion and eloquence for subjective/objective psychoanalysis. In Stephen LaBerge, Sigmund Freud meets Anton Mesmer and passion and eloquence sleep together.

The earliest recorded experience of lucid dreaming is that of Hervey de Saint-Denys, an aristocrat who lived in 19th century Paris, where he was such a respected scholar of Chinese culture that he was made a member of the Académie Française. His sideline interest in dreams derived from his moral quandary regarding unsavory aspects of his own oneiric life. He resolved to restore conscious will and moral conscience to his sleep and, by becoming lucid in his dreams, he claimed to have succeeded in doing so. Meanwhile he described and illustrated his nightly experience in a 23 volume dream journal which, sadly, has never been found. Fortunately, he published a book entitled "Les Rêves et Comment Les Diriger" (1867), which is now available in English translation as "Dreams and How to Direct Them."

A more succinct and enlightened instructional text is *Studies in Dreams* (1921) by Mary Arnold-Forster, the niece of the famous English novelist, E. M. Forster. She wanted to test psychological hypotheses about dreaming that were then popular and, when she became lucid, to explore the experiential aspects of her newfound state. Flying to and fro about her house was as exciting to Mary Arnold-Forster as it is for most lucid dreamers. Other amusements are possible. A faculty colleague of ours enjoyed lecturing to his students about lucid dreaming as he circled, weightless, overhead. A facsimile copy of *Studies in Dreams* is now available.

The technique of lucid dream induction is quite simple. You place a recording device on the bedside table to physically instantiate the task. When the bedside light is turned off, hypnotic autosuggestion tells the participant that awareness of dreaming will be triggered by the recognition of bizarreness (cognitive discontinuity or incongruity). Self-induced awakening is followed by detailed mental rehearsal of the dream and/or the physical recording of a report. Ten trials and a time investment of about 5 minutes per trial should suffice. Young students have more luck with lucid dream hunting than older professors (but both Saint-Denys and Arnold-Forster were more middle-aged than adolescent/young adults when they made their successful experiments).

The phenomenological features of dream lucidity include: awareness of awareness replacing the usual uninsightful single-mindedness of dreaming; dream interruption with partial arousal facilitating recall; plot selection or redirection; return to the interrupted dream; voluntary abandonment of control, allowing the dream to run free; self-treatment for relief of anxiety, fear, aggression, or submissiveness. The scientific importance will be discussed below but, besides home entertainment, dream lucidity permits self-observation of a unique sort: one part of the brain-mind watches and influences another part of the brain-mind. One of us enjoyed the full panoply of dream lucidity on reading Mary Arnold-Forster's wonderful book.

Laboratory Studies of Lucid Dreaming

While still a graduate student in the Stanford University sleep and dream laboratory of William Dement, Stephen LaBerge recorded brain waves, eye movements, and muscle tone from lucid dreamers and provided the necessary third-party, objective observations that helped to bring the fringe (and therefore suspect) phenomenon of lucid dreaming into the mainstream of consciousness science. The subsequent study of lucid dreaming has more than repaid LaBerge's charismatic zeal and conviction. It is our opinion that lucid dreaming now offers fundamental and signal scientific opportunities.

According to LaBerge, lucid dreaming always arises out of REM sleep. This claim has two important meanings. One is that dreaming is indeed a REM sleep correlate. The other is that REM sleep subjectivity is both experimentally observable and manipulable. Even dream consciousness may be rendered causal. Dream consciousness appears to be able to control brain physiology if it is given some of the power of waking. This is self-hypnosis. But we are getting ahead of ourselves.

Skeptics who thought that LaBerge's lucid dream subjects were awake were partially quieted by his report that they could "signal out" of REM by making a preplanned sequence of voluntary eye movements (e.g. left-right-left-right-left-right) that were interspersed with the more chaotic eye movement sequences of REM. This meant the lucid dreaming was the product of a combination of waking and REM sleep physiology. Lucid dreaming does not occur in *either* REM sleep *or* waking but in *both* REM sleep *and* waking. The controversy that drove Stephen LaBerge away from the center of dream science is resolved by this shift in logic.

Our formulation is that lucid dreaming is an experiment of nature that lends itself to the experimental study of consciousness. The rarity and evanescence of the state can be countered (as LaBerge did) by using lucidity adepts as expert subjects. That their exceptional ease at becoming lucid in the laboratory is not qualitatively abnormal is affirmed by the epidemiological evidence to be presented in a subsequent section of this chapter.

Ursula Voss has conducted quantitative EEG studies which suggest that specific regional brain activation differences distinguish lucidity from both waking and non-lucid dreaming. Even the fMRI scanner, a hostile environment for any delicate human behavior or conscious state, has already shown itself to be capable of hosting lucid dreamers. The bottom line is that myriad conscious states are determined by myriad brain states, just as dual-aspect monism predicts.

The Early Development of Lucid Dreaming

It is generally agreed that dreaming, as we adults know, cannot be reliably identified in reports of children before the age of five or six. David Foulkes, a psychologist who trained with Nathaniel Kleitman, performed extensive laboratory research at the University of Wyoming in the 1960s, and concluded that dreaming, like language itself, was a cognitive skill that was a product of prolonged brain development. It is understandable that the onset of dreaming (or at least reporting about it) is contemporaneous with grade school readiness, when other cognitive skills ripen and separation from caregivers is possible.

The early occurrence of reports of lucid dreaming by children came much later and was surprising. From the scientific literature and hearsay, it was thought that lucid dreaming was a young adult–middle age phenomenon. Two interpretations may explain this misperception. One is that children, like older people, often do not talk about their dreams if they are not asked. The other is that dream recall, at any age, is fleeting. Even subjects who are faithful dream recorders have no recognition that a dream, recorded by them in detail, ever really occurred. How many of you have any recollection of dreaming in childhood, let alone of dreaming lucidly? Possible exceptions to this amnesia rule are repetitive traumatic dreams but this subject has not been studied either systematically or prospectively.

Ursula Voss (1947–) is a psychology professor working in Frankfurt, Germany. She pioneered the study of the brain basis of lucid dreaming. To accomplish her success, she mastered quantitative EEG measurement from multiple electrodes; the brain's electrical activity was then frequency-analyzed and statistically compared so that the signatures of the conscious states could be compared. Her husband, a physicist, collaborated in this work. I wish that Hans Berger could see these fruits of his 1928 discovery: the brain is an electrical organ. Is thought electrical?

In an epidemiological, prospective study of three thousand German school children, ages 6 through 19, the University of Frankfurt psychologist, Ursula Voss, developed a questionnaire which revealed that claims of dreaming lucidly began at age six and peaked at age nine. The incidence of lucidity was low but, because of the large sample, the peak age results were significant when compared with pre- and post-peak data. Could the age of the brain, especially the maturation of the frontal lobes, determine this peak?

We can offer only a speculative answer for future scientists to confirm or deny. Age 9 is school grade three in the US system. Children are then beginning to become self-aware and to show an increasing capacity for self-reflection when awake. They are taught the reading of chapter stories and can perform rote mathematical calculations. Cortical development almost certainly underlies these changes and could contribute to the increase in lucidity claimed by the children.

These correlations may explain the age 9 peak in the Voss epidemiological data but the subsequent decline remains a mystery. Naps are lost as night sleep consolidates and shortens between ages 9 and 19, indicating that deeper sleep may interfere with both lucidity and its recall. Only an even more searching study using imaging methodology and more quantitative psychology can decide this issue, but the feasibility of lucid dream science in children has been unequivocally demonstrated.

Lucid Dream Psychology

Some lucid dreamers declare that they are able to let their mind run free for a while and watch the dream (as from outside a window) but the more freely they dream, the less they watch and the tendency is then to lose the divided brain-mind that defines lucidity. On the other hand, the more they watch the greater is their tendency to wake up and lose dream consciousness altogether. In other words, the degree to which they could divide consciousness in two was dynamic and variable as if the brain-mind wanted to be in one state (dreaming) or the other (waking) but not both at once. This led to the hypothesis, later confirmed, that lucid dreaming was at the sleep-wake interface.

Subjects who become adept at lucid dreaming assert that lucidity is the gateway to a variety of unique subjective experiences. They can fly (such as the case of Mary Arnold-Forster), have physical intimacy with a chosen partner – or not (à la Hervey de Saint-Denys), or engage in a series of Harry Potter-like adventures (walking through a wall, for example).

The pioneer self-observers, Jay and Janice Vogelsong, have described home-based experiments along these lines. Again the scientific hypothesis of a unique hybrid state of consciousness springs to mind. Subjects learn to voluntarily wake themselves up (to enhance recall), go back to the same dream, recommence it (as if the brain were in pause mode), change the dream plot at will, and delight in their innocent psychedelic prowess. Their prowess is innocent and psychedelic because no drugs, no danger, and no invasion of privacy are involved.

Home-based studies using simple, portable monitoring systems have confirmed LaBerge's finding that lucid dreaming always emerges out of REM sleep and further demonstrates that, if the awakening occurs early in REM, the same REM period may be resumed and re-entered with a second bout of lucid dreaming. LaBerge has developed his *Dream Light* device to signal when home sleepers are in REM. This facilitates awareness of dreaming. Experimentation on consciousness has thus been put in the hands of every man, woman, and child (over nine), including high school students, collegians, and many curious people outside of academic institutions. Most people will do this for fun but systematic, hypothesis-testing science is not out of reach.

Frontal Lobe Activation and Lucid Dreaming

A technique akin to brain imaging called quantitative or qEEG allows consciousness scientists to quantify the degree of activation in regional recordings of brain waves. The electrical power of the brain sensed by multiple electrodes can be averaged and displayed in pseudo 3-dimensional color-coded arrays. When this was done, again by Ursula Voss and her colleagues, dramatic and highly significant differences between waking, lucid, and non-lucid dreaming were observed. Activation of the frontal brain was highest in waking and lowest in non-lucid (normal) dreaming. Lucid dreaming was associated with frontal lobe activation that was intermediate between the waking and non lucid dreaming extremes. This result confirms the hypothesis that lucid dreaming is a state of consciousness at the interface of waking and sleep. It is also compatible with the idea that lucid dreaming is a hybrid state of consciousness with elements of both waking and dreaming.

The importance of the frontal lobe localization of the qEEG findings is its corroboration of classical psychology. A major difference between normal waking and normal dreaming consists of differences in the working memory functions of: self-reflective awareness, short-term memory, and volition. All three of these cognitive features of waking cognition are lost in dreaming. Furthermore, as previously mentioned, they are all supposed by neuropsychologists to be frontal lobe functions. In fact, they probably constitute working memory, which is normally defective in non-lucid REM.

We may tentatively conclude that the activation state of the frontal lobe critically determines our state of consciousness. We assume that the activation of the frontal lobe is, in part, a response to brain stem signals but must now assume that the frontal lobe is also responsive to self-activation via the mind-brain loop that is part of the new causal paradigm of dual-aspect monism.

Figure 19.1. Quantitative electroencephalographic (EEG) studies comparing brain activity during waking, lucid dreaming and REM sleep. Frontal areas are highly activated during waking but show deactivation during REM sleep. During lucid dreaming there is an increase in 40 Hz power and coherence in frontal areas compared with non-lucid REM sleep. In lucid dreaming, additional electrical activation of the brain is needed to activate the dreamer's forebrain enough to recognize the true state without causing waking and thus terminating the dream. Differentiated regional activation may underlie the phenomenological distinction between the states of REM sleep, lucid dreaming, and waking. Scale bars indicate standardized power based on scale potentials (0.50% to 1.50% power).

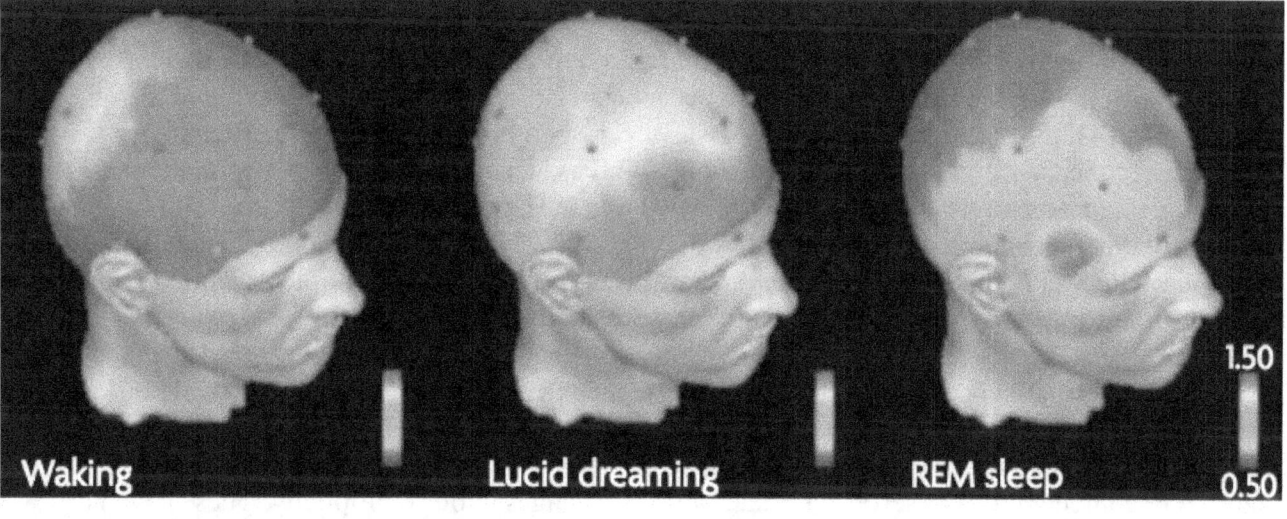

Waking   Lucid dreaming   REM sleep

1.50

0.50

A critical test of these hypotheses is the induction of frontal lobe activation and lucid dreaming by electrical stimulation by Clemens Frenzel and Jana Speth. Human subjects can be fitted out with scalp electrodes in addition to monitoring sensors. Low-voltage stimuli delivered to the scalp overlying the frontal brain during REM sleep surprisingly do produce immediate activation of the regional EEG and the electrically-induced frontal lobe activation is associated with dream lucidity.

The frontal lobe is not the only cortical region activated concomitantly with dream lucidity. A parieto-temporal network has been identified by the Washington University neuroscientist, Marcus Raichle, which becomes active when experimental cognition subjects are not on task, that is to say when they are resting quietly in the fMRI recording apparatus instead of performing tests. We do not know if these subjects are thinking, fantasizing, dreaming, or dozing, but the same network has been shown to be activated in lucid dreaming. This discovery was made by Martin Dresler and other members of the team led by physicist, Michael Czsich, at the Max Planck Institute for Psychiatry in Munich, Germany. Despite the obstacles of sleeping in the magnet, subjects can and do fall asleep, and can and do dream both normally and lucidly in an fMRI magnet. This is just the beginning of a brave new age in consciousness science.

The Cartesian Theater Revisited

The spirit of Descartes may not be entirely dead after all. With dualism, he was on to something and now that we know a great deal more about the brain than he did, his metaphor of the theater comes back to us. In lucid dreams, a virtual self is seated, we now have reason to suppose, in the frontal brain and, suddenly self-aware, he watches his very own dream played out on a stage, or screen, in the posterior brain (or all over the cerebrum including the watcher's seat). This interpretation is not easily accommodated to dual-aspect monism because there are two dual-aspect selves in play. *Pace* Descartes, neither of the two selves is a perfectly synchronized watch set in motion by God, but still there are two of us so our theory must allow at least for multiple selves.

A word which, throughout the twentieth century, was almost as vehemently censored as "consciousness" was "homunculus." The idea of a little man, seated in the head and watching a show, was said to explain nothing and we must agree with the essence of this critique. But how are we otherwise to account for the lucid dreaming evidence? We must accept the multiplicity of selves and assume that each self is the subjective awareness of a regional brain state. The brain regions can communicate with each other much as persons do at cocktail or dinner parties, at conventions and committee meetings, and, indeed, in our fantasies when we "talk to ourselves."

The split brain-mind of lucid dreaming bears comparison to the split brain of epileptic patients whose corpus callosum has been surgically cut to prevent seizure spread from one hemisphere to the other. These patients were studied by the embryologist, Roger Sperry, and the cognitive psychologist, Michael Gazzaniga, at the California Institute of Technology in Pasadena, California. Split-brain humans have two cerebral hemispheres each with its own isolated mind. They can sense and reason with both their left and right brains but cannot integrate the two. They can only speak and report out from their left hemisphere. This experiment confirms the theory of multiple selves and multiple psyches. Unlike lucid dreamers, their split is unfortunately structural and irreversible.

The rarity and evanescence of lucid dreaming is evidence that nature tries her best to avoid this sort of confusion. She wants us to believe that we are one and only one person in one and only one state, but nature is not perfect and she makes mistakes, which scientists must recognize and utilize. The truth is that we are confederations of selves and states with advantages to our multiplicity. These advantages relate to the adaptational value of consciousness itself and hence to its function.

An animal that has a sense of itself in relation to its conspecifics is aware of its competitors. If some of its competitors are not conspecifics, they are likely to be, at best, limited to primary consciousness. While wary and instinctually gifted escape artists, they are no match for conscious humans. With the aid of secondary consciousness, humans can design traps that give them food and warmth for survival purposes, and bedroom decoration useful to procreation. Moreover, these basic functions constitute a virtual self or multiple virtual selves which can be used to construct social strategies of an infinite variety. This advantage culminates in man's creativity, the ultimate product of consciousness.

René Descartes needed no such insights to construct his dualistic philosophy, and just as we cannot prove him wrong in his theological assumptions, we can keep him respectfully in mind as we try to philosophize without them. We assume that Descartes was a lucid dreamer and regret that he was unaware of his own split dualism. It is perhaps a little too early to tear down the Cartesian theater.

Dialogue 19. Virtual Sex

TH: In the 19th and early 20th century, many lucid dreamers wanted to expurgate their dreams. They were at pains to bring their private desires into register with public convention.

AH: Since about 1960, thanks to birth control, humans have been considerably less squeamish about sex.

TH: Still, even licentious people need to protect their privacy.

AH: Fantasy and dreaming are as private as normal waking. Now sexual pleasure can be achieved without fear of exposure.

TH: We ought to have a consciousness consolation prize for lucid dreamers. We might call it the Hervey de Saint-Denys award.

AH: Let's do a thought experiment to a description of the challenge. How does lucid out-of-body sex sound to you?

TH: Great! As long as we teachers can practice OBE sex, too.

AH: Like education, sex is wasted on the young. The first step is to open your eyes when you make love.

TH: It helps to make love in daylight or with electric illumination. In my dreams, the lights are always on.

AH: In waking, you can install mirrors above, beside, and aft of the bed.

TH: You are always a first person lover but you can add third person sensation to the mix.

AH: Voyeurism has long been practiced as an aphrodisiac.

TH: In a hall of mirrors, making love is multiplied and integrated. In our thought experiment, one could practice voyeurism on one's self.

AH: Why might men take to this more easily than women?

TH: Women are often committed to motherhood and the cult of virginity. This may make self-indulgent sensuality difficult.

AH: Keeping their eyes closed in the dark helps them concentrate on the task at hand. At the same time that they are going for it, they cannot watch.

TH: That helps them to deny the obvious.

AH: The visual enhancement of eroticism indicates that brain activation can be intensified, possibly by the recruitment of more neurons.

TH: It is theoretically possible to leave your body with your partner on the bed, float up to the ceiling, and watch the two of you having fun down below.

AH. The brain of the dreamer would presumably evince a specific activation pattern.

TH: That's a testable hypothesis. All you have to do is make love in a scanner.

AH: Be sure that neither you nor your partner moves your head.

TH: I think I will do this for fun. Screw the science.

AH: Amen. Ah women!

Chapter 20. Sleep Disorders

We think it significant that the clinical problems of sleep and dreaming are called disorders and not diseases. Thus insomnia is usually a reversible change in psychophysiological state, not a permanent alteration of structure or function. Some sleep disorders, like narcolepsy and sleep apnea, have genetic underpinnings and thus warrant consideration of disease formulations, but their reversibility also warrants the disorders nomenclature. We will try to do justice to both nomenclatures while admittedly favoring the disorders concept, which we hope will spread further into medicine and clinical psychology. In this chapter, we will base our necessarily limited discussion upon the AIM model.

Pathophysiology and AIM

Pathophysiology may be defined as the dysfunction caused by an exaggeration and/or the diminution of a normal function. Thus insomnia may be a dysfunction caused by an exaggeration of anxiety, which in moderation is a normal emotion. Sleep disorders are of three general types: 1) too much sleep (the hypersomnias) or too little sleep (the insomnias); 2) the wrong kind of sleep (the parasomnias, motor disturbances like sleep walking, sleep talking, or bedwetting) and; 3) the exaggeration of intrinsic sleep dysfunction, like narcolepsy and sleep apnea. These disorders are caused respectively by: 1) *set point errors* — resulting in too little or too much wake-state activation at the expense of sleep; 2) *timing errors* — resulting in the activation of motor pattern generators before REM sleep motor inhibition is instantiated; and 3) *genetically determined state overlaps* — such as falling asleep at unwanted times (narcolepsy) and reduction in breathing efforts during sleep (central sleep apnea).

The brainstem can be viewed as set of coupled oscillators which must interact in concert to guarantee smooth and harmonious state changes. The circadian oscillator is in the hypothalamus where its 24-hour rhythm governs the cyclic alternation of rest and activity. The circadian oscillator times the activation and deactivation of the pontine NREM-REM sleep oscillator in the pontine tegmentum. The respiratory oscillator in the medulla controls the frequency and strength of breathing efforts. All three oscillators interact as conscious states change.

Consider insomnia, the most common sleep disorder and one for which millions of medical prescriptions are issued each year. Too much wake-state activation opposes sleep, and is epitomized by the expression "don't take your troubles to bed with you." Waking life is normally conflictual; competition and loss are inevitable in life and it is difficult to stop dwelling upon the problems of waking. The first line of defense is philosophical resignation. This mind reset is reinforced by systematic relaxation, an integral element in meditation. Muscle relaxation decreases arousal signals from the spinal cord. Reading and bed partner soothing are imperfect adjuncts which should be maximized before turning to medication. Two simple rules are relevant to the naturalistic approach advocated here: Good sleep normally follows bad. As they say in New England, "If you don't like the weather, wait a minute." All soporific drugs are habit-forming. Think twice before taking a sleeping pill.

The failed task of synchronizing far-flung and diverse neuronal generators in the production of state transitions is one of the causes of the *parasomnias*. The spinal cord and the cortex are widely separated brain structures and linked by very few direct synaptic connections. State changes are time-consuming and involve the polysynaptic activation of one brain subsystem and the inactivation of others. No wonder that their coordination in brain-mind state changes is imperfect. To avoid walking in one's sleep or running around the bedroom in one's dreams, it is necessary to install the motor inhibition of the spinal motorneurons before turning on the motor pattern generators of the cortex and subcortex in the transition from NREM to REM sleep.

At sleep onset, the widespread deactivation that is necessary to facilitate sleep diminishes the drive on both the medullary respiratory oscillator and the mechanisms regulating muscle tone. Thus, cessation of breathing in sleep is normal, especially in males. At the same time that breathing efforts become less frequent, the airway may be narrowed by muscular relaxation. Descent into NREM sleep stages II–IV passes via Stage I with micro-dreams normally greeting the sleeper. For reasons not yet entirely clear, some people get stuck there and have very unpleasant and unwelcome sleep onset dreams. The take-home message is that sleep is both necessary and hazardous to human bodily and mental health.

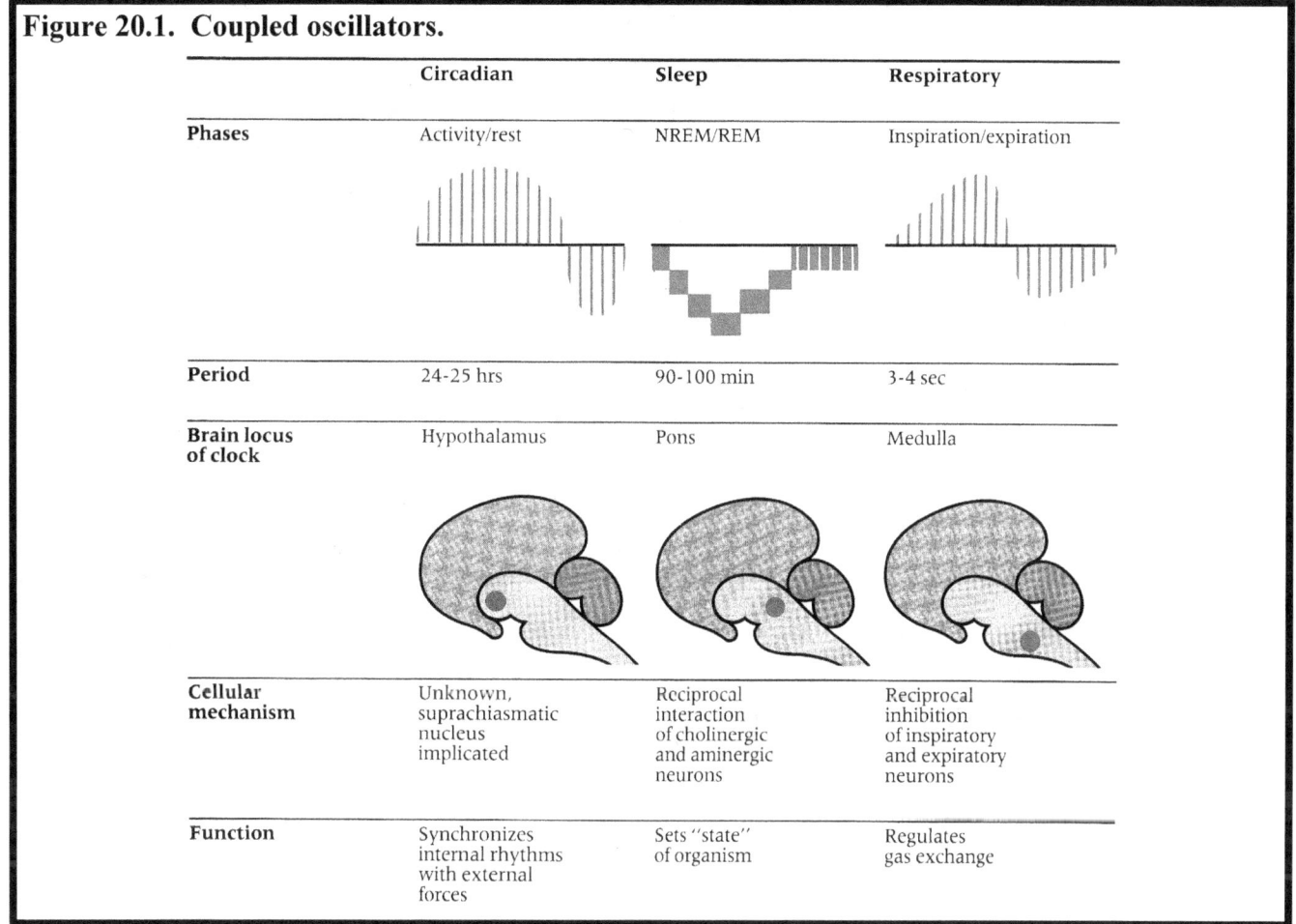

Figure 20.1. Coupled oscillators.

	Circadian	Sleep	Respiratory
Phases	Activity/rest	NREM/REM	Inspiration/expiration
Period	24-25 hrs	90-100 min	3-4 sec
Brain locus of clock	Hypothalamus	Pons	Medulla
Cellular mechanism	Unknown, suprachiasmatic nucleus implicated	Reciprocal interaction of cholinergic and aminergic neurons	Reciprocal inhibition of inspiratory and expiratory neurons
Function	Synchronizes internal rhythms with external forces	Sets "state" of organism	Regulates gas exchange

The Parasomnias

When we say that the parasomnias are sleep of the wrong kind, we mean that sleep may be coexistent with behaviors normally seen only in waking. Walking, talking, and urinating are three graphic examples. The minimalist approach advocated above for insomnia is even more appropriate for the parasomnias. Roger Broughton, who worked in Ottawa, Canada, has called the parasomnias "disorders of arousal" by which he means a too-forceful Stage IV NREM sleep. This concept is related to sleep inertia. Young people are prone to very deep unconsciousness early in the night and sleepwalkers tend to be young persons whose cortex remains in Stage IV throughout the sleepwalking episode. That's the bad news. The good news for sleepwalkers is that Stage IV sleep disappears between ages 30 and 40. Children may also "grow out" of night terrors, parasomnic awakenings associated with psychosis. As with lucid dreaming, parasomnias may also respond to hypnotic suggestion.

The discovery that sleepwalkers ambulate while only partially awake tells us that conscious states can be dissociated, as we have noted in discussing lucid dreaming. In this case, the sleepwalker is partially awake but from the neck down. The brain waves of sleepwalkers can be recorded and shown to be in Stage IV as they wander out the door of the recording room and up the hall. In this state, they might mumble incoherent answers to questions posed to them, avoid obstacles in their path, and urinate in a toilet if directed. When they are led back to bed, they rapidly return to unresponsive deep sleep and, unless intentionally waked up, have no recollection of sleepwalking on awakening the next morning.

Figure 20.2. Ferdinand Hodler painting, "Night," of a man awakening in fright.

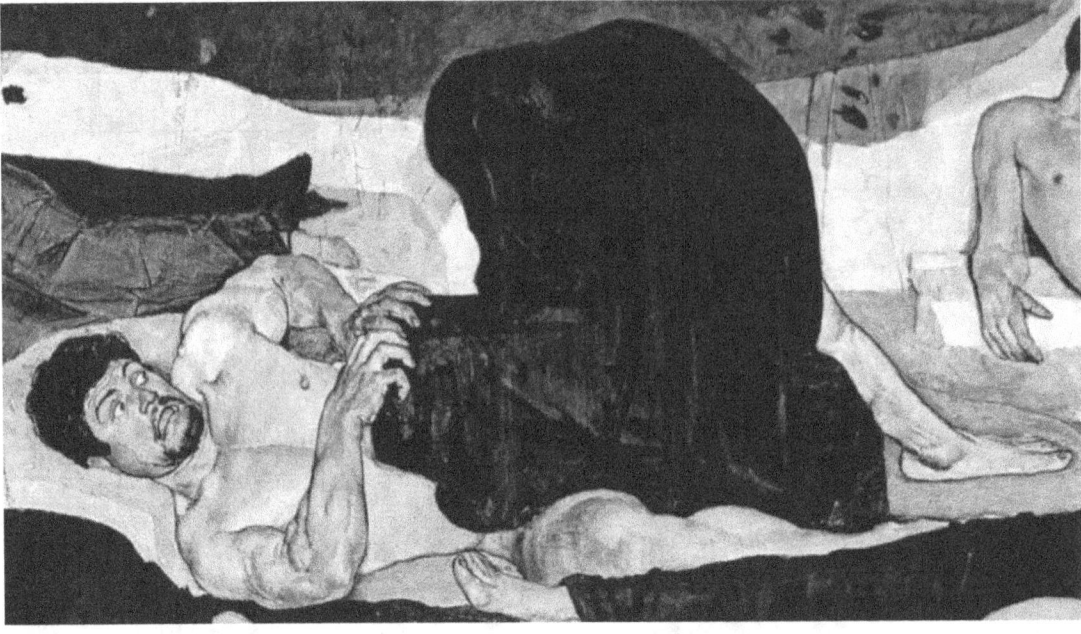

Sleepwalking and related parasomnias are *dissociations* in that the motor quiescence of normal sleep is replaced by the motility of waking. To be more precise and more provocative, sleepwalkers are functionally split-brained: the top of the brain mind (the forebrain) doesn't know what the bottom (the brain stem and spinal cord) is doing. The split in this case is transverse rather than paramedian; it is functional rather than structural; and it is reversible rather than permanent. Sleepwalking is unusual but not statistically rare. It should be considered innocent but potentially dangerous. Sleepwalkers have only to be reassured that "this too will pass" and be protected from harm by preventing falls from a roof or a stairway.

When children between the ages of 5 and 10 wake up from deep sleep, they may evince dissociation as their sleep inertia holds them in its night terror thrall. They hallucinate in fear as their anxious parent tries unsuccessfully to interrupt their functional psychosis. The children do not say that they are hearing or seeing things but the parental inference that they are is possibly correct. The good news again is that watchful waiting is rewarded. We are not suggesting that night terrors are normal or easy to witness. But they are natural dysfunctions, not diseases, and care providers can reassure children and their parents by informing them about the understandably natural process.

Nightmares

The definition of a nightmare might vary from person to person. A dream has to be more than merely unpleasant to be a nightmare. When people are awakened by nightmares, they have strong physiological reactions including racing heart and perspiration. A nightmare, by definition, provokes a strong negative emotional response. People may have difficulty in getting back to sleep after waking from a nightmare. Some people are more prone to nightmares than others, with high-anxiety subjects being particularly vulnerable. Teenagers are more likely than older adults to suffer; indeed the frequency of nightmares (or at least our ability to be awakened by them and remember them) declines with age, possibly for the same reason as the decline in the parasomnias that comes with the normal disappearance of Stage IV sleep between the ages of 30 to 40.

Surprisingly little is known about the causes of nightmares. Of course in as much as dreams are intrinsically anxious and because they often process the conflictual aspects of waking life, we would expect some strong negative emotional dreams occasionally, but this consideration does not explain why some people are more susceptible than others. Bodily pain, stress, food before sleeping, and drugs have all been linked with nightmare occurrence. People suffering from post traumatic stress disorder (PTSD) have recurrent nightmares linked to their trauma.

Those who suffer from frequent nightmares are at risk of disrupted sleep, which as we have seen often leads to physical and cognitive impairment. A number of treatments are available to such people. Various talking cures, including hypnosis, appear to be effective. More recently the drug Prazosin has been used, particularly with those suffering nightmares following PTSD. Prazosin reduces the vividness of dreams.

Figure 20.3. A Pathophysiological model of sleep disorders.

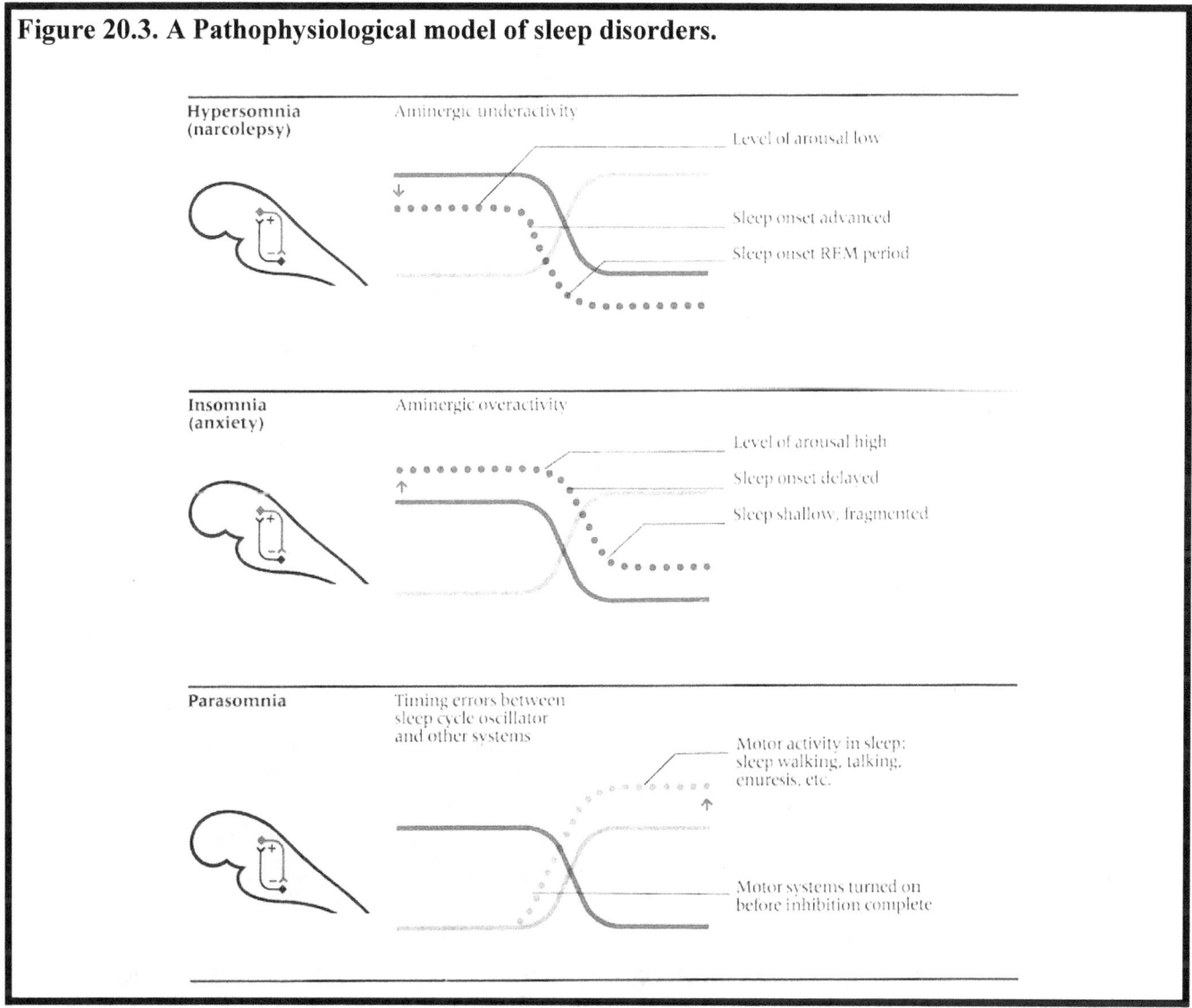

Narcolepsy

The most informative of all sleep disorder/diseases is narcolepsy. The French physician, Jean-Baptiste-Édouard Gélineau (1828–1928), described four signs and symptoms, all of which are caused by the loss of the hypothalamic inhibition of REM sleep in waking. Daytime sleep attacks with or without loss of muscle tone (cataplexy), sleep onset hallucinations, and paralysis, with hallucinations on awakening from sleep, may occur. Narcolepsy illustrates all of the pathophysiological features enumerated in the introduction to this chapter: uncoupling of the circadian and NREM-REM sleep oscillators, emergence in waking of REM sleep signs, and the loss of normal boundaries between waking and dreaming. Laboratory study shows that the daytime sleep attacks and cataplexy of narcolepsy are REM sleep events occurring at the wrong time of day; the sleep onset, post-awakening hallucinations, and the sleep paralysis are caused by REM events that are normally held in check at sleep/wake boundaries by hypothalamic inhibition. According to AIM, narcolepsy is triggered by a combination of axis M defects: weakened aminergic inhibition and reciprocally strengthened cholinergic excitation. The amelioration of narcolepsy by amine reuptake blockers and amphetamine-like drugs complements this formulation.

Narcolepsy is not only a human sleep disorder/disease. In fact, its change of status from disorder to disease derives from its occurrence in Doberman Pinscher dogs (which constitute a valuable animal model for the Stanford University team of scientists under the direction of William Dement). Genetic scientists have shown that there is a hereditary propensity and an immune signature underlying that propensity. It is now thought that narcolepsy is an autoimmune disease caused by the degeneration of hypocretin (orexin)-secreting neurons in the hypothalamus. As instructive as narcolepsy is as a sleep disorder, so is its enshrinement as a classical medical disease.

Sleep Apnea

Snoring is the widely heard indication of throat muscle relaxation at sleep onset. In bedrooms the world over, flabby throats rustle in the respiratory breeze. Partners move to adjacent bedrooms so that they can sleep and avoid the ominous silence when the snoring stops, often for 15 seconds, before a snort is heard and the snoring begins again. Older, overweight men may arouse their insomniac wives up to 50 times a night. The pathophysiology is clear: reticular activation of the medullary respiratory oscillator and neck muscle tone normally decrease in tandem with the cortical deactivation of sleep onset and NREM sleep. If the cessation of respiratory efforts exceeds 50 times per night and the post-pause grunts become noisier and more prolonged as the sleeper struggles to move oxygen across a now forcibly closed airway, the normally snoring sleeper moves into pathological sleep apnea territory. The continuum of normal and abnormal sleep is both seamless and sinister. If the wife is sleeping soundly in the next room, her husband may have 300 or more obstructive sleep apneas per night, never waking up enough himself to realize the danger he is in. However, by day, he knows he has not slept soundly because he always feels fatigued and because, like Charles Dickens' Mr. Pickwick, he nods off repeatedly as he tries to recover from the apnea-induced sleep deprivation of the preceding nights. An obstructive sleep apnea sufferer is never either wide-awake or deeply asleep.

When the breathing efforts cease, the life-giving blood oxygen level falls and the brain responds by creating enough waking to resuscitate the brain stem. The apneic sleeper then gasps to save his life. But he is not out of harm's way. His already compressed airway may become completely shut by his exaggerated recovery breathing effort. Again, the victim is unaware of his peril. An alarmed wife may turn on the bedside light and see that her pink-cheeked hubby has turned blue. It is then time for a sleep medicine laboratory evaluation.

The life-threatening respiratory obstruction can be obviated by applying continuous positive airway pressure (CPAP) via a mask and connection to an electrically powered respiratory pump. If they can tolerate the mask, sleep apnea sufferers then sleep soundly, lose their excessive daytime sleepiness, and repair the cognitive defects that are caused by a sleep-starved brain. Without treatment, consciousness may be compromised by irreversible dementia. The brain depends upon oxygen to sustain itself and normal consciousness.

Figure 20.4. Pathophysiology of the sleep apnea syndrome.

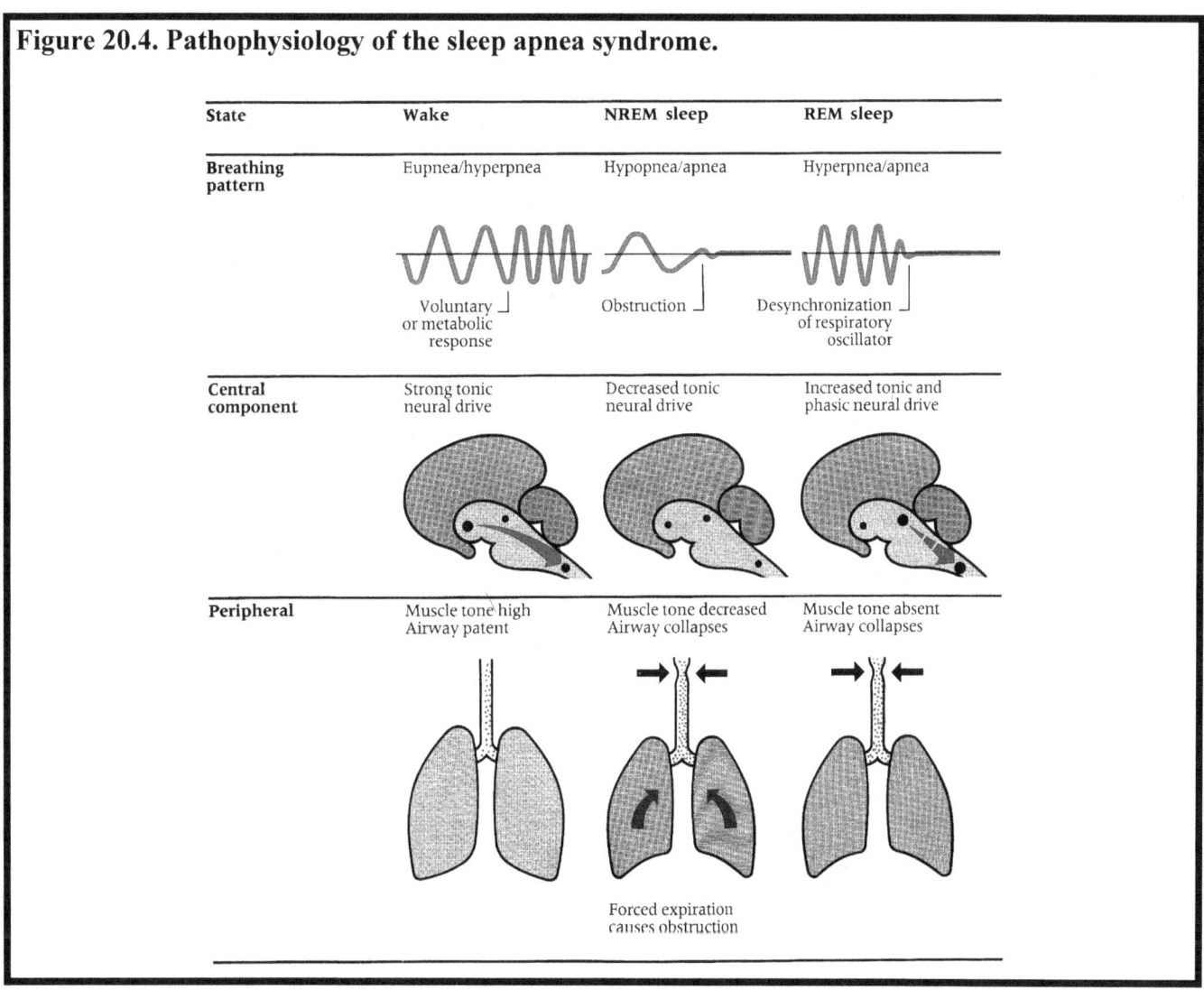

Obese people are counseled to lose weight for this and other good health reasons. Adipose tissue expands inward as well as outward. It thereby narrows the airway in the back of the mouth and in the upper neck. The most serious medical threat is to the heart. Heart failure may result from increased lung blood vessel pressure. Excessive body fat requires oxygen as much as the normally fatty brain and the heart can supply only so much before it fails. The take-home message is: be born a woman and (whether you are a man or a woman) don't eat too much.

REM Sleep Behavior Disorder

Scientists first believed that sleepwalking was the only dramatic parasomnia and had good reason to doubt that dreams were the cause. Sleepwalking, as we have stressed, arises out of NREM sleep and NREM sleep may persist throughout the sleepwalking episode. Now we know that the animation of dreams accompanies brain motor-pattern generator activation and that real movement can be the result. REM Sleep Behavior Disorder (RBD) results when the inhibition of muscle tone of REM is too weak to quench motor commands. Like sleepwalking, dreaming may continue throughout the behavioral episode of RBD. We are thus confronted with still another example of conscious state component dissociation.

RBD persons are in two states at once. As if awake, they literally act out their dreams. They dive into imaginary swimming pools (supposing themselves to be Olympic springboard champions), tackle chests of drawers (whom they take to be the opposing football team's running backs), and swerve violently in bed (to avoid oncoming cars threatening head-on dream collisions). This oneiric motor behavior is comical to consider but it incurs real and hazardous risk. Steps to avoid risk are thus necessary. Fortunately such steps exist but the long-term prognosis is nonetheless dire.

Like sleep apnea, RBD afflicts principally middle-aged men, but women may also enact their dreams. The cause is unknown but an imbalance of AIM factor M brain chemicals is widely suspected. An important and worrisome clinical finding is that RBD always heralds Parkinson's disease. Parkinsonism is known to be caused by a deficiency of dopamine, the do-everything aminergic modulator manufactured by the substantia nigra of the midbrain. Another fact supporting the dopamine deficiency theory is the observation of RBD in persons treated for depression and narcolepsy with amine reuptake blockers and dopamine enhancers.

A French internal medicine specialist who moved to California, **Christian Guilleminault** led the scientific recognition of sleep apnea, the tendency to stop breathing due to reticular deactivation in sleep (See again Figure 20.1). As a result of his work with William Dement, sleep disorder medicine is now a worldwide specialty. The realistic threat to life and the disability secondary to untreated sleep apnea could be said to be the most practical consequence of sleep science. Beyond that, Guilleminault and Dement have also contributed to the science of consciousness emphasized here by documenting the impressive state dependency of self-awareness. Sleep apnea sufferers do not know that their lives are in peril because they are not conscious of their true state.

Treatment of RBD is usually immediately and completely effective with the benzodiazepine drug, clonazepam. The benzodiazepines are not ideal drugs because they have side effects (like hangover drowsiness) and because they are habit-forming (if they are not frankly addicting). They are popular but problematical sedatives.

For the student of consciousness, it is important to notice that both natural and manufactured chemicals often affect both motion and mentation. Since the goal of all directed behavior is motoric, we include movement in our list of consciousness components. We also recognize programmatic movement in the theory of protoconsciousness and virtual reality. For us, even thought is a kind of motor act and dreaming is a conscious state that is always animated.

Self Appraisal of Sleep and Dreams

Sleep and dreams both have rich subjective content which should be combined with objective study. If the mind and the brain are two aspects of a unified process, we should be able to learn about the one by studying the other in the clinic as well as in the laboratory. Our philosophy is didactic: active investigation is a mutual and cooperative collaboration. Teaching and treatment both profit from shared responsibility. This section introduces self-observation instruments which can be used by students in the home-based laboratories of their bedrooms. If they have a bed partner to observe and be observed by, so much the better. We have already made many of these suggestions in Chapter 13. Because they are so relevant to effective evaluation and treatment, we repeat them here.

Before beginning these exercises, self-observe the clear visibility of the corneal bulge as the eye moves beneath the closed lid. This can be seen in a bathroom mirror by holding one lid down and watching with the open eye. In the early morning light, telltale REMs can be seen in bed partners or other accomplices. REM sleep is often associated with palpable twitches of the bodily musculature and facial expressions of dream emotion can be seen. This is especially true of newborn infants in whom the motor inhibition of REM is still weak. In utero, fetal REM movement is even more uninhibited. The fetus is afloat and mothers can feel its kicking, a welcome sign of vital energy.

Sleep charting is simple. Each 24-hour day is represented by a line. Seven lines make up a graphic week. The sleep-wake graph reveals, at a glance, the diurnal pattern of rest and activity. The clock times of awakening, daytime activities, going to bed, and relevant details are recorded, allowing the data to be quantified. Students often sleep-deprive themselves on weekdays and recover lost sleep on weekends. But they rarely recover what they have lost, since many drink or take drugs before they go to bed on late weekend nights. Our goal is not to police sleep. We know that boys will be boys and that girls will be girls but we do wish to help young people to protect the organ which they are spending so much time and money to educate.

A desktop computer or laptop is easily programmed to chart sleep. It can also serve as a consciousness journal allowing one to record recollection and thereby to increase dream recall. Dreams may not reveal a repressed self as Freud supposed but they certainly do underline what is on our minds. Dream harvesting and storage are the first steps in lucidity incubation and are key to the recognition that dreams and other products of consciousness may be recognized en route to self-understanding. Your dream journal may not be great literature but it may be the most important book that you will ever read.

Objective adjuncts to self-observation should be investigated. If your university has a sleep lab, you should visit it. Volunteer to be a subject and see how it feels. Inexpensive video equipment can be programmed to perform time-lapse studies of your sleep at home. A good bioelectrical engineer can help you rig up a bedroom recorder so you can perform studies on your own sleep and dreams as well as on guinea pig friends. You may not want to publish — or even to reveal — your findings but you will never forget this introduction to yourself for the rest of your life.

Figure 20.5. Sleep Charting.

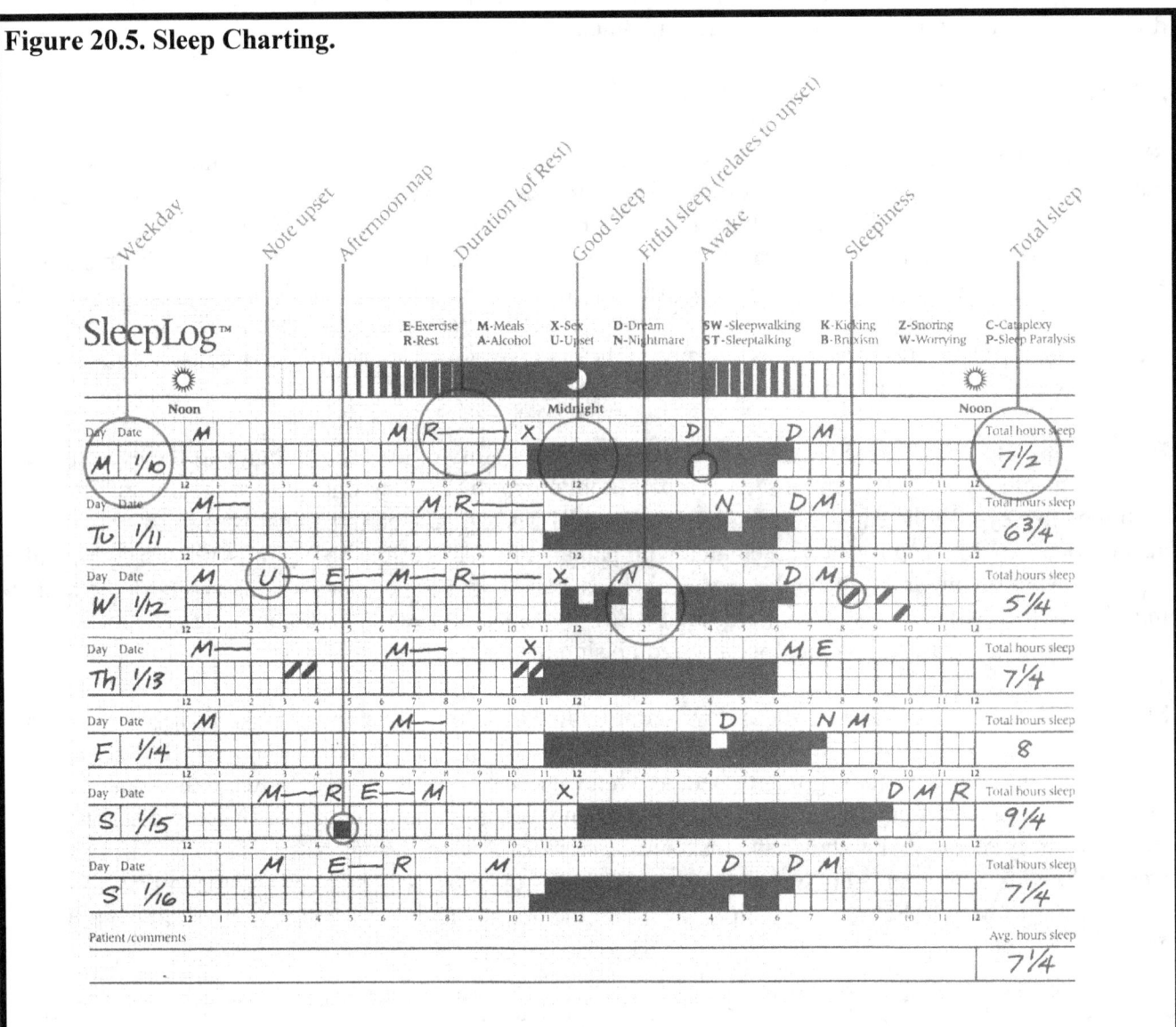

Dialogue 20. The Long and Short of It

TH: Are you a long or a short sleeper?

AH: Now I sleep ten to twelve hours. When I was younger, I got away with 6 and a half.

TH: 10 would be considered long but 4 or 5 is indeed short.

AH: Sleep length is as biologically variable as height and weight. All biological variables change with age. I was never a really short sleeper (4–6 hours) but now I am a truly long one.

TH: Tall men are adept at basketball while short ones may become champion jockeys.

AH: The point is that sleep length is a moving average. Every biological function is variable.

TH: Doesn't sleep length correlate with personality?

AH: Yes. Short sleepers tend be outgoing over-achievers.

TH: They have more time to get work done and, being active and non-anxious, are socially adept. To what can we attribute these assets?

AH: The answer, I am afraid, is genetics. All people are not created equal.

TH: Long sleepers tend to be depressed and introverted. What is the biological reason for this?

AH: We aren't yet sure. AIM predicts that short sleepers are high M (aminergic dominance) and long sleepers are low M (cholinergic dominance).

TH: This fits with the data showing that depression is low M and can be treated by raising M. Biogenic amine reuptake blockers, like Prozac, are effective antidepressants because they elevate M.

AH: But we would like to adjust our sleep length and our mood by more natural means.

TH: I hate exercise but I must admit that it peps me up and facilitates sleep.

AH: I don't like being old and immobile, but I do appreciate my longer sleep. And in retirement I have time for pleasures like writing this book.

TH: Drinking less and going to bed at a regular time of day are also helpful.

AH: Nonetheless a "white night," however unwelcome, is inevitable. I drink much less now but occasionally suffer from insomnia, which never bothered me when I was a young, drunken buff.

TH: Like a late airplane flight, I try to take advantage of the opportunity to read, reflect, plan and philosophize in the dark.

AH: That sounds like meditation. Do you use your mind to soothe and enrich your mind?

TH: Yes. Consciousness treats itself in keeping with the causality assumptions of dual-aspect monism.

AH: My father didn't know anything about consciousness science but he practiced good sleep hygiene.

TH: And he gave you half your genes!

Chapter 21. Psychosis

If we define as psychotic any brain-mind state characterized by hallucinations and delusions, dreaming is psychotic by definition. When we dream, we see things that are not real and we mistakenly believe things that could not possibly be true. All three pathological psychoses (schizophrenia, affective disorder, and organic delirium) share these two defining traits.

Dreaming as Sleep Psychosis

That these traits occur in normal dreaming must mean that the brain-mind is capable in normal sleep of the same symptoms that are considered as mental illness if they occur in waking. We are not suggesting that there is anything pathological about dreaming. We only assert that understanding the brain basis of the hallucinations and delusions of dreams may help us better to understand the hallucinations and delusions of mental illness.

We freely admit that although we do not wish to pathologize normal dreaming, we do wish to promote the recognition of a formal continuum between the normal (dreaming) and the abnormal (mental illness). We also admit to promoting a more tolerant attitude toward so-called patients who differ from us less sharply than we would like to believe. It is indeed frightening to imagine going out of our minds, because the loss of mind control is so debilitating, but also because we are terrified to realize that the normal and the abnormal are separated by so very fine a line, one that we normally cross several times every night of our lives.

Dream hallucinations are predominantly visual: we see things with vivid clarity although our eyes are closed in the dark. Colors are perceived and described in dream reports. We hear less well in dreams (unless we are musicians), and touch may be only inferred. The sense of movement is continuous but taste, smell, and pain are rare. Why? We will try to answer these questions but we must admit that we aren't really sure. What do you think?

We feel emotion in dreams, often with nightmarishly intense fear, embarrassingly strong rage, and unreservedly high elation. These emotions trouble mental patients, too, and our cognition crosses the line of abnormality in our dream disorientation, memory loss, and single-mindedness. We are certifiably wacko when we sleep. What a relief it is to wake up and realize that our craziness was "only a dream."

Hallucinations and delusions are thus not the only normally abnormal aspects of dream psychology. By now, we trust that you have learned enough neuroscience to answer the how question. Your brain is activated off line and modulated in a manner distinctly different from waking. We think that the how question is now quite well understood but exactly why this happens is still a mystery, as we will later discuss when we consider functional theories in Chapters 24 and 25.

Psychosis as a Waking Dream

The analogy between dreaming and psychosis long antedates the sleep lab era. Carl Jung, among other distinguished colleagues said, "Let the dreamer awake and you will see psychosis." While we accept the spirit of this epithet, we prefer a more critical examination to a sweeping endorsement of it. There are two reasons for our skepticism. One is that psychosis is more difficult to access than dreaming. The clouding of consciousness, which makes access difficult, is intrinsic to psychosis. Psychotic persons often lack insight (they do not awaken) supposing themselves (like dreamers when dreaming) to be normal when awake. When psychotic persons recognize that others regard them as abnormal, they often become further withdrawn and refuse to divulge their subjective experience.

Figure 21.1. Dream Caused by the Flight of a Bee around a Pomegranate a Second Before Awakening (Salvador Dalí). In this painting, Salvador Dalí depicts a loosely connected set of dream images in the space above the sleeping body of his wife, Gala. Dalí's surrealist vision of dreaming anticipates modern dream science findings in several ways. First and foremost is the representation of intense and vivid visual imagery, which arises within the head of the inert sleeper. Second is the implicit abrogation of perceptions of external stimuli by the dream hallucinations and the implied takeover of critical judgment that deludes the dreamer into believing that she is awake. Third is the loosely associated linkage of the bizarrely discordant images: an exploding pomegranate emits a fish that belches out a ferocious tiger which is in turn transformed into a bayonet pointed aggressively at Gala's head. An incongruously long-legged elephant roams behind this image sequence. Fourth is the emotional salience of the bizarre sequence, which helps to bind the images together in a coherent but probably fleetingly remembered narrative of Gala's experience. Given the intensity of dream experiences such as this, it is surprising how few are remembered. Gala may have remembered such a dream but many others were almost certainly lost to her after waking. It was Dalí's idea that this dream was caused by the sting of the bee shown buzzing behind Gala's left ear. Thus, Dalí himself was probably not aware that dreams with this intensity may occur with no external stimulation whatever. With the help of modern neuroscience, many of Gala's dream features can begin to be explained naturalistically without recourse to questionable psychological speculation. The emerging picture suggests that Gala's dream consciousness state may have helped her subsequent waking consciousness to be more accurately perceptive, more orderly in thought, more temperate emotionally and better able to remember her experience.

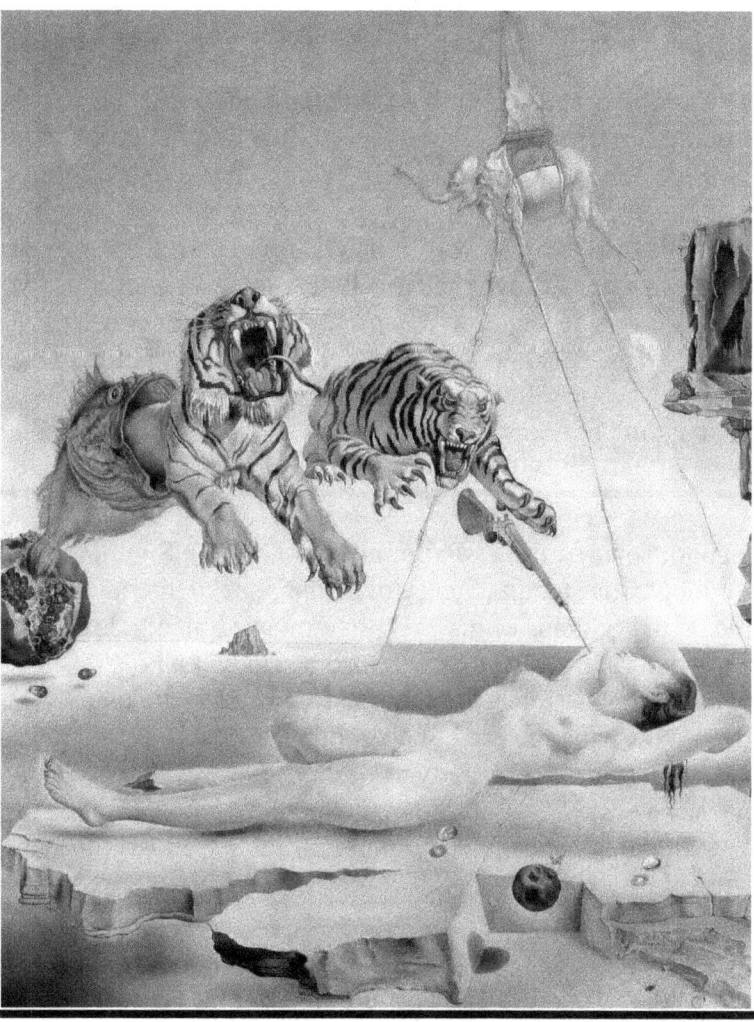

Carl Jung (1875–1961) was a Swiss psychiatrist who collaborated with Sigmund Freud but argued against the theory of infantile sexuality and other concepts that he considered to be excessively reductive. He argued for a "collective unconscious" by which he meant a store of universal ancestral memories that fed dreams and artistic sensibility. At first committed to the scientific study of associations and, following his break with Freud, Jung championed religious experience, intimate interpersonal relations, and artistic self-development. Jung's own theories thus overlap with those of modern dream and consciousness science more commodiously than do those of Sigmund Freud. Jung's theory of a collective unconscious is related to Immanuel Kant's postulation of *a priori* knowledge and Helmholtz's predictive formulations. While Carl Jung did not acknowledge those theories and, without endorsing the idiosyncratic nature of many of his views, he is clearly a notable pioneer.

Carl Jung was the Swiss-born son of a Protestant minister who, with Sigmund Freud, founded psychoanalysis. Because Jung contested the theory of infantile sexuality and criticized Freud openly, he was rejected as a disciple on the grounds of his excessive mysticism. No doubt, Carl Jung was a mystic enamored especially of oriental religion. But he was also a scientist who studied word associations with Eugene Bleuler and a painter who produced the famous Red Book, who courted and bedded many women, some of whom were his patients. Freud had good reason to jettison Carl Jung but with the bathwater out went the baby — the Jungian analysis of dreams as revelatory products of the creative mind — which is a view that we want to promote in this book. Another of Jung's prophetic pronouncements was, "Let the dreamer awake and you will see psychosis." We are all mad in our dreams and mad is not necessarily bad.

This withdrawal into the self is particularly pronounced in schizophrenia, the most common and most debilitating psychotic mental illness. We return to this condition and the obstacles to its scientific investigation below but wish now to summarize the phenomenological differences that we perceive to be limiting to the dream-psychosis analogy. Schizophrenic persons are so withdrawn as to be loath to talk confidentially about anything, let alone their subjective experience. Some normal dreamers are hesitant to reveal the intimate details of their dreams but freely relate most of what they remember because they know that dreaming is normal and that it is normally abnormal. Having a hallucination that stops on awakening is quite different from having a hallucination that begins with waking in the morning, continues off and on throughout the day, and may become even more invasive in the dark.

Dreaming may normally be more social even than waking. One is rarely alone in normal dreams and the boundaries between the self and other dream characters are both fluid and porous. This dissolution of interpersonal boundaries is not restricted to dreaming and some psychotic persons have more than their share of it when awake. Manic and hypomanic persons invade the personal space of others with impunity, and normal dreaming shares this trait, only to have it disappear upon awakening. In waking, we work alone even when surrounded by other persons. We feel comfortable with the conventional niceties of intermittent communication with co-workers but concentrate on our work, often alone. Psychotic persons have no such ease as their psychosis leads them to imagine (or even hear the voices of) people talking about them. To our knowledge, this very common psychotic paranoia never occurs in normal dreaming.

Our reservations about the psychosis/dreaming analogy lead to a clear conclusion: while there are similarities between the abnormal and normal states, there are also pronounced differences. In seeking understanding, we must stop short of expecting definitive explanations especially with respect to schizophrenia and affective disorders. The psychosis with which dreaming bears the closest resemblance is organic delirium, now rarely seen because of improvement in its treatment.

Organic Delirium

When alcoholics stop drinking — or amphetamine abusers crash — they may be subject to the psychosis of organic delirium. This disorder is increasingly problematical in the post-operative course of many senior citizens. Organic delirium is defined as disorientation, hallucinations which are predominantly visual, memory loss, and confabulation (the making up of stories to cover up the cognitive memory deficits). The attentive reader will realize that these are among the formal features of normal dreaming. Because we understand its brain basis, we consider dreaming to be an organic psychosis.

That dreaming is a psychosis (defined by the presence of hallucinations and delusions) is beyond doubt. It is organic because it is biologically determined. Furthermore, it is biologically determined by exactly those very same factor M alterations that are involved in organic delirium, especially aminergic demodulation, but also by cholinergic hyper-modulation.

Alcoholics incur organic delirium because alcohol suppresses REM sleep. Over the long term, the alcoholic's REM deprivation leads to an increase in REM pressure. When the alcohol REM deprivation suddenly ceases, the REM rebound is so intense as to break through into waking. Subjects are then awake and in dream psychosis often lasting for several days.

Amphetamine is also a potent REM suppressant. Amphetamine mimics the action of norepinephrine in the brain causing a desired hyper-vigilance, elation, and an increased sexual potency. No wonder that amphetamine and its first cousin, cocaine, are so popular. The high that is sought by amphetamine abusers incurs the same REM debt that is accumulated by alcoholics. The REM breakthrough in withdrawal causes their psychosis, too. They then become visually hallucinatory, with many of the formal features of dreaming already mentioned.

Needless to say, alcohol and amphetamine addicts become increasingly desperate to maintain their habit because the toxic delirium of withdrawal is so unpleasant. Worse than that, withdrawal can be either permanently disabling or fatal. Before the days of benzodiazepine treatment, it was not unusual to see the delirium tremens victims of alcohol withdrawal die of temperature dyscontrol, the same fatal defect as that of Allan Rechtschaffen's REM sleep-deprived rats. These unfortunate patients had lost body temperature control when they lost REM.

Figure 21.2. Amphetamine molecule (a) and Norepinephrine molecule (b).

a.

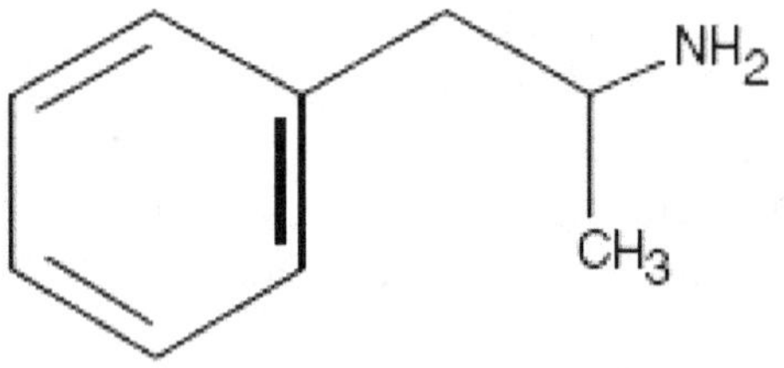

b.

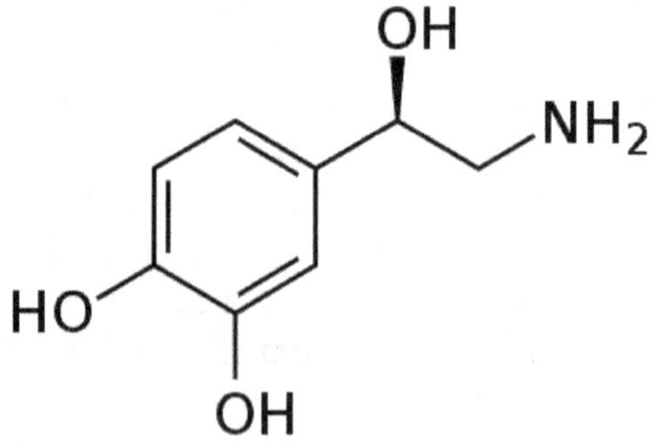

Is there any good news to be derived from these conditions? We need to be critical of our own enthusiasm. Nothing is perfect and the fit between toxic delirium and dreaming is not perfect. The visual hallucinations of addicts in withdrawal are typically stereotyped involving insects or reptiles. Such creatures are not usually beheld by normal dreamers or by other mental patients. The conversations between alcoholics and their hallucinated drunk cronies are stereotyped in ways not found in dreams. The delirious states of addicts are much longer lasting. As with dreaming, normal awakening ultimately occurs and, with abstinence, normal dreaming and mental health can once again be established.

Schizophrenia

The "primary" symptoms of schizophrenia have nothing to do with dreaming. This observation should have dampened the ardor of the REM sleep/dreaming pioneers who vainly hoped that this most severe of mental illnesses would have a simple explanation and remedy. In the early 1960s, experiments with "dream deprivation" were undertaken with the expectation that they would lead to psychosis. They often did but it was not because dreaming was deprived; in fact, it was shown that NREM sleep deprivation was every bit as deleterious to mental health as REM deprivation. Subsequent work has shown that REM sleep deprivation may be life-threatening because of its disruption of energy metabolism.

The primary symptoms of schizophrenia (withdrawal, isolation, and negativity) often antedate by years the outbreak of the secondary symptoms (hallucinations and delusions) that lead to hospitalization and which do bear some relationship to dreaming. The primary symptoms are thus far more likely to be genetically determined traits than physiologically determined states (and it is in the latter spirit that we pursue the dream/psychosis analogy). Why the secondary symptoms should emerge in late adolescence rather than in childhood is no clearer than why 9 year olds are lucid dreamers. Although the explanatory hypothesis of neuronal pruning is not unreasonable, it is not obvious why such a process should unleash hallucinations and delusions often requiring hospitalization.

The link between D2 dopamine receptor affinity and effective antipsychotic medication brings us closer to home. The same Mark Solms who showed that reports of cessation of dreaming followed certain strokes has suggested that dopamine, another aminergic neuromodulator, is a dream/psychosis instigator (while norepinephrine, serotonin, and histamine are instead wake state instigators). In REM sleep, dopamine is released as it is in normal waking but, in REM, dopamine acts in the absence of norepinephrine, histamine, and serotonin. It could be this imbalance which is the psychotogenic culprit in dreaming. A unified theory is thus standing in the wings of our renovated Cartesian theater. The absent recording of identified dopamine neurons in cats or rats is the only obstacle to scientific progress here. Pilot studies suggest that, indeed, dopamine-containing neurons do continue to fire in REM, suggesting that the Solms dopamine hypothesis of dream psychosis may be valid.

Novel antipsychotic drugs (like clozapine) have the theoretically undesirable propensity to affect the serotonergic and noradrenergic neuromodulators of waking (whose inhibition is essential to the release of REM dream hallucinations and delusions) as well as affecting dopamine. Perhaps it is this very messy pharmacological non-specificity which makes clozapine so effective an antipsychotic in recalcitrant patients. We speculate that clozapine pushes all the right molecular buttons at once and thus relieves the disabling secondary symptoms, allowing chronically schizophrenic patients to live outside hospitals.

Allen Braun is the NIH neurologist who performed the PET imaging study of the human brain in its three cardinal conscious states of waking, sleeping, and dreaming. See again Figures 1.1 – 1.4. His results are compatible with those of animal studies, suggesting that the biological basis of subjectivity was a widely shared property of all mammals and that a common mechanism underlies the alternation of states. Recently he has shown that jazz musicians' brains shift into a REM-like state when they improvise. When we say, "That was a very dreamy solo," our statement of appreciation may be more than metaphorical.

Bipolar Affective Disorder

Closer to dreaming are the affective disorders, especially mania, with its wild elation, its associatively loose mentation, and its uncontrollable driven quality. Depression, with its sadness, its hopelessness, and its suicidal ideation fits poorly from an experiential point of view but its REM sleep dynamics are a biologically important link. The dream ≡ psychosis analogy here is stronger than in schizophrenia, though the equivalence is far from as tight as it is in organic delirium.

Manic persons amuse others because they think and talk fast, often humorously so. "All the world loves a manic," psychiatrists like to say when they are teaching medical students. Stand-up comics are often manic (or at least hypomanic). Dreamers love the wild elation of their "good" dreams and are frequently amused both in the dream itself and when delivering a report of its comical content. The feeling of omnipotence (no fear of flying) is another common point. A "flight of ideas" is also shared by dreamers and manic patients. They both jump from adventure to adventure, from one part of the world to another, from one companion to the next and without wondering who, where, or how this could all be possible.

That's the good news. The bad news is that manic hallucinations are more commonly auditory than visual and manic delusions are frequently paranoid. These are the reverse of trends in dreaming. Good or "big" dreams are not uncommon but dreams of fear (and anxiety) or aggression are equally likely for normal young adults. Rapid cycling, from one affective state to another, is more common in dreaming than in affective disorders but it does occur, often in the middle of the night, which helps to explain why scientist-clinicians have always considered affective disorders and sleep to be bedfellows.

Among the many reasons for consciousness scientists to be interested in the shared mechanisms of dreaming and affective disorder are the findings of shortened REM latency and intensified first REM periods in depression. REM latency, the time elapsed between sleep onset and REM onset, may be halved, from 90 to 45 minutes, in depression: the first REM period may be twice as long and have a doubled REM density. These dramatic findings are not surprising, given the diminished aminergic and increased cholinergic synaptic efficacy in depression. Indeed, they are among the most cogent and practical facts of sleep and dream science.

The facts are cogent because they increase the reductionistic power of AIM. Mood is down and REM is up when factor M is low. Moreover, the extent to which M is down predicts the clinical efficacy of antidepressant medication when depression is treated with aminergic reuptake blockers like Prozac. Mood goes back up when factor M is elevated.

The paradox still needing to be explained is why dream content does not become depressive when factor M decreases in the course of the normal REM/NREM cycle. An answer could be that REM only changes mood over many days (rather than in many minutes) and this decoupling is a function of second messenger physiology, a subtlety beyond the scope of this book. Suffice it to say, the causal link between sleep and energy is one of the most important ties in consciousness science and one that we will make much of when we discuss sleep and dream function later.

State Stabilization

The healthy brain-mind has an orderly progression of discrete conscious states. Sleep and waking succeed each other with almost seamless continuity. Within waking, fantasy is so effectively kept in the background of consciousness as to be difficult to recognize, yet it provides data that is used in foreground decision-making. Within sleep, dreaming is also so evanescent as to escape detection but it may play an important role in preparing waking consciousness for efficient action. When fantasy and dreaming escape from the background and invade waking consciousness, mental illness is the result. We have seen that sleep disorders are caused by the same sort of interstate instability. It is the task of treatment of both psychosis and sleep disorders to restore stability.

Psychotherapy restores state stabilization by strengthening insight, a presumably cortical process involving, almost certainly, the frontal lobes. The frontal lobes are thought to be the seat of executive ego and of conscious will. Elsewhere we argue that the mind is causal and the conscious will is, at least in significant part, free. The role in lucid dreaming is an experimentally demonstrated expression of consciously willed, frontal-lobe power. Whether the force of psychotherapy is strengthened by an exclusive focus on personal history is, we think, dubious, and more efficient ways of achieving the insight necessary for state derive from the new science of consciousness.

Pharmaceutical intervention is rational when it is targeted on state stabilization. Despite its drawbacks, the prescription of some drugs is finally in scientific harmony with sleep and dream science. Unfortunately, the U.S. pharmaceutical industry, the Federal Drug Administration, and academia are still locked into the application of the medical model of disorders of the brain-mind. A good example is the labeling of the new "antidepressants." The advent of the biogenic amine reuptake blockers was a missed opportunity to recognize that these state stabilizers have a wide variety of uses. Yes, they are often effective antidepressants but they are also good at treating narcolepsy whether or not depression is in the clinical picture.

These drugs work, we submit, by stabilizing state. The natural order and discreteness of states is restored by them. For the same reason, antiepileptic drugs are useful in reducing psychosis. They stabilize neuronal excitability. And many so-called antidepressants are also good sedatives whether or not the sleep loss is related to depression (or even anxiety for that matter). They ease and make seamless natural state change.

In our view, nothing less than a revolution of theory and practice is now feasible and desirable. We hope to imbue students with a passion for scientific research and the changes it already mandates. Like an oncoming REM period, these trends will increase exponentially as time goes on.

Kay Redfield Jamieson (1946–) is a psychologist who has written eloquently about the bipolar affective disorder from which she suffered. Her testimony is a demonstration of the art of self-observation and the success of the psychotherapy of one's own ego. It may be no accident that Jamieson was the wife of a sleep scientist and psychiatrist, the late Richard Jed Wyatt, who studied the biological basis of mental illness at the National Institute of Mental Health. Progress in this field is painfully slow because the brain is (by far) the most complex organ in the body. The facts that the brain is responsible for both dreams and psychosis and that they share formal properties justify their continued conjoint research.

Dialogue 21. A Mind To Go Out Of

TH: Non-human animals don't go crazy because they lack the wherewithal to do so.

AH: Secondary consciousness (and the brain that gives rise to it) must be essential to the madness of severe mental illness.

TH: We are skeptical about dreaming in our fellow mammals even though they all have REM sleep.

AH: It may be more important to stay warm than to write poetry, although most poets would dispute that point.

TH: Most poets don't even know they are kept warm by their dreaming brains.

AH: Many of them celebrate madness as a sign of poetic inspiration.

TH: The English poet Sylvia Plath, and the American poet Robert Lowell, come to mind in this regard. They both had minds to go out of and did so periodically.

AH: When depressed, their psychological pain was so intense that they both committed suicide.

TH: Non-human animals do suffer loss and depression but they are never (as far as we know) suicidal.

AH: To commit suicide, one must realize that one is alive and that one can take one's own life.

TH: The center of virtual reality consciousness is the sense of self-as-agent.

AH: The sense of self is what holds dreams and life together.

TH: No wonder it is such an important part of secondary consciousness.

AH: We all need to take our selves more seriously even as we laugh at our predicament. Our selves are precious.

TH: Schizophrenia is a clear example of the failure of the sense of self.

AH: My late colleague, Dr. Richard Wyatt, told me that schizophrenic patients had a weakened sense of self, even in their dreams.

TH: Wyatt's widow, Kay Redfield Jamiesen, has written about her attacks of mania in a telling way. Her reports of manic excitement make it seem dream-like, as if the barriers between waking and dreaming had broken down.

AH: The main point here is that strongly differentiated and rich waking and dreaming consciousness are both essential to mental health.

TH: You might even say that rich and strongly differentiated states of consciousness are the essence of mental health.

AH: A strong mind is both precious and costly. Handle it with care. You might go out of it.

Chapter 22. Epilepsy and Migraine*

When Edward Evarts proposed the first neuronal theory of REM sleep dreaming in 1962, he thought that the bursts of discharge by cortical motor neurons that he observed might be the result of a decrease in feedback inhibition from Renshaw cells which quenched the neuronal fire and thereby guaranteed excitability control. Renshaw cells are interneurons which are excited whenever a cortical motorneuron fires and which limit the firing in waking to one and only one neuronal discharge. This is one of several mechanisms by which the brain-mind achieves excitability equilibrium and prevents the excessive and uncontrolled discharge that results in epilepsy.

Excitability Control and Consciousness

Epileptic seizures are takeovers of the brain which can invalidate the mind, distort consciousness, or even subvert it altogether. This chapter explores some fascinating aspects of this disorder, its causes, and its effects. The central idea is that we are always on the edge of epilepsy, especially when we sleep and dream.

Mircea Steriade later showed that Evarts' theory was wrong because cortical interneurons, including the Renshaw cells, fired more frequently in REM, not less, as Evarts had suggested. Steriade's improved recording methodology detected other excitability changes in sleep, particularly those that result in the EEG sleep spindles and slow waves that may underlie the suspension of consciousness in the NREM phase of sleep. Early in the night, it would appear that we undergo normal seizures, and it therefore comes as no surprise that patients who are seizure-prone by day reveal their pathology more clearly by night.

In discussing factor I of the AIM model, we have already sensitized our readers to the importance of understanding how the brain-mind shifts its excitability from one conscious state to another via the shift in activation from one brain region to another. Nowhere is this principle made more clear than by the startle response and the PGO waves that are concomitant with it. PGO waves, you will remember, are high-voltage EEG deflections which look exactly like the spike and wave complexes that appear in the EEG of many epileptic patients. PGO waves are normal in REM sleep, when they are disinhibited. They would cause a motor seizure were the spinal motor neurons not rendered less excitable by their active inhibition.

We are not exaggerating when we say that normal dreaming is our conscious experience of the brain's normal expression of paroxysmal neuronal discharge akin to that of epileptic seizure. Seizure, as the name implies, takes over our brain-minds when we dream and causes us to be deluded and hallucinated, disoriented and demented, and emotionally driven by anxiety, aggression, and elation. Fortunately, these nightly fits are rarely remembered. Critics have attacked this pathological analogy saying that nothing normal can be abnormal. We take this point seriously but counter that the distinction between the normal and the abnormal is statistical, not absolute. The brain-mind is normally abnormal because shared normal and abnormal mechanisms and states are the rule and not the exception in the control of consciousness.

It is difficult, we know, to learn about all these facts but consider yourselves lucky: your brain-mind keeps you out of harm's way whether you understand how it all works or not. Consciousness is a privilege which is granted to you whether you study hard or not. Of course we hope that you will share our commitment to and pleasure in learning more about your selves than meets the conventional eye. You are a self that is the elaboration of your brain. Understanding how it all works, largely without your supervision, is a privilege as well as an onerous assignment.

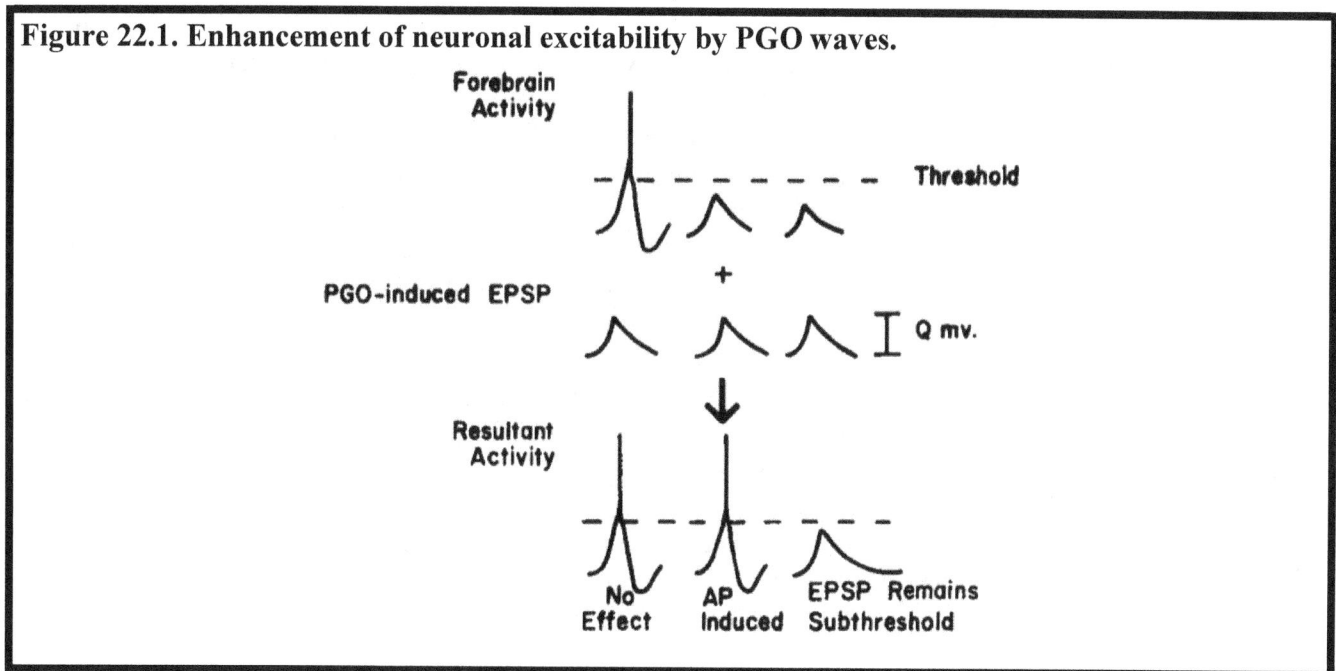

Figure 22.1. Enhancement of neuronal excitability by PGO waves.

Seizure Disorders

Clinical epilepsy comprises a wide variety of disorders, from Grand Mal at the most extreme end of the spectrum, through Petit Mal in the middle, to Pseudo Seizures at the low end of the spectrum. We do not entirely endorse this way of looking at epilepsy but wish to describe it.

Grand mal seizures are characterized by gross motor phenomena and comatose loss of consciousness that are easily observable as "fits." The motor signs are called tonic-clonic because the patient becomes tense (tonic) and exhibits bilateral muscular flexion and extension (clonic) activity as he falls to the ground. Grand mal epilepsy was once called "The Falling Sickness" because of the inability of the afflicted person to maintain an erect posture. The interruption of consciousness is caused by the replacement of integrated cortical electrical activity with the spike and wave discharges alluded to in the previous section of this chapter. Grand mal seizure unconsciousness may last for three hours but is usually over within minutes.

Grand mal seizures can be induced by transcranial electrical stimulation for the treatment of mental illness, especially vegetative depression. Induced seizures are typically "modified" by precipitating unconsciousness with intravenous injection of the same barbiturate sedatives used in surgical procedures. Blockade of the motor components can be effected by pharmacological antagonism of the acetylcholine excitation of the motor end plate of skeletal muscles. In this case, the cessation of breathing movements must be countered with artificial respiration. Despite the life threatening invasiveness of Electro Convulsive Therapy (ECT), it may be preferred to drug treatment by some patients and practitioners because its effects are immediate rather than delayed. Memory for both spontaneous and induced seizures is as nonexistent as memory for deep NREM sleep.

Petit mal seizures are much less severe with respect to both their motoric effects and their consciousness disruption but they may be socially debilitating by virtue of their very subtlety. Short of fugue (the prolonged automaticity of behavior, with loss of orientation and insight that may affect patients with seizures of the temporal lobe), petit mal seizures are characterized by momentary losses of awareness (or "*absences*" in French) with only lip smacking (or no motor signs). In the fugue states of temporal lobe epilepsy (TLE), the subject may literally travel and behave effectively for weeks before "waking up" and returning to normal consciousness. The special case of TLE will be treated in more detail in a separate section. The take-home message now is that even relatively minor epilepsies demonstrate the fragility of consciousness and its liability to alteration by brain state change and disorder.

The good news is that all forms of epilepsy are now treatable with medication and that effective antiepileptic medications are being found useful in the treatment of so-called mental illness. It has long been known that epilepsy and mental illness were "co-morbid" conditions. It could just be that while the one begets the other, the other begets the one. It is also possible that epilepsy and mental illness are not really discontinuous states. We will assert that when we REM-dream, we experience both normal epilepsy and normal madness.

When it comes to "pseudo seizures," we cross the line between the domains of neurology and psychiatry as traditionally defined. As the term implies, the "pseudo" seizure is thought to be motivationally mediated. Here disease and disorder meet and confound each other. Our view is that the two medical specialties have more to gain by acknowledgement of shared principles and practices than by an exaggeration of nosological and conceptual differences. For us and for our dual-aspect monism philosophy, there is no brain anatomy and physiology without psychology and philosophy, and vice versa.

Electrical Stimulation of the Waking Conscious Brain

Much has been revealed about the timing characteristics of attentional consciousness by the experiments of the Canadian neurosurgeon, Wilder Penfield, interestingly dubbed in his time "the greatest living Canadian." Penfield took the opportunity of operating on the brain as a treatment for epilepsy to study the effects of mild electrical stimulation of very specific points on the surface of the cerebrum using a fine electrical probe, (a technique known as *direct electrical stimulation*); as the brain has no pain receptors, such a technique didn't hurt the patient, who was operated upon under local anaesthetic. The technique also told Penfield exactly where to cut, and which regions he could remove while minimizing damage. In this way, he built up a map of the cortex of the brain; he created the sensory-motor homunculus, showing that specific regions of the motor cortex controlled precise regions of the body.

Penfield found that electrical stimulation of regions of the temporal lobe provoked the apparent recall of very specific memories (as, for example, the smell of toast). However, the probe had to stimulate the somato-sensory cortex for a minimum period of time — close to half a second, which is similar to Libet's half-second delay. These results taken together suggest that consciousness (in the sense of awareness of a stimulus) takes half a second to build up; we say that there is a neuronal adequacy for consciousness — we need the persistence of 0.5 of a second in the brain. This does not mean the stimulus has to last 0.5 second, as we can immediately see from the literature on subliminal processing, where very brief presentations of material of a few tens of milliseconds can lead to awareness (and possibly influence behavior without conscious awareness). It would seem that if we are able to be conscious of a stimulus, there has to be a fairly stable pattern of activation in the brain for about half a second. A stimulus can be brief but its conscious elaboration takes time.

Another important aspect of the Penfield findings is that the temporal lobe seat of memory suggested in his experiments is spontaneously and selectively activated in REM sleep dreaming. This could be related to the richness of remote memory access in that conscious state as well as to the enhanced emotionality of dreaming consciousness. These observations also reinforce that the REM sleep dream activation of the temporal lobe is epileptiform. When we dream, our brains are spasmodically activated as if we were a patient on Penfield's operating table.

Wilder Penfield (1891–1976) was a neurosurgeon working at McGill Medical School in Montréal where he founded a world famous institute for the study of the brain basis of mind. Among his many collaborators were Brenda Milner (who described the memory defects caused by temporal lobe damage of HM), Herbert Jasper (who confirmed and extended the reticular activating effects described by Giuseppe Moruzzi and Horace Magoun), and Krešimir Krnjević (who described the cortical neurophysiology of acetylcholine). Penfield himself was larger than life and persuaded his surgical patients to undergo experimental electrical stimulation of their brains while he removed their cortical epileptic foci. Among many other important contributions, Penfield elaborated sensory and motor "homunculi," the maps of our selves in our brains. There are lots of little men (and women in our heads). They are us.

Temporal Lobe Epilepsy and Dreaming

The 'dreamy states' of patients with Temporal Lobe Epilepsy (TLE) are enlightened by understanding the psychophysiology of REM sleep. Activation of the forebrain includes the temporal lobe, which receives a strong and direct projection from the brain stem where the PGO waves have been shown to be associated with eye movement direction. Moreover, the work of the late José Calvo at the National Institute of Mental Health in Mexico City revealed that the projection pathway conducted the epileptiform PGO wave signals from the pontine reticular formation directly to the temporal lobe where they could initiate seizure-like neuronal discharge. Calvo's working hypothesis was that dreaming was a modified TLE seizure and that, conversely, a TLE seizure was a modified REM sleep dream.

The temporal lobe is the brain's most sensitive region to the "kindling" of seizures. By kindling, the experimental psychologist Graham Goddard of the University of Otago, New Zealand, meant that the second of two electrical stimuli caused seizure activity more easily than the first stimulus. The temporal lobe might thus be said to remember its prior experience, an appropriate metaphor for that brain region's potent mnemonic capacity. It is germane to note that the brain itself was insensitive to Penfield's and

Libet's surgical procedures. This fact makes possible the scientific study of the unanaesthetized human brain.

The fugue states of some TLE patients are dream-like in that they appear to be automatic, to occur outside of awareness, and to be forgotten when they end. A major difference in fugue is that the subjects are awake, the behavior is sensible and appropriate, and it can go on for days or weeks. Fugue states bear some resemblance to hypnosis, which also occurs outside normal waking consciousness but is much more short-lived. In all likelihood, we are confronted here with a multiplicity of conscious/non-conscious states, each with its own distinct brain mediator in the temporal lobe and its connections.

The 'dreamy states' of TLE have been investigated by the neurologist Arthur Epstein, of Tulane University in New Orleans, Louisiana. The patients describe an aura (or *warning prodrome)*, sometimes associated with an olfactory hallucination (such as a burnt almond odor). These are never features of normal dreams. Nor are the paranoid delusions of TLE patients dream-like. Dreams are rarely, if ever, paranoid. Auditory hallucinations are more frequent in schizophrenia than the predominantly visual ones of dreams, while cognitive discontinuity and incongruity have neither been noted nor quantified. Thus the 'dreamy states' of TLE would appear to resemble schizophrenia more than normal dreaming. Schizophrenia is a disorder of consciousness, as we have suggested in the preceding chapter on psychosis.

Returning to the neurophysiology, Calvo reported that electrical stimulation of the amygdala could precipitate a REM sleep episode with an associated increase in the high-amplitude individual PGO waves that are typical of NREM-REM transitions. These waves followed the electrical stimuli of the amygdala, indicating that there was a two-way connection between the pons and the temporal lobe and that increasing the excitability of this pathway in the pre-REM phase could tip the balance in favor of REM. The amygdala has a well-documented role in anxiety, the most common dream emotion. One conclusion seems clear: REM sleep facilitates internal communication between two parts of the brain that are normally disconnected. This process may be common to other conscious state shifts.

Fyodor Dostoyevsky and Temporal Lobe Epilepsy

The Russian romantic novelist, Fyodor Dostoyevsky, has been retrospectively diagnosed as a temporal lobe epileptic by none other than the neurologist-turned-psychiatrist Sigmund Freud. The reasons are circumstantial and entirely historical but conceptually compelling: Dostovevsky was impulsive; he was an habitual gambler; as such, he was given to fugue-like flights to Baden-Baden and other gambling spas where he squandered the money he had just received from writing the thrilling installment of still another novel; he was a liberal in political rebellion against the tsar, who staged a mock execution to threaten him and then exiled him to Siberia; he was a religious fanatic and mystic who yearned throughout his life to communicate with God; he was a hypergraphic writer whose very literary talent suggested abnormality. For Dostoyevsky, life was a seizure-like dream and it is in that sense that we discuss his "case."

For Sigmund Freud and other scholars, it was the novelistic works themselves which provided the most persuasive evidence of his hypothetical brain affliction. His novels teemed with mirrors of his purportedly epileptic self: Raskolnikov, the hero-murderer of *Crime and Punishment*, hacks an old woman to death to silence her witness to his robbery; Prince Myshkin of *The Idiot* is torn between naïve innocence and mad passion; *The Demons* (or *The Possessed*) are criminals who have lost their faith; and *The Brothers Karamazov* represent all of these unbridled passionate traits. Their father is an irascible quasi-epileptic whom they conspire to kill (thus Freud's famous essay, *Dostoyevsky and the Parricide*); Dmitri is impulsively sensual and chases women; Ivan is a cynical intellectual who is disenchanted by politics; and Alyosha repairs to a monastery to find a spiritual father.

Whether all of this adds up to temporal lobe epilepsy is beside the point. Only an EEG, a brain scan, or an MRI could clinch the clinical case and Dostoyevsky lived before any such tests existed. We speculate that the results might probably have been negative anyway. There can be little question, however, that authorial vision and writing are the products of a dream-like waking state involving the temporal lobe. Many neuroscientists have speculated along these lines in discussing Marcel Proust's memory of his madeleine cake. Just as the frontal lobe lights up in dreamers who become lucid at the sleep-wake margin and just as the occipital lobe lights up in those who see or imagine the world in waking, so the temporal lobe lights up in dreaming, and we hypothetically propose, in waking imagination.

To demonstrate how such hypotheses can be tested, we mention Allen Braun's recent MRI studies of improvisational jazz musicians. When these artists create solos, they close their eyes and enter a brain-mind state which resembles REM on the MRI. Improvisational jazz typically is played and listened to at night in dark nightclubs where drugs and alcohol conspire to influence the brains of REM-ready musicians and their audience.

Fyodor Dostoyevsky (1821–1881) was a Russian novelist whose own murderous rage was attributed by Sigmund Freud to unconscious hostility toward his father. Rage motivated characters in *The Brothers Karamazov, Crime and Punishment*, and *The Possessed*, but it now appears that Dostovesky may have suffered from temporal lobe epilepsy, a brain disorder characterized by brief lapses of consciousness and by "dreamy states" as well as impulsive and compulsive spending sprees, uncontrolled writing and speaking and emotional instability. Whether or not Dostoyevsky was an epileptic and/or a father hater, he was certainly a waking dreamer whose literary genius reflects the breakthrough of fiction into conventional reality. In this light, it is exciting to realize that dreams which are normally experienced in sleep are physiologically enhanced by seizure-like temporal lobe discharge.

Normal dreams are certainly created and they may be creative but they are not reliably mnemonic. Most of their content actually bears no relationship to events remembered in waking except in a generic sense. And most dreams do not have the coherence of a Dostoyevsky novel. In the chapter on memory, we discussed alternatives to the theory that dreams replay real or symbolically transformed recollections. It may be the function of the temporal lobe and other forebrain structures to prepare us for the future as much as to relive the past.

The epileptiform nature of the seizure-like discharges that take over and cause fits in our brain during REM sleep may constitute an evolutionarily proven way of preparing us for tomorrow as learning. REM sleep dreaming may be a modified temporal lobe seizure.

Religion and the Temporal Lobe

The American neuroscientist Michael Persinger (now living in Canada) argues controversially that religious states can be artificially induced by magnetic stimulation of the temporal lobes of the brain, and that natural religious experiences arise from spontaneous activation of the same areas. The whole field has been given the name neurotheology (a term coined by the writer Aldous Huxley). Persinger used a device known colloquially as the "god helmet" to produce weak magnetic fields over the temporal lobes. Some participants reported religious experiences and a sense of oneness and mysticism, although these results have been difficult to replicate. It has also been proposed that the results that Persinger and colleagues have obtained are due to expectancy effects on the part of the participants, as the experiments were not carried out double-blind — people taking part knew what the expected results should be. However, other evidence also points to some role for the temporal lobes in religious experiences. Norman Geschwind noticed that epileptic seizures with a focus in the temporal lobes (TLE) were often associated with intense religious and mystical feelings.

It has further been suggested that some mystics and visionaries (examples include St. Paul and Ste. Bernadette) suffered from TLE. Persinger (1983) argued that full-blown TLE might not be necessary to experience religiosity; some people might be prone to transient deep temporal lobe "micro-seizures," leading them to perceive life and events as more meaningful than others, distorting space and time, and enhancing religious feelings. An admirably calm discussion of these and other cases is to be found in William James's book *Varieties of Religious Experience* (1902). Although much of this work is both speculative and controversial, it does show how altered states of consciousness, and intense religious experiences, can arise from the malfunction, even if just temporary, of an otherwise normal brain.

Evolution and Devolution of the Brain-Mind

John Hughlings Jackson was one of the pioneers of modern brain-mind integration in pursuit of consciousness. Inspired by Charles Darwin, Jackson was the superintendent of the West Riding Lunatic Asylum in Yorkshire in the days before the schismatic split between neurology and psychiatry. Jackson realized that many of his insane inmates suffered from incurable brain disease, but his most memorable writings were testimonies to the dynamic unraveling of the mind as the brain 'devolved.'

We see this today in normal humans who undergo devolution of their minds as the brain reverses its evolution every night in sleep. We, too, are inspired by Darwin and we hope to be both neurologists and psychologists at the same time. As you can appreciate, this is not easy to do.

Jackson was convinced that temporal lobe epilepsy (and epilepsy in general) could teach us much about consciousness. He defined the "march" of grand mal seizures as the paroxysmal hyperexcitation moved across the cortex. He also made more than a nod to Darwin when he formulated his tripartite functional division of the central nervous system: the oldest parts were the spinal cord and lower brain stem; the newest parts were the forebrain, especially the cortex; and the parts in between regulated the whole system. Today we can add specific mechanisms to this germinal formulation.

It was the analysis of the "psychomotor" seizures of TLE that brought Jackson his lasting neurological fame. His descriptions of the symptomatology are unsurpassed in their deductions about the anatomy and physiology of the temporal lobe in relation to abnormal consciousness (including the 'dreamy states' — his terminology — which were treated in the previous section). Jackson was also the first to separate the primary symptoms of schizophrenia (e.g. withdrawal and apathy) from the secondary symptoms (e.g.

delusions and hallucinations). The link between TLE and schizophrenia, which was made via his dynamic and structural concepts, is now the subject of the scientific research of Robert McCarley, the pioneer of sleep and dream science to whom this book is dedicated.

According to the protoconsciousness/virtual reality model which we espouse, sleep and dreaming are a back-to-basics return to evolutionary primitives (or priors in the Bayesian sense of the word). The primary consciousness elements that we experience in our dreams include: the sense of self, the sense of space, the Darwinian survival emotions, and sensorimotor integration; the Jacksonian secondary consciousness elements are the cognitive and perceptual abnormalities (compared to the waking consciousness achieved by evolutionary adaptation). This analysis, which is made for didactic purposes, blurs the distinction we modernists like to make between primary consciousness (which includes hallucinations and delusions) and secondary consciousness (which is reserved for more abstract cognitive skills such as calculation and language).

Marcel Proust (1871–1922) was a French novelist who worked at the turn of the century when society was undergoing radical change and the scientific study of consciousness was in its infancy. His eight-volume novel *The Remembrance of Things Past* is celebrated for its poetic evocation of childhood experience and for its portrayal of changes in the adult social world. This masterpiece was made possible by the detailed introspection of Proust's life abed. He is justifiably renowned for his careful description of the states of consciousness that arise in sleep and dreaming and, most impressively, the mentality of the borderland between sleep and waking. Most of us cannot afford to lie abed as Proust did but we can thank him for exercising the luxury of laziness for us.

In talking with Hughlings Jackson, we are bridging 150 years of time in search of scientific commonalities. We find many.

Migraine Headache

Migraine bears some similarity to epilepsy: both disorders show abnormal EEGs, although the EEG characteristics of migraine are subtler and more variable than those of epilepsy. Obviously the defining characteristic of migraine is the severe headache, often accompanied by nausea to the point of vomiting, accompanied by photosensitivity, increased sensitivity to sound, and even hypersensitivity to smell. The headache is often unilateral. Migraines may be preceded by an *aura*, a transient disturbance of psychological processing, and are often marked by visual disturbances.

The aura indicates that an attack is imminent. Prior to the aura, however, there might be other changes in awareness, sensitivity to particular stimuli, and even personality alterations in what is known as the *prodromal phase*. In some cases, consciousness can be lost during the migraine attack. Migraine headaches are usually recurrent. Migraines run in families and affect women more then men. The prevalence of migraine is estimated to be around 15%.

In basilar migraine, lower regions of the brain are affected, and the symptoms can be even more severe. One of the symptoms of basilar migraine is confusion — consciousness is affected to the extent that the person might become disoriented, unaware of where she is or what is happening to her.

These disorders show that changes in consciousness can be induced in some people by specific environmental triggers (e.g. certain foods, lack of sleep, or stress). Consciousness might be an internal state and a hard problem, but it can be affected by something so simple as a crumb of cheese or piece of chocolate. The importance of a healthy brain to healthy consciousness is further proof of the significance of the basic principle of this book and indicates both our ignorance and the opportunity to replace it with scientific knowledge.

Dialogue 22. Dream Fits

TH: I suffer from migraine and now you tell me that my dreams are quasi epileptic? What a downer!

AH: Science tells us that the brain resorts to elaborate guises to ensure neuronal excitability control.

TH: And the brain sometimes makes mistakes? I would rather think of myself as perfect (or at least free of disease), but my migraine headaches prove me wrong in this conceit.

AH: Or right, if you accept the conscious states mantra. Creating and maintaining a healthy brain-mind is a big deal.

TH: I guess I should be grateful that my mental heath is not worse. But I freely admit that I am too old to learn neurobiology.

AH: Today's students are treated to vast scientific progress in neurobiology. But many fail to appreciate the psychological import of those discoveries.

TH: I thought you must be kidding when you suggested that my dreams were like the delusions of madmen and epileptics.

AH: You can have a modified seizure as you sleep. The brain is activated but inputs and outputs are blocked so you don't really have a fit.

TH: Only its central benefits are the result. I wish Wilder Penfield were here to hear this story. He would be thrilled.

AH: Maybe so, maybe no. The medical disease model dies hard. It has been very successful, after all.

TH: You are courageous to take on the medical establishment.

AH: The reviewers of a draft of this book were no readier for it than the students they teach.

TH: Most scientists do not remember or write off their dreams as scientific nonsense.

AH: Just the opposite is the case. Dreams make sense only when viewed through the lens of the conscious states paradigm.

Chapter 23. Altered States of Consciousness*

Most of our lives, we inhabit a narrow range of consciousness that we can call *normal consciousness*. We have ups and downs, are sometimes tired and sometimes exhilarated, sometimes focused on a task at hand, other times not, but the variations are quite small, and rarely involve any change in perceptual quality. Certain states however can stray from this normal range, and we call these *altered states of consciousness*, or ASC for short.

The most familiar ASC is of course dreaming: as we have seen, dreams are often bizarre, but have some internally-generated narrative. Their perceptual qualities are often distorted or selective or both. This chapter looks at other states that differ from our baseline of normal waking consciousness.

Drugs

The easiest — but by no means safest, or often entirely legal — means of experiencing an ASC is through certain mind-altering drugs. Such drugs are called *psychoactive* or *psychotropic*, and there are many different types with many different effects. All have their effects by affecting brain neurotransmitter systems, and all are dangerous in large amounts. Even the mildest of them cause some reaction after we stop taking them. *Caffeine*, for example, makes us feel more alert and gives us a buzz, but if we stop taking tea, coffee, and pop drinks with caffeine added, we may experience caffeine withdrawal, with headaches, tiredness, lethargy, constipation, and irritability heading the list of symptoms.

The intensity of caffeine withdrawal depends on just how many cups of coffee were regularly consumed before a subject stops caffeine use. The lethal dose of caffeine is around 10 grams, depending on body weight; the average cup of coffee contains 100–200 mg, so you would have to drink 100 cups of coffee in short succession to be at risk. Caffeine is the most widely used psychoactive drug in the world today. It has a number of effects in the brain, but the primary one is that it is an adenosine receptor antagonist. Because caffeine's shape is similar to that of adenosine, it occupies some of the adenosine receptor sites in the brain. Because adenosine is a *neuromodulator* involved in regulating sleep, the effect is to make us more alert.

Amphetamine ("speed") is an even more effective stimulant and performance enhancer, but it has numerous side effects, can be addictive if used recreationally, and has severe long-term consequences if abused, particularly *amphetamine psychosis*.

Alcohol has marked psychoactive effects. In small doses, it helps people to relax and become socially more disinhibited, increases confidence, and leads to mild euphoria; in larger doses, it interferes with memory and sleep, and may cause stupor. A liter of spirits or more than three bottles of wine ingested within a short time will kill most people. Many more people are killed by accidents, particularly drunk driving, where alcohol has decreased alertness and motor control. Many suffer short-term health impairments (the dreaded hangover) and long-term brain damage.

Although widely consumed across the world and across history, the social cost of alcohol ingestion is enormous. The degree of alteration to the state of consciousness depends on the dose. It is widely thought that alcohol helps you to sleep: this belief might be true in very small doses. While alcohol facilitates falling asleep and increases the amount of slow-wave sleep early in the night, it has many effects throughout the night, resulting in an overall decline in sleep quality. Alcohol has many effects on brain, neurotransmitter, and hormone systems: it enhances the effects of the neurotransmitters glutamate and GABA, resulting in the mood effects at low doses and memory impairments at higher doses. Alcohol also interferes with the action of the brain's reward system.

Cannabis (marijuana or "weed") is legal in some countries and states of the USA. It is prescribed for pain relief. Its primary psychoactive constituent is *tetrahydrocannabinol* (THC), a member of the *cannabinoid* group of chemicals. The brain has cannabinoid receptors, and THC produces its effects by binding to these receptors, which ordinarily prevent the operation of neurotransmitter systems, particularly those involving dopamine and norepinephrine.

THC induces euphoria, increased relaxation, an enhancement of all senses, increased enjoyment of music, and distortions of time and space. In larger doses, THC can produce hallucinations and increase anxiety. There are no known cases of a fatal overdose resulting from cannabis use, although there are physiological costs to smoking cannabis (similar to the harm done by smoking tobacco) and possible psychological costs. Some researchers believe that prolonged cannabis use can increase the risk of mental illness, particularly schizophrenia and depression, but the evidence is equivocal. It could be, for example, that people more prone to schizophrenia are more likely to smoke weed. This type of research is also very difficult to carry out given ethical considerations: you can't force people to take a potentially harmful, illegal drug.

The *opiates* are a large class of drugs resembling morphine. They act on the opioid receptors of the central nervous system, many of which are located on REM-off cells, which could be why they have such potent effects on waking and why they alter consciousness. Action at these receptors mimics the action of endorphins, often described as the brain's natural painkillers, but as well as these pronounced analgesic effects (which explains why morphine is used to relieve extreme pain), they lead to profound euphoria and extreme relaxation.

Morphine is derived from opium, the extract of the opium poppy, *Papaver somniferum*. Morphine can be processed in the synthesis of heroin, which has similar but even more pronounced psychoactive effects, being about three times more powerful than natural morphine and acting more quickly. Most of these opioids are highly addictive; furthermore, the body quickly adapts to become tolerant, so that ever higher doses are necessary to achieve the same effect. As a consequence, lethal overdoses are common for drugs such as heroin.

Hallucinogens

From the point of view of consciousness, perhaps the most interesting drugs are the hallucinogens; these are drugs that distort perception, or give rise to hallucinations of all types. They also distort a person's sense of reality, giving the user the feeling that he has great power, special knowledge, or insight into the nature of the universe. It is these properties that have made hallucinogens popular for studying or enhancing consciousness across time and cultures. Many cultures use naturally occurring hallucinogens in shamanistic rituals: examples include *mescaline*, found in the peyote cactus, and *psilocybin*, found in so-called "magic mushrooms" (actually a number of types of mushroom, but of which *Psilocybe cubensis* is the most common).

The most notorious hallucinogen is *lysergic acid diethylamide*, or LSD. LSD was first synthesized in 1938 by the Swiss chemist Albert Hoffman, who was searching for new types of drugs to act as stimulants to the circulatory system. LSD is very potent, and was inadvertently absorbed through Hoffman's fingertips. The drug was first used in the late forties for psychiatric purposes, but LSD was a clear case of the treatment being worse than the disease. In the fifties, the CIA experimented with the drug as a potential means of mind control, or even as a weapon. However, after it was popularized in the early nineteen sixties by people such as the Harvard psychologist Timothy Leary, LSD entered into popular use, playing some role in the development of the counter-culture. The naïve idea was that LSD would liberate people's minds. It is now clear that enslavement is the price of LSD mind liberation.

The British writer **Aldous Huxley (1894–1963)** famously took mescaline, hoping that it would free his mind from everyday constraints and expand his creativity. He wrote a book about his experience called *The Doors of Perception,* a title taken from a couplet from William Blake's poem, *The marriage of heaven and hell*:

"If the doors of perception were cleansed every thing would appear to man as it is, Infinite. For man has closed himself up, till he sees all things thro' narrow chinks of his cavern."

Drugs that give people the feeling that they have a divine purpose, or an insight into the divine, or are used in religious rituals, are called *entheogens*. Sadly for most of us, the insights given by hallucinogens are spurious: there is no "real reality" beyond the present one, and the profound insights given by the drugs usually turn out to be utter rubbish in more sober light. Even worse, insights scribbled down during the drug trip often prove to be illegible the next day (The same is true of deep sleep insights, by the way).

More seriously, there are many adverse effects associated with LSD use, including the "bad trip," wherein the person has a psychonoxious experience; panic and extreme anxiety, self-harm, and even suicide. Furthermore, some people report later "flashbacks" to the hallucinogenic experience. There are claims that LSD use can lead to psychosis, but this sort of claim is difficult to evaluate.
Some researchers distinguish different types of hallucinogens: psychedelic drugs such as LSD, which alter the quality of consciousness by changing perception and cognition; and deliriants, found in several plants such as Jimson weed, but not widely used recreationally because the symptomatic confusion and stupor is usually found to be unpleasantly dissociative. Ketamine also produces feelings of dissociation. These types of hallucinogen all derive their effects by affecting the neurotransmitter systems, which mediate normal consciousness.

We have barely scratched the surface of the myriad psychoactive drugs. Some people are tempted to "raise" or "explore" their consciousness by experimenting with these drugs. Potential users should remember that many of these drugs are illegal, and most have possible or potential consequences, some of which are fatal.

Albert Hoffmann (1906–2008) was a research biochemist who discovered LSD accidentally while testing synthetic ergot alkaloids of possible use in the treatment of arterial disease. His discovery led to the recognition that minute doses of a chemical can cause dramatic alterations in conscious state and that normal and abnormal hallucinations can be induced chemically (as is recognized and explained by the AIM model of consciousness). That Hoffmann became visually psychotic by trace amounts of LSD gave impetus to the still ongoing search for intrinsic molecules that might mediate mental illness and lead to widespread recreational abuse of LSD and other psychedelic agents. The science of consciousness, amateur experimenters, and the criminal justice system all owe a large debt to Hoffmann's discovery.

Hypnosis

A common image of the hypnotist is a man with a moustache, staring into a person's eyes, swinging a pocket watch in front of him. "Look into my eyes..." We associate this image with the German physician Franz Anton Mesmer (1734–1815), who treated patients in late eighteenth-century Paris (and who gave us the noun 'mesmerism' and the widely used verb 'mesmerize'). We see stage hypnotists select susceptible individuals, and then influence them so that they act out of character, performing scenes such as acting like pigs or thinking that people are objects. *Post-hypnotic suggestions* are made during trance. Subjects are given instructions that they will later carry out a particular action when they come out of trance. They then perform the action (saying something or scratching their ear), and when quizzed, cannot explain why they so acted.

Anton Mesmer was way ahead of his time. He postulated that the mind was a form of magnetic energy which could be influenced by controlling attention (see again Chapter 11). Indeed, human subjects can be "mesmerized" by a hypnotist who coopts their consciousness and substitutes his own will for theirs. We now recognize that the self is dynamic, plastic, and constantly reformulated. The popularity and power led Freud and his contemporaries to assert that psychoanalytic access to what they called the unconscious was in no way hypnotic, a difficult and highly unlikely claim. Mesmer has an undeserved reputation as a charlatan and a fake as people are loath to admit the obvious: their will is certainly not free in hypnosis and hypnosis can be a force for evil as well as for good.

Hypnosis involves susceptible, suggestible individuals who are put into a trance, through a process known as hypnotic induction, wherein they become hyper-suggestible. Attention is focused on the hypnotist, and there appears to be reduced awareness of other peripheral stimuli. However, frustratingly little is known about what exactly happens during hypnosis, and there is no real agreement about how suggestibility is enhanced. Neurophysiological studies show that participants display more alpha waves in a trance, but then they are likely to be highly relaxed. One controversial proposal is that hypnotized people are role-playing: they are playing the role that is expected of them. It is not clear how social role-playing theory explains the frequent successes of hypnotherapy.

Hypnosis is widely used as a therapy for some psychopathologies, particularly anxiety-related disorders, and for unwanted habits such as smoking. Freud began his clinical career using hypnotherapy, and authoring *Studies on Hysteria* with Joseph Breuer (1895). He concluded that hypnotherapy was not getting to the core problems of his patients, so abandoned it in favor of the techniques of free association and dream interpretation of psychoanalysis. It is quite difficult to evaluate the success of hypnotherapy because of the difficulty of finding appropriate controls, but studies generally show a benefit of hypnotherapy over no therapy. For example, people wishing to quit smoking are likely to be twice as successful in doing so with hypnotherapy than without hypnosis.

Franz Anton Mesmer (1734–1815) was a pseudoscientific pioneer of consciousness science. He should be remembered for his institutionalization of the power of suggestion (still poorly understood) even as we ridicule his particular theory of invisible fluid force. Indeed his humbug should make us cautious about the postulation of any unobservable force of mind. We insist on experimental proof of a such hypotheses (including our own ionic theory of mind). And we must be equally resistant to the denial of hypnotic effect simply because we regard Mesmer as a charlatan (which he was) or because we wish to substitute our selves for him or our theories for his. His story is educational for the debunking effect of the French government's warning of charlatanism, via its review panel on the scientific basis of Mesmerism. A similar warning might have headed off modern abuse of trust by contemporary charlatans.

Whatever the scientific basis of hypnosis might be, there can be no doubt as to its importance to the study of consciousness. Whether the subjects are merely role-playing or are genuinely dissociated hardly matters. We are all suggestible and unaware of the myriad influences upon our minds. The power of positive thinking is hypnotic. Only the assiduous application of neuroscientific methodology has any hope of objectifying this important psychological process.

In the chapter on lucid dreaming we emphasize our view that sleep science offers objective evidence of the brain basis of hypnotic suggestion. The frontal lobe is activated by a sleeper's autosuggestion, giving him access to something like free will. If suggestion can alter consciousness via selective frontal lobe activation, then we have at least a reasonable chance of understanding the mind in terms of fluid-like energy and Anton Mesmer, like Sigmund Freud, may achieve their rightful places in the history of science. They were both discredited, like Stephen La Berge, for their excessive hucksterism.

Figure 23.1. Cajal as Hynpotist.

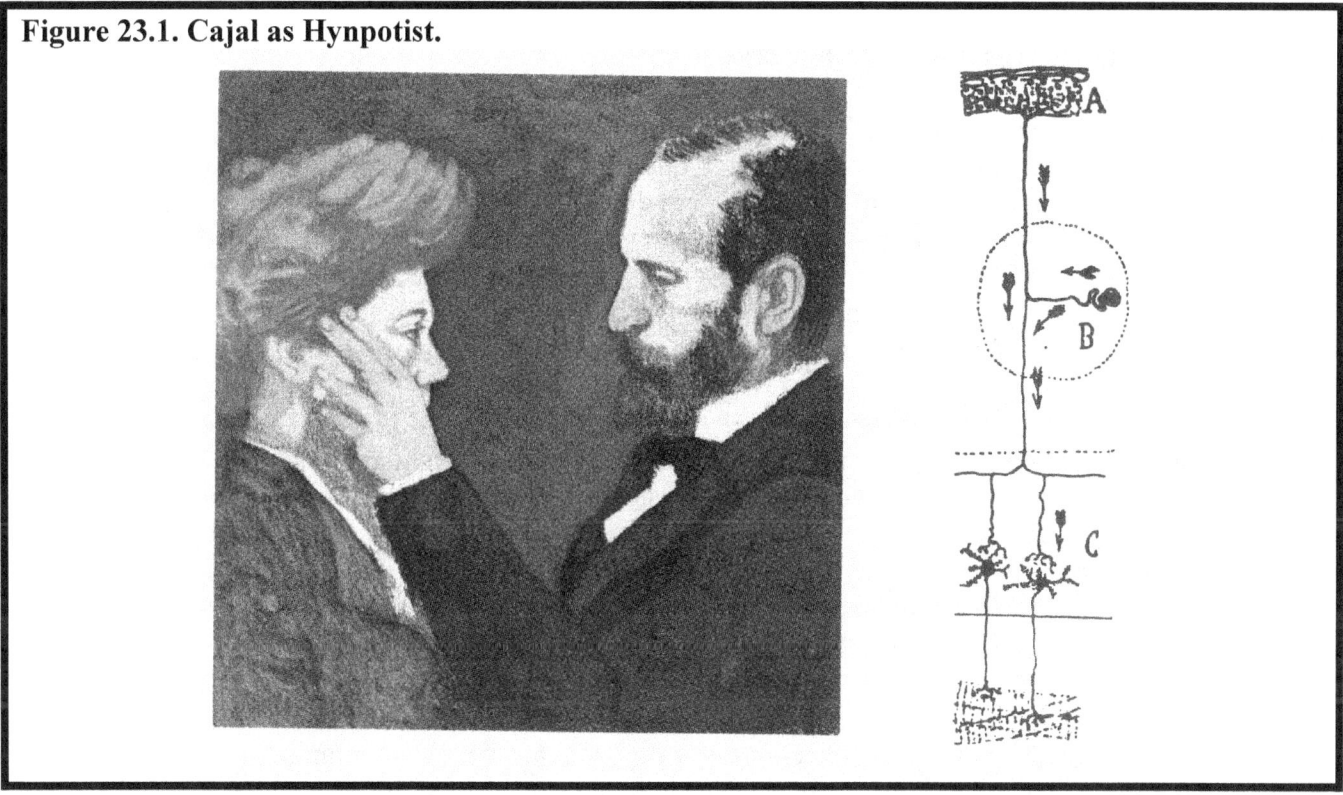

Meditation, Prayer, and Mindfulness

Meditation has been in the news. There is now considerable evidence that meditation has substantial beneficial effects for both mental and physical health.

Perhaps the best known technique is transcendental meditation (or TM), although all types of meditation involve focusing attention on a simple, easily accessible stimulus. In TM, it is a simple word or phrase called a *mantra*; in breath-based meditation, it is respiration, either focusing on breathing efforts and/or counting them. The practitioner typically meditates for 15 to 20 minutes once or twice a day. If attention drifts from the object of concentration, the meditator is instructed to bring attention back into focus; there is meant to be no striving and no recrimination. However, there are many other types of meditative practice. We can take prayer to be no more than a particular type of meditation that is practiced in a religious setting with particular words and intention.

In all cases, meditation leads in the short term to a state of deep sleep-like relaxation, characterized by increased alpha and theta EEG wave activity. fMRI imaging shows increased activity in the prefrontal and frontal cortex, and in the anterior cingulate cortex. There are suggestions that repeated meditation can lead to long-term changes in EEG, although these findings are controversial. Similarly, there are some long-term structural changes revealed by brain imaging, such as activation levels in the frontal and anterior cingulate cortex, suggesting that regular meditators are better able to self-regulate their mental states, and have enhanced meta-awareness. There is now considerable evidence that regular practice of meditation can lead to the reduction of the symptoms of stress, and may even slow cognitive aging.

Whereas meditation is actively directed towards focusing attention and excluding extraneous stimuli, *mindfulness* involves trying to be as open as possible. For example, in being mindful of the world around you, you would try to observe your perception, without evaluation or judgment, and simply notice your mental states. It has been argued that mindfulness training can be as effective as if not more so than SSRIs in the treatment of depression.

Trance and "Higher" States of Consciousness

Some cultures, including our own, make use of drugs to induce shamanistic, religious states. Others use forms of meditation and associated techniques to induce trances, leading people to believe they have access to the future, the ability to cure themselves or others, or endure pain and physical suffering without analgesics. Trances can be induced by hypnosis, self-hypnosis, or prolonged repetitive movement. Dervishes are Sufi Muslims who choose to live a life of extreme hardship and self-deprivation. Sufi whirling ("whirling dervishes") is a form of repetitive dancing involving the dancer spinning around, and is a form of moving meditation where the dancer hopes to obtain insight into the nature of existence.

Buddhists strive to obtain satori (understanding of the nature of being and the world) and ultimately nirvana (liberation from the cycle and demands of existence) through the prolonged practice of meditation.

Trance involves states similar to meditation and hypnosis. It is not for us to comment on the religious content associated with these states. They all involve forms of consciousness different from the normal waking state, but it is not scientifically justified to think of them as being "higher" in the sense of better.

Some people occasionally have "peak experiences" or *epiphanies*, which are moments of exhilarating euphoria where everyday scenes of objects may suddenly take on great significance or appear to be exceptionally beautiful. The Victorian poets, particularly Rossetti, Tennyson, and Wordsworth, wrote about them; Wordsworth called them "moments in time." Maslow believed that peak experiences were something to strive for, and were more likely to be experienced by self-actualized individuals. We have probably all experienced such moments, when the everyday suddenly becomes almost a religious experience filling us with great joy.

The writer Colin Wilson argued that we should all strive to achieve more epiphanies, and that we are not achieving our full potential unless we do so. We see that the notions of self-awareness and mindfulness are intertwined with raising our awareness and peak experiences. However, very little is known about what triggers peak experiences, or what is happening in the brain while they are occurring subjectively. To our knowledge, there has been very little neuroscience research on higher states of consciousness.

Sensory Deprivation and Habituation

What happens if we deprive the brain of sensory input? This process of deliberately depriving the brain of external data is known as *sensory deprivation*. The typical setup involves a person being immersed in an *isolation tank* full of water maintained at body temperature, with enough salt dissolved in it for the person to float so that they do not feel the effects of gravity. The tank is isolated so that it is completely dark and silent inside. In this way, there is no input to any of the senses of the participant; at the very least, sensory activation is greatly minimized. After a while most people become very calm and extremely relaxed; hence sensory deprivation results in an altered state of consciousness.

Giulio Tononi studied neurophysiology with Ottavio Pompeiano in Pisa before collaborating with Gerald Edelman, then at Rockefeller University in New York. He is the author of *PHI*, the theory that consciousness can be defined and measured as the net amount of information processed by any system. Tononi now directs a research program at the University of Wisconsin at Madison. Among other important experimental findings, Tononi has shown that isolated parts of the brain can be awakened during sleep. Conscious states are thus not necessarily global, an idea also advanced by Tore Nielsen. As AIM predicts, consciousness varies in degree and kind according to the intensity of brain activation.

People's responses to sensory deprivation depend mostly on the length of time that they are isolated. For short periods of deprivation, many people find the experience pleasant and relaxing, and they may use the tanks to meditate and relax. Indeed, commercial chains now sell periods in these tanks for these purposes as *flotation therapy*. They make suggestive claims that flotation therapy reduces the effects of stress, promotes physical healing, and enhances creativity.

People often find longer periods of sensory deprivation to be unpleasant. The experience of deprivation progresses from relaxation and stress reduction to anxiety and an increase in unpleasant consciousness. Many participants may even become paranoid. After a while, the person in the sensory deprivation tank starts to hallucinate: it is as though the brain cannot bear to be without sensory input, and after a period of deprivation of input, starts to fabricate its own sensations. This sequence bears comparison to normal dreaming caused by the brain's elaboration of pseudosensory stimuli.

Performance on cognitive tasks is significantly lower than on pre-tests after 48 hours of sensory deprivation. Indeed, so unpleasant is prolonged deprivation for many people that it has been used as a means of psychological torture ("enhanced interrogation techniques").

A related procedure is *sensory habituation*, known as the Ganzfeld technique. Sensory habituation is easier to accomplish than sensory deprivation because a flotation tank is not needed. People generally find the experience very pleasant. Because they're not trapped in a tank, subjects are less likely to become claustrophobic or paranoid. The Ganzfeld is easily set up: all that is needed is a comfortable chair, couch, or bed; headphones and a recording or source of pink noise (similar to random noise but more pleasant to the ear), and a means of projecting a uniform visual field.

The classic, and rather comical, method is to tape halved Ping-Pong balls to the eyes, suitably padded out with cotton wool to avoid sharp edges, and to shine a red light through the plastic. You can now buy ready-made "Ganzfeld goggles." People find the Ganzfeld to be very relaxing, and move into a very relaxed, comfortable state of consciousness within half an hour. The Ganzfeld has been one of the most frequently used and controversial techniques in parapsychology research (the idea being that any weak signal is more likely to be detected when other external stimuli are absent or reduced).

Eye Movement Directed psychotherapy is one the more specific ASCs which suggest wake-state activation of the brain's own REM dream motor. As is typical of therapeutic novelties (like hypnosis and psychoanalysis), practitioners are enthusiastic but fail to produce efficacy or mechanistic evidence. This leaves us with a sense of promissory notes which we would like to cash in but can't find a bank that will risk its credit.

In conclusion, these variations on the theme of suggestion should be taken as grist for the mill of understanding consciousness. They all reflect the investment that people are willing to make in order to achieve peace of mind and are testimony to motivation and persistence in the quest for comfort. From the analytical viewpoint, they celebrate the mind's capacity to examine and control itself. As such, they would seem to vouch for free will but this implication must be critically evaluated lest it, too, become a plaything of our suggestibility.

Dialogue 23. Good and Bad Trips

TH: When I was a student, the drug craze was just getting started and I cut my psychedelic teeth on pot.

AH: Pot? That's Cambridge, UK for you. Beer was the only consciousness destroyer available when I was young. We spoke of getting high on pitchers of brew at Uncle's bar.

TH: Then you got up at 8 AM to go to biochemistry class because you wanted to go to medical school.

AH: No wonder I never learned a thing (even though I passed all the tests).

TH: And no wonder they say that education is wasted on the young.

AH: Getting "wasted" was my pride and joy. Now the stakes are higher. With recreational drugs, it is possible to blow your mind at will in the privacy of your own bedroom.

TH: The effects of LSD can be psychotic. Visual illusions and hallucinations are often sought and easily found.

AH: This confirms the predictions of dual-aspect monism philosophy and virtual reality dream theory (as if we needed more proof).

TH: We always need proof. I suppose that young people have two motives: they want to test the limits of psychological experience and they want to rebel against people like us who warn them about the risks.

AH: We might do better if we promoted rebellion. Students might then rebel against the rebellion that we seem to promote. This trick is called paradoxical intention.

TH: In my day, gurus like Timothy Leary and RD Laing led as many youth into the river as the Pied Piper of Hamelin.

AH: Nothing that we say is likely to change their minds but, like parents, we need to issue warnings anyway.

TH: Some mind trips are bad and others are one-way (with no return to base).

AH: The puritan in me advocates the amateur study of dreams. I now have ready and natural access to psychosis in my own bed in my own bedroom.

TH: The only chemicals you really need are your own brain juices. Natural mind trips are not only safe, they are also free.

AH: But I miss the hangover that I nursed through biochemistry class. It convinced me that I was free to be sick.

TH: I'm still quite partial to a tipple of port. Alcohol is a venerable conscious state alterer and now we know why. It puts parts of the brain to sleep.

Part IV. PHILOSOPHY

Introduction and Summary, Part IV.

I have come a long way but I still have far to go. If it took evolution millions of years to tinker up our brain-minds, we can hardly expect to understand them in little more than a century of research. History teaches us to be patient but most people ignore history. They want answers now. Nowhere is that impatience more manifest than in the desire to understand dreaming. No one wants to be told that his dreams are evidence of a built-in system designed to make his waking mind function better.

The digital revolution may help understanding by providing people with brain-like gadgets. We don't have to believe that our computers can think and feel in order to be impressed with their cognitive power. And the iPhone operator of today is in a better position to appreciate the importance of brain hardware than a telephone operator or a typist was in the mid-twentieth century. It is now possible for every high school student in the world to communicate with one another over Facebook.

In this final section of our book I blue-sky unabashedly as I speculate about why we sleep and dream. I adopt the concept of virtual reality that the digital revolution has made popular. Whatever critical science peers think about my conviction that the mind-brain problem is already solved, the mind is clearly a brain function. The attribution of cause to one part of the brain, the mind, is, in my view, unexceptional and unexceptionable. Thus I do not apologize for my enthusiasm. Rather I invite you to share it with me.

We are, indeed, such stuff as dreams are made on. We can now take Shakespeare to bed with us and write plays of our own devising in our private Cartesian theaters.

Chapter 24. Virtual Reality

Virtual reality may be defined as an artificial (virtual) copy of the external world (reality) that is taken for the world itself. Video games work, and are justifiably popular, because they are as effective illusions as dreams and yet they occur in waking. One does not have to be drowsy to succumb to these illusions. In the computer age in which we live, it is easily understood as the purposeful exposure of human subjects to sets of elaborate stimuli of video games that are designed to fool the subject into believing that he is really shooting an imaginary enemy, really driving at high speed in an imaginary vehicle, or really flying through space as if he were weightless. This is virtual reality.

Virtual Reality and Subjective Experience

Short of video games, it is the amazing talent of the brain to create an entirely virtual reality. There is no real world within the head and yet the image of the world that the brain creates is such a faithful copy of the real world as to convince us that our subjective experience is itself real. It really is real but it really is an illusion. Dreaming illustrates this illusion. After the fact, we realize that dreaming must be illusory but, on reflection, we can easily convince ourselves that waking consciousness is every bit as illusory as dreaming. Waking consciousness is as virtual as dreaming in that it is entirely artificially constructed in spite of its use of real physical stimuli. How and why this is done and how can dreaming can help us understand the similarities and differences of these two virtual realities is the subject of this chapter.

The why question is easy to answer for waking. Our survival depends on the fidelity of our virtual image of the world. But why are our dreams accepted as real when they are so clearly virtual and so unfaithful to the real world (whatever that world "really" is)? While this sort of conjecture may further soften the problem of the why of both waking and dreaming, it is the how question that is still elusive. This again is the "hard problem" to which we have often alluded. It is our contention that the scientific investigation of REM sleep dreaming has already softened the hardness of the how problem and that an extension of this experimental approach can be expected to add further softening (and even resolution) of the problem.

Our confidence is based on the following observations. Dreaming, despite its infidelity to external reality, is nonetheless a surprisingly good simulacrum of the world. That means that waking perception (and several other cognitive faculties) might profit from a model of the world that could be used to predict external reality. Instead of merely reading the world, the brain encounters that world with expectations about it. In other words, a possible function of dreaming is to facilitate the predictive capacity of waking consciousness. That's more "why?" but does not yet tell us much about "how?" The how question is beginning to be answered by the scientific study of REM sleep dreaming in animals, which permits detailed neurobiological mechanisms to be determined.

To illustrate these and related themes we present reports of some relevant dreams:

"Sperry's slice" illustrates several important aspects of the virtual reality concept:

1. It is derived from the dreamer's experience. He is a real as well as a dream golfer.
2. The dreamer's brain is activated and subjected to the convincing experience of really playing golf.
3. The dream golf is animated in a distinctive way which the dreamer depicts as curving and circular trajectories.
4. The dreamer wishes that bad golf shots, slices like his and Sperry's, could be magically corrected.

Figure 24.1. The golf dream of LL Buchanan from The Dream Journal of the Engine Man.

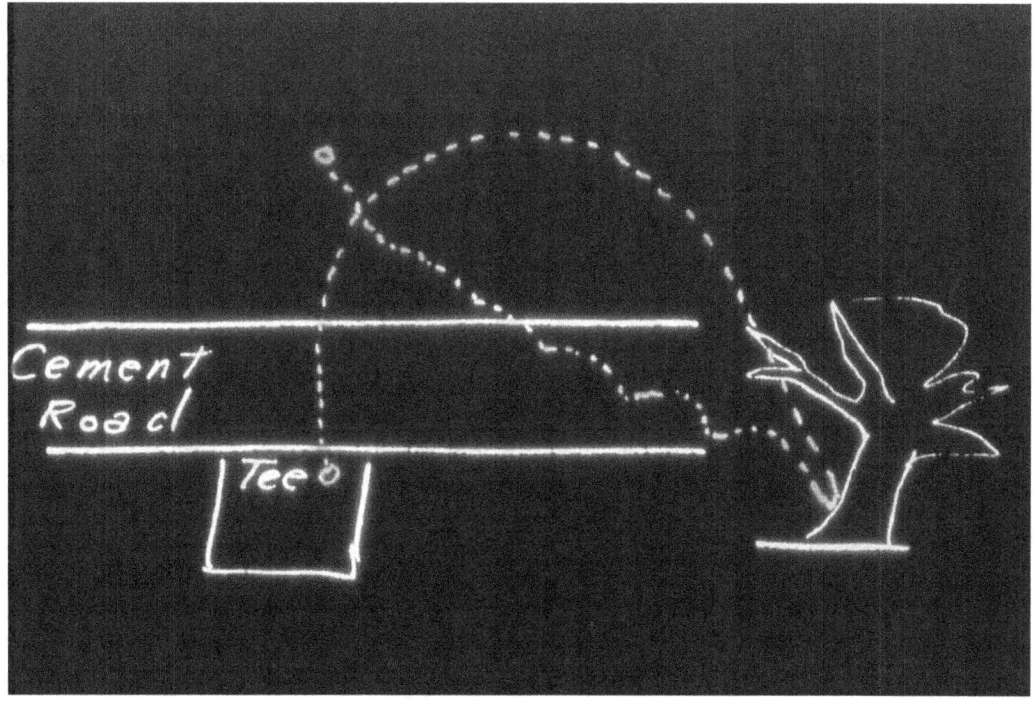

The study of REM sleep dreaming makes clear that the physiological basis of virtual reality is a material entity whose subjective experience is physical but immaterial. All physical systems have both a visible and invisible immaterial aspect. If the brain, in either waking or REM sleep, is at once visibly and invisibly material, it may create virtual worlds that we experience subjectively as real. In the case of waking, the fit is good. In the case of dreaming, it is good enough. It could hardly be expected to be perfect. Reproduction of the world is not perfect even in waking, and in dreaming it does not have unadulterated real world data to work with.

The virtual reality of which we speak is a poor man's video game. Your brain plays tricks on you free of charge and your dreams provide a nightly set of golf games, imaginary enemies, imaginary high-speed vehicles, and imaginary weightlessness.

Protoconsciousness Theory

Consciousness develops slowly both in evolutionary time and in human developmental time. It stands to reason that animals which become even partially conscious pass through a preceding phase that we call protoconsciousness to signal both its precedence and its incompleteness. Protoconsciousness is a theory that has been developed to help us understand the "how?" question posed in the preceding section. By examining immature or primitive brains, we hope to get a fix on those brain parts and functions that must be in place for consciousness to supervene. Before playing dream golf, a sport must have been learned. But the pre-golf baby experiences proto-golf via excitation of his developing oculomotor and vestibular brain in REM sleep.

To further comprehend protoconsciousness theory, it is helpful to reiterate Gerald Edelman's distinction between primary and secondary consciousness. Recall that primary consciousness was defined as sensation, perception, and emotion in distinction to secondary consciousness, which consists of awareness of awareness, along with other capacities for abstract thought, including the sense of self as named agent, the ability to comprehend and create mathematics, speech, and literature, all of which we call "knowledge." It is these abstract capacities of the mind that we assume to be uniquely human.

According to our virtual reality theory, protoconsciousness precedes primary consciousness (with which it is fused) and both precede secondary consciousness (with which they are fused). By strategic design, experiments can be designed to test and learn from the theory even if it is unlikely that protoconsciousness theory can ever be definitively proved or disproved. In this sense, it is like the Darwinian theory of evolution, which explains aspects of nature while remaining unproved and probably unprovable because it is historical and history can never be repeated. It is nonetheless scientific because it leads to testable hypotheses which might otherwise escape notice.

At the risk of circularity, we suggest that REM sleep/dream consciousness is protoconscious in that it is abundant in early life and arises in utero as the brain is being formed and elaborated. By studying dreaming, we can gain insight into an earlier state of our existence. To the end of life, dreaming is single-minded, that is to say, its consciousness is primary rather than secondary. For this reason, many scientists prefer to assign dreaming to the unconscious and restrict the definition of consciousness to the secondary state described by Edelman. We are unhappy with this decision for both semantic and scientific reasons. Our objections are simple and strong.

Antti Revonsuo is a Finnish philosopher turned dream scientist. His functional theory is that dreaming reinforces threat avoidance behavior in waking. This idea is attractive because: a) it is evolutionarily Darwinian, b) it is based on the solid evidence that anxiety and flight are common dream experiences and c) it is philosophically innovative. Revonsuo thus represents the highly desirable integration of psychology, philosophy, and physiology championed by William James over a century ago. The human data of Revonsuo can now be further integrated with the cellular and molecular findings deriving from animal studies, featured here.

Dreaming is not so much unconscious as it is consciousness which is inaccessible to waking memory. Memory is only one of several important components of consciousness and its absence during and after dreaming cannot therefore be definitional. Our scientific reasons are functional. Perhaps the weakening of memory within and after dreaming is in favor of strengthening and enriching waking memory. This would mean that dream amnesia is misleading. We hold that waking and dreaming are two states of consciousness, each in the service of the other and add that the wall that normally separates them can be broken down in the interest of both entertainment and scientific enlightenment.

Sensorimotor Integration and Modeling of the External World

A step on the way to protoconsciousness is the development of sensorimotor integration. The brain must coordinate and unify its inputs and outputs before its possessor can go into the world with a chance of survival. According to protoconsciousness theory, it does this offline, at first uninterruptedly in utero, then intermittently around the clock in infancy, and finally only in sleep during childhood and adult life. With each successive phase, the brain utilizes real-world information increasingly as if to augment its genetically limited instructions. This is trial-and-error learning and is at the root of studies of the sleep enhancement of skills that were reviewed in Chapter 17.

As Immanuel Kant insisted, *a priori* information arises sui generis. Protoconsciousness theory ascribes *a priori* information to a genetic program for sensorimotor integration and further assumes that animals which succeed in the prenatal self-organization task of sensorimotor integration will have a better chance for survival than those who do not. A good example is locomotion as stressed by the physiologist, Rodolfo Llinás. Of course we must learn to walk, but our brains are equipped with gait generators which can be activated off-line. We practice walking, running, and even flying in our sleep and it is no surprise to learn that the circuits for gait selection and motor program activation are located cheek by jowl in the pontine brain stem, adjacent to the neural machinery for REM sleep control.

No wonder that REM sleep dreams are so intensely animated. There is not one sentence of a REM sleep dream report that does not contain an action verb. By running motor programs offline, the organism is accorded the opportunity to unify the necessary sensory systems involved in actual movement in the world. These motor actions can be virtual without being either dangerous or ineffectual. Millions of synaptic connections may thus be made off-line while we are floating in amniotic fluid or snugly asleep in our cribs and beds.

To better appreciate the safety of this paradigm, recall that even in adult life we are protected from harm by the active inhibition of motor output in REM sleep. A further insight is provided by the claimed utility of visualization by Olympic-level gymnasts, acrobatic skiers, and champion divers. Television coverage reveals the movement rehearsal sequences by these athletes as they are waiting to perform. They often close their eyes as if to suppress external visual inputs which are not relevant to their performance routines. These fellow humans are waking dreamers, are they not?

Self-organization is a process that characterizes all complex systems, and the protoconscious brain is one such system. An easily understood and relevant example is the simple case of water at various temperature ranges. Below 0 degrees C., between 0 and 100 degrees C., and above 100 degrees C., water changes state from ice to liquid and then to steam. These are three states (or phases) of water.

The brain and its mind are far more complex than water and yet the sequence of self-organized states is tripartite and its three phases are distinct. In this section, we have emphasized the role of genetic control but once the system is set in motion the rules of self-organization apply.

Remarkably, order arises out of chaos and chemical oscillators achieve rhythmicity by virtue of their tendency to synchronize the activity of distributed components. At one extreme of a rhythmic continuum, waking, the brain-mind is environmentally constrained but it nonetheless exhibits consciousness. In sleep, the "temperature" drops and the system first reprocesses environmental data, sometimes with perseverative thought consciousness, only to "heat up" again and become single-mindedly conscious. It can then compare recent environmental inputs to its intrinsic store of information. The point of invoking the self-organization paradigm is to show that the force of genetics is continued with the unsupervised freedom and order of the system itself.

Construction of the Self and its Virtual World

Atop the sensorimotor brain-mind must sit a virtual self who sees, senses, and feels the virtual world of protoconsciousness. As we have repeatedly noted, the German philosopher Thomas Metzinger has explored the functional development of the self in his books *On Being No One* and *The Ego Tunnel*. According to Metzinger, the self is a state, not a structure. As such, it is constantly recreated and is subject to both dissolution and positive reconstruction. These are important and clinically far-reaching assertions to which we will return.

As far as protoconscious virtuality is concerned, Metzinger's theory is compatible with our conception of the spontaneous construction of the self as an agent who is embedded in an *a priori* virtual world. Protoconsciousness theory adopts the philosophical functional self model and embeds it within the sleep-dream framework. The protoself inhabits a proto-space, navigates within it, manifests and feels emotion as it "warms up" or rehearses for life on the real-life waking stage. The protoself does not need to be self-aware or to remember its feelings or actions because they are instantly reinvented via functional iteration. An actor no longer needs a script to remember his lines. They are built in by automatic and spontaneous iteration. Dream golf is one of many possible scripts.

Figure 24.2. Portrait of Mary Arnold-Forster.

The first protoconscious state is the REM sleep primordium of the third trimester of life in utero. This state is shared by all mammals and accords to all such animals a proto-self which continues to operate throughout their lives together with primary consciousness. Pet owners confer an identity (and even a name) on their non-verbal companions. This could be seen as meaningless projection, but no less a scientist than Ivan Pavlov noted the distinctive character of many of his canine experimental subjects. We see no reason to deny primary consciousness to non-human or to human animals.

In human dreaming, we see evidence of persistent pre-consciousness in the near universal iteration of the self. At least 99% of REM sleep dream reports are narrated in the first person. Dreamers almost always perceive themselves to be at the center of dream action. This makes sense if we take dreaming to be an automatic self-model iteration and updating process. Short of that interpretation, it is clear that the self never sleeps. Even in the thought-like perseverative dreams of NREM sleep, it is always "I," never "he, she or it" that has the unrealistic obligation or the grandiose self appraisal. Thus Gottfried Leibniz again prevails over John Locke in their famous controversy: the mind is not entirely dependent upon external stimulation for its creation or its maintenance.

The protoself perceives itself to live in a protospace. REM sleep dreamers perceive themselves to be awake and behaving in a virtual surround which they take to be real as would be the case if they were awake. The fact that they do not normally recognize that they are not awake is testimony that the veridical formal aspects of the dreamed space closely match those of the real space of the world. Only after awakening does the dreamer realize that his dream space was not (and could not be) any space that he has ever inhabited. Formal similarity is all that is required to suspend disbelief. The theater analogy again springs irresistibly to mind. At the very least, we may conclude with confidence that sensation and perception can be constructed artificially.

In the preceding section, we have alluded to the illusion of movement in protoconsciousness and dreams. *Emotion*, too, is simulated virtually. The protoconscious animal, like the adult human dreamer, feels, and displays emotion. Preverbal human infants smile, frown, and grimace in REM sleep. Puppies twitch, squeal, try to escape, or virtually give chase. Such outward displays of motion and emotion add unmistakable weight to the rehearsal aspect of the protoconsciousness-virtual reality dream theory. The predominance of anxiety/fear, elation, and fight-or-flight aggression over the social emotions of sadness, shame, or guilt in adult human dream reports is in favor of the adaptive value of virtual reality.

Fantasy, Protoconsciousness, and Virtual Reality

If protoconsciousness precedes dreaming, and dreaming precedes waking, then what does fantasy do? Undoubtedly, fantasy parallels waking and many psychologists think it is continuous with dreaming. If so, this means that fantasy is continuous with protoconsciousness. Whatever the true sequentiality, fantasy is certainly virtual reality and it is certainly functionally significant in its parallelism with waking consciousness. Fantasy affords a meaningful escape from reality but also provides alternative strategies for dealing with the problems presented to waking consciousness. Empirical study reveals fantasy to be shot through with behavioral rehearsal and review.

Fantasy may be defined as the *background* processing of information in waking, where input-output information processing is *foreground*. So faint and subtle may be fantasy as to make it all but imperceptible. Careful introspection, however, reveals fantasy to be omnipresent, hence a constant silent partner of waking consciousness. It is this very evanescence which invites us to regard fantasy as an unconscious mental process. We may decide to decline this invitation in order to avoid being drawn into the swamp of wish-fulfillment associated with the psychoanalytic theory of the unconscious.

When trained self-observers pay attention to fantasy they tell us that, indeed, fantasy is difficult to "trap" and, even when trapped, it may be as elusive to reporting as dream recall. When recorded verbally, reports of fantasy share with dreams many formal properties but the bizarreness of fantasy is never so intense as in dreaming, and one category, character transmogrification (one person turning into another), was never reported by the fantasists so far studied. Far more common than in dreaming was the close relationship of fantasy to ongoing life events. This realistic feature is more typical of NREM sleep thinking than of REM sleep dreaming. Fantasy reports commonly described what could be construed as review and rehearsal of life strategies.

Typical review fantasies took the form of what did I do right (or wrong) in this or that social interaction (with my boss, spouse, friend, lover or colleague). Rehearsal fantasies projected plans for future social interaction. These two categories were far more common than the elaborate and seductively imaginary escape scenarios made famous by James Thurber in *The Secret Life of Walter Mitty*. Such grandiose adventures are decidedly rare in fantasy, however engaging they may be as literature. They capture the wish-fulfillment motive more dramatically than is our commonplace lot in dreaming.

Frank Sulloway (1947–) is the author of *Freud, Biologist of the Mind* which was the first scholarly critique of psychoanalysis launching a tradition later pursued by the philosopher of science Adolf Grünbaum and the literary scholar Frederick Crews. All three consider Freud to be a pseudoscientist whose ideas were romantically self-serving philosophical speculation. Sulloway has gone on to the empirical study of *Birth Order and Scientific Revolution* and returned to his interest in scientific history via his research on Charles Darwin in the Galapagos Islands. Sulloway demonstrates strong independence of mind and scholarship. He has never had a job, preferring to pursue his intellectual curiosity in parallel with academia.

The human mind seems to be deeply rooted in the here and now. It occasionally does fly, like Peter Pan, to the never-never land of fairies and sprites but this trope also captures a folk-psychology view of fantasy more successfully than the prosaic view of science. You can guess which commands the interest of readers and publishers. Escape fantasies make good stories and we all wish that wishing would make it so. But our life of quiet desperation is often lived very close to the ground.

Dialogue 24. Do It Yourself

TH: The Virtual Reality Model of Consciousness can be practiced as a tool for pleasure, self-analysis, and scientific understanding.

AH: It took me a long time to realize that truth. I used to think that science was what other people did. My job was to learn from them.

TH: You didn't realize that you could be a self-respecting psychologist until you embraced physiology. I didn't become a psychologist until I took my own mind seriously.

AH: It was much easier for me to relate to literary than to scientific writing when I was young. I think that my literary bent helped me get into Harvard Medical School.

TH: In college I devoured novel after novel but never cracked a scientific journal or treatise. Yet I earned a PhD in psychology at Cambridge.

AH: Novels are more clearly and directly about us. Proust, Joyce, and Mann speak to us with passion. Scientists are trained to regard life as an object to be described rather than a process to be experienced.

TH: I found science dry and irrelevant until I began to study sleep and dreaming.

AH: This book provides a chance for us to communicate our shared conviction that psychology can, indeed, become a science.

TH: Mind and brain are still thought to be hopelessly incompatible.

AH: Consciousness is not a will-of-the-wisp. It is clearly a functional state of the brain.

TH: We know we have a brain in our head even though it doesn't feel that way.

AH: We may be convinced by the neurobiological facts but don't see how to use them in our work and in our lives.

TH: It is widely supposed that subjectivity can never be objectified. Dream science challenges that assumption.

AH: I was excited to learn that you kept a dream journal. What have you learned from it?

TH: More about myself than I care to tell you. Like Sigmund Freud I have conducted a self-analysis.

AH: Your dreams are yours and mine are mine. We both share the formal features like bizarreness, disorientation and hallucinations, but the stories are quite different.

TH: You liked the journalistic approach so much that you incorporated descriptions of your waking life. Didn't you risk discovery of your secret self?

AH: My first wife, who correctly suspected me of excessive adventurism, said that my journals read like novels. She was an English major.

TH: The novelistic and scientific approaches come together in the study of dreams.

AH: I argue that the science of consciousness should be considered one of the humanities.

TH: Humanists will be threatened by the invasion of their turf. They want to preserve, at all costs, the two-culture myth promulgated by C.P. Snow in the 1950s and 60s.

AH: The work of humanists like Frank Sulloway, Adolf Grünbaum and Frederick Crews already shows that science can benefit from blurring the two-culture boundary distinction.

TH: Territoriality is a biological fact that works against integration. Only the young can cross the demilitarized zone in which integrators risk their academic lives.

AH: The mind and brain are already secretly married. Its high time they came out of the closet as the indissoluble couple that they really are.

TH: Some marriages fall apart but this one will have an endless future.

Chapter 25. Subjectivity

Consciousness cannot expect to be defined until its nature is more fully and accurately described by conscious persons. We hold that this seemingly impossible task can best be pursued within the context of a scientific investigation of conscious states and that a promising start has already been made in recording and analyzing reports of waking, sleeping, and dreaming consciousness. To illustrate my point, I share a recent dream report of my own. It concerns the writing of this book.

Choral Dreams — Friday, April 8, 2016

Last night I went to bed late after watching The Masters golf tournament from Atlanta. I was more taken with the spring flowers than the golf. I thought that the azaleas and rhododendra, which I love, were a sure sign of American decadence. So were the pastel-colored pants of many of the coddled young players.

I thought I must have forgotten to take my sedative pills. Why else did I lie so long wide awake. I nonetheless must have dozed off because I dreamed that Chris Gates was extolling the virtues of Ming Tsuang, my former boss whom I consider criminally corrupt. At one point Gates announced approvingly that Tsuang and his wife were members of a church choir. "He is even worse than I supposed," I said to myself, implying that Tsuang thought himself to be one of the chosen. At 3:30 AM, I got up and hobbled to my desk for chemical help.

I slept deeply until 9 AM when I awakened from a long, vivid dream which must have been REM engendered. I was about to be visited by a choral group, perhaps an American contingent visiting Italy. My job, as host, was to prepare for the visit. The setting was rural. Distressed by the lack of order indoors I went outdoors into the woods to gather logs for a fire. I found that dead tree branches could be made into fireplace-length pieces by hitting them on live trees or stones. I was only faintly aware that this activity was relevant to the task of preparation for the chorus visit and enjoyed my surprising ingenuity and success until I woke up. I would estimate that this dream lasted five or ten minutes.

When I saw how late it was I got up, had breakfast, and worked long and hard on Reciprocal Interaction, a chapter of my new book on States of Consciousness. I had originally truncated the REM neurophysiology and melded it with dream psychology. Today it was easy to separate and begin to expand the two subjects. Tomorrow I will tackle the next chapter on dreaming in relation to neurophysiology.

A part of me asks: what do the choral dreams have to do with my work life? The honest answer is that I do not know but the dream clearly links my current work life, my perennial function as host and my past history as a choir boy/adventurer. As such the dream is transparently meaningful and its integrity is facilitated by sleep physiology. It is in this sense that I conclude that the mind-brain problem is solved.

Subjectivity

From the solid scientific base already established, more detailed and hypothesis-testing studies of subjectivity can proceed. We are delighted that our young colleagues, Clemens Frenzel and Jana Speth, are embarked on a systematic research program at the University of Dundee in Scotland. Frenzel and Speth have chosen to study motor activity at the levels of subjective consciousness and brain activity. Several other examples will be discussed in the course of this section, the main point of which is to insist that subjective experience is the very essence of consciousness science. This position runs counter to the dominant trend advocated by reductionist neuroscience.

An easily understood example is the so-called stream of consciousness. We still do not know the mean and standard deviation of the duration of a segment of the hypothetical stream of consciousness in any state. Nor do we know whether the segments are continuous or discontinuous or whether the segments flow one from another (as suggested by the stream metaphor) or whether they instead jump from one streambed to another. Is there a difference in these parameters when conscious states are compared with one another? The psychologist John Antrobus claims that waking consciousness is just as choppy as dreaming consciousness. This controversy remains to be resolved, perhaps by a neutral doctoral student. A problem is that there is no such thing as neutrality, any more than associations are ever free.

Do you, as a reader, expect your teacher to provide your mind with an easily followed sequence of conscious state segments? The question is rhetorical. Of course you do. When we ask you this question, we are asking you to consider whether classroom learning is an artificially constructed stream of consciousness.

Vernon Benjamin Mountcastle (1918–2015) is best known for his physiological work on the brain basis of extracorporeal space orientation. By means of this mechanism we unconsciously know where our bodies are when our brains are activated in waking (real orientation) and in REM sleep (virtual orientation). A student of Philip Bard, he had a lifelong interest in sleep and consciousness. This interest led to his recruitment by Francis O. Schmidt to discuss (together with Gerald Edelman) the brain basis of awareness in the early days of the MIT Neuroscience Research Program. Throughout his distinguished career, Vernon Mountcastle supported the sleep and dream research of younger scientists.

When you daydream or doze either in class or over this book, does your mind wander along a set path or does it leap from one path to another? In other words, does the background stream of consciousness have the same temporal dimensions as the foreground stream is supposed to have if your teacher has planned the lesson carefully? Please don't tell us that you really don't care about the answers to these questions because all you want to do is pass the course and get on with your life.

Your life will be successful or not according to your ability to control these aspects of your consciousness. To control them, you need to be aware that they exist even if questions about them persist. We hope to recruit some of you to the intellectual effort to answer questions that define subjective parameters of consciousness, but we hope to excite all of you to their importance to your life.

When you prepare a paper or study for an exam, you will benefit from a strategy to organize your attention, to keep your mind on track, to cope with distractors (be they internal or external). Since your mind is causal, you can use it to help you master its operation in your work or domestic life.

Subjectivity is not an illusion that can be dispensed with once its neural correlates are understood. Subjectivity is the physical mind and the physical mind is causal. The role of psychology and philosophy is to keep subjectivity in focus. All people have the right and the obligation to take their own subjectivity seriously. The goal of conscious state science is not to eliminate subjectivity by explaining its neural underpinnings. It is rather to reveal that the conscious brain-mind is at once the most interesting entity in all of nature and the most promising tool in understanding itself and meeting life challenges.

Karl Friston (1959–) is an English psychiatrist who has created a statistical package for the analysis of fMRI and other brain imaging data pertaining to consciousness and related cognitive studies. He brings to this task an unusual math-physics perspective. His deep interest in the Free Energy and Sensorimotor Prediction theories of Hermann von Helmholtz gives his work a seamless interface with protoconsciousness dream theory. As such, he is at the cutting edge of consciousness science. Professor Friston is not only a world leader in computational neuroscience but a masterful writer and thinker with a clear vision of the relationship of brain to mind. For Friston, the brain and the mind are both physical but while the former is physical but invisible, they interact with each other in such a way that consciousness is causal and the will is free.

Mind as Physical and Causal

That the mind must be causal is evidenced by human experience and by philosophical reasoning; how can an entity be physical but invisible? When William James announced, as the first principle of his doctrine, pragmatism, that he would believe in free will because it suited and served him to do so, he was advancing an article of faith. The fact that scientists have not yet found certain proof that James's pragmatic pronouncement is valid and instead found evidence that such freedom may be illusory leaves us in limbo with respect to this very central issue.

The position taken by consciousness science is that the mind is invisible but material. This seeming paradox needs concluding discussion. How can an entity be physical but immaterial? Are we saying that the mind is extracorporeal, i.e. spiritual in a Cartesian dualistic sense? No. We are saying that the mind is to the brain as gravity is to mass. Mind is thus a force, an energy state as dependent upon the brain as gravity is upon mass. Yet mind is no more substantial, and no more directly observable, than is gravity — and no less physical. The mind/gravity analogy is heuristic but important in spite of its manifest inadequacy. Another analogy may be found in magnetism. Mind is as dependent upon the brain as magnetism is dependent upon moving electrical charges. Much as a magnetic field generates forces of magnetic attraction and repulsion, mind generates thoughts that can exert causal influence on the world. In other words, mind is to (living) neurons what a magnetic field is to (moving) electrons.

The analogy is still inadequate because we cannot imagine that gravity or magnetism thinks and feels as the brain-mind certainly does. The only logical way out of this dilemma is to consider the mind itself — not just free will — to be illusory and to embrace monism pure and simple. There is brain and nothing else that is not reducible to brain. This is the position of hardheaded neuroscience, at least Monday-to-Friday neuroscience. On Saturday and Sunday many hardheaded neuroscientists become soft-hearted dualists.

We reject Cartesian dualism but embrace dual-aspect monism as a seven-day-a-week credo. The inadequacy of the mind/gravity or mind/magnetism analogy stems from the persistently "hard problem" of imagining how qualia could emerge from any substantial object. In our opinion, this "hard" problem has already gotten softer thanks to conscious state research and we anticipate further softening of the hard problem in the near future.

Since 1928, we have known that electrical energy radiating from the brain is detectable on the surface of the scalp. Does that mean that we equate mind with the EEG? Certainly not. It only means that brain energy, like weight-as-gravity, is now directly measurable. No one supposes that the EEG itself is thought but it is a scientific fact that both waking and dreaming consciousness are correlated with 40 Hz activity in the EEG. This is knowledge in progress. The progress is agonizingly slow, but it is progress nonetheless. Current enthusiasm for a science of consciousness may shortly fade, only to brighten again in the distant future. When it brightens anew, we predict that brain energy (or force) will be better understood and that it will be shown to better explain qualia than we are now able to do.

If mind is one of two aspects of a unified entity, does that guarantee causality as in volition and free will? No. But it helps to explain both mental causality and free will, at least theoretically. This formulation resonates with the common-sense pronouncement of William James. If consciousness emerges from the brain (as we think it does), then it is not far-fetched to suppose that it is both caused by and causal of changes in the brain. A physical force (akin to gravity, or magnetism) changes the course of an object (akin to a mass) as it hurtles through space.

In 1912, William James was still sitting on the fence between science and religion. James was born when Charles Darwin was 31 and about to complete his famous Beagle voyage. *The Origin of Species* was published in 1859, when James was 17. At that time, God-fearing Christians still went to church on Sunday. Creationism is still believed even now by many laypersons but scientists now favor the theory of evolution.

Free Energy and Prediction

The free-energy principle derives from the theoretical treatment of thermodynamics by the Austrian scientist, Hermann von Helmholtz, who was a pioneer in the physics of consciousness. Its current application, by psychiatrist-physicist Karl Friston, takes sleep and dream research into account by postulating that dreaming serves to reduce perceptual surprise (in waking) by revising the brain-mind's model of sensory input expectation. The theory is important because it links information hypotheses to energy dynamics and helps us understand why mammals take such risks as the abandonment of temperature control in REM sleep. Why is REM sleep so important to survival, and what cognitive purpose does dreaming itself serve?

William James (1842–1910) was an American philosopher who was educated at home and abroad together with his brother, Henry, by their father. While Henry moved to Europe and became a famous novelist, William lived in Cambridge, went to the Amazon with Agassiz and attended medical school before becoming an experimental psychologist at Harvard. William James was notoriously broad-minded. He befriended spiritualist mediums, psychoanalysts, and unorthodox students like Gertrude Stein. He founded the philosophical school of pragmatism which condoned the acceptance of a belief, like free will, because it worked for him. His Gifford Lectures at Saint Andrews University in Scotland were published as *The Varieties of Religious Experience*. As in his earlier *Principles of Psychology*, James integrated subjective experience with neuroscience in his philosophy. It is disappointing that he said so little about dreams; he must have been a poor recaller.

Free energy theory is not the only attempt to develop a physics of consciousness. By far the most widespread approach is that taken by the large contingent of quantum physicists who ascribe consciousness to "collapse of the wave function." Leading exponents of the quantum approach are physicist Roger Penrose of Oxford University, and anaesthesiologist Stuart Hameroff of the University of Arizona. Their theory proposes that the microtubules of neurons constitute the physical basis for the quantal origin of consciousness. This theory is laudable in its attempt to embed quantum theory in the brain but has been criticized as unrealistic by other physicists. Consciousness science is thus still in search of a realistic link between the brain-mind and subjectivity. Enter free energy theory.

Hermann von Helmholtz theorized that all thermodynamic systems (such as the brain-mind) must tolerate energy that is free to do the work required to maintain or restore thermal equilibrium when the system is perturbed by unexpected challenges. In other words, free energy serves to minimize surprise and improve prediction. There must be enough free energy to facilitate this function but not so much that the system is inefficient or, worse, erroneous in its strategy. According to Karl Friston, the function of

REM sleep dreaming is to minimize surprise and reduce prediction error at the same time that it promotes thermal equilibrium.

To understand this theory, consider your tendency to misinterpret stimuli when there is ambiguity (as in the case of a scientist-colleague's hat atop a rack outside his office which was mistaken for the colleague himself) or a weakening of intensity (as in the case of indecipherable stimuli at night which give rise to anxious speculation). In the first case, the scientist must correct his instantaneous misperception that his colleague is standing there. In the second case, a person must realize that the rustle of the wind is not a burglar. These examples invoke consciousness itself as a discriminator between erroneous and accurate readings of the environment whereas free energy predictions may be entirely non-conscious since the system corrects itself spontaneously via self-organization.

Our point is not to insist that free energy is the only or even the best way to think about the physics of consciousness. But it is the only theory which takes the science of waking, sleeping, and dreaming into account. It is therefore the best theory that we have today. Furthermore, it is a theory which lends itself to experimental test. REM sleep deprivation has two consequences which affirm the theory; prediction error is increased by sleep deprivation (subjects "see things" that are not there; and they cannot regulate body temperature).

Protoconsciousness Precedes and Follows Waking Consciousness

The evolutionary, developmental, and diurnal cycles that we experience and now have begun to understand scientifically tell us that we must broaden our minds and deepen our questions about our reality. Until recently, we have tried to understand consciousness only as it exists in waking. Consciousness, we are still told, is composed of awareness and vigilance. It is communication to and from the outside world. Even if we accept this narrow definition, we must ask how consciousness came to be awareness and vigilance, how it evolved, how it developed, and how it is maintained. This book is dedicated to answering these difficult and important questions. However challenging our theories might be, we offer them in the hope that they will inspire the experimental work of the future as well as summarize the state of the art of today.

Protoconsciousness theory posits a primordial state which predates mammalian speciation, predates the postnatal life of mammals and precedes and follows each day of our individual lives. The most direct access to protoconsciousness that we have as adult mammalian persons is via our dreams and we have had scientific access to them only for the last sixty-three years. Attention to dreaming can be expected to reveal the way in which the brain creates and is created by the mind. This is the shared goal of psychology, philosophy, and physiology. The recognition that consciousness has such a long prehistory in evolution and development is a difficult fact to keep in mind. We are hampered by our topicality, our near-sightedness. It is a stretch for us to imagine that our dreaming may contain something meaningful about yesterday or our life as children, much less our lives as fetuses, apes, or even our own lives' tomorrows.

To overcome our topicality, three tactics are helpful: the *first* is direct psychological observation and includes trained self-observation; the *second* is physiology, which permits third-party identification of states in ourselves and in other humans as well as in other animals; and the *third* is philosophy, which provides the framework in which to frame our inquiry and the logic we need to analyze our observations. Using these tools, we can recognize protoconsciousness in ourselves at psychological, philosophical, and physiological levels. According to protoconsciousness theory, we are conscious selves in waking; we dream every night; we dreamed in utero before we were born and our biological ancestors dreamed before the dawn of human history. Awareness and vigilance are only part of the story and a superficial part at that.

The physiology underlying psychology and philosophy is studied by scientists like Ralph Lydic, now working at the University of Tennessee in Knoxville. While still a Harvard colleague of Allan Hobson, Lydic pioneered the long-term recording of REM-off cells and subjected the data to cycle averaging in the objectification of the reciprocal-interaction model. Recognizing the formal resemblance of sleep and anaesthesia, Lydic currently attempts to elucidate the deep brain effects of sleep and chemical agents. In his experimental paradigm, a drug (say one affecting acetylcholine or serotonin metabolism) is administered and the effects on sleep measured. Needless to say, the effects are profound, indicating that whether we go to sleep in our beds or are put to sleep on the operating table our brains are changed such that we lose our most precious skill, waking consciousness.

Figure 25.1. Ralph Lydic's collage homage to Allan Hobson and Robert W. McCarley.

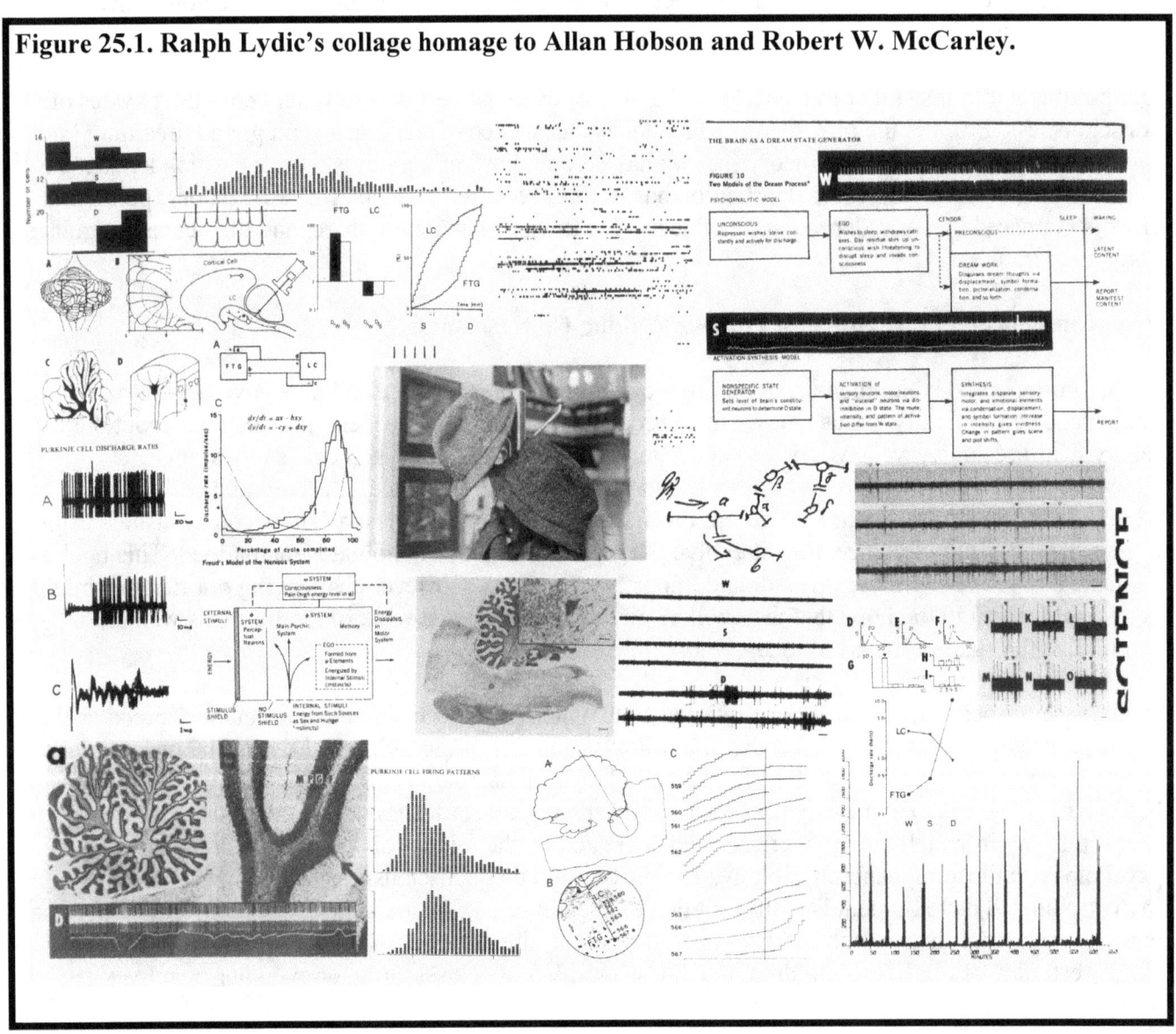

Virtual Reality (Again)

Consciousness is virtual in any conscious state. The brain is a physical object which has the amazing power to create a model of the world so accurate as to fool us into thinking that the model is the world and not just a model of it. That dream content is virtual reality is an easier idea to grasp. While the virtual reality of dreaming is similar enough to the virtual reality of waking to beg comparison, it occurs in sleep, when the brain-mind is offline from the external world, and dream consciousness is cognitively different from waking consciousness. The bottom line of this section is to insist that we humans are both models and models of models. Our bodies and our brains are real but our perceptions of them and all other physical objects are virtual.

It is easy to imagine why we have a model of the world but difficult to see why we need a model of that model. The answer must be that it is difficult to conceive how the model of the world that we have in waking evolved. It was difficult to make that model and is difficult to maintain that model. Eons of time were required to evolve that model and 5 years of an average human lifetime are devoted to its maintenance and repair so that it can function in the face of intrinsic changes within our being and our bodies and extrinsic changes in our physical and psychological surroundings.

The virtual reality of dream consciousness cannot be dismissed as an accident. The virtual reality model of the world that we so easily recognize in dreaming is not simply a pale copy of the world. The virtual reality of dreaming is a blueprint (or map) for the model of the world that we use as builder in (or navigator through) that world. It is frightening to realize that this is so, but the evidence for such a bewildering idea is overwhelming. The evidence is physical as well as psychological and philosophical. No one questions that we sleep and most people acknowledge that we dream. How and why we do both are the subjects of this book, and if we have failed to convince you of these facts, we apologize.

Whether or not you are convinced by the scientific story, we trust that you will realize the logical necessity of the virtual reality principle. Whether or not you understand or accept the implications of protoconsciousness theory, we expect you to acknowledge that sleep precedes as much as it follows waking. To suppose that sleep only follows waking is a colossal mistake. Sleep precedes and follows waking in exactly 50% (precisely half) of all instances. This is a fact, not a theory. The theory attempts to explain how and why the sequence might be useful.

Education as Brain-Mind Training

One view of education is the rote learning of facts. Quite another view is that education is the training of conscious awareness. Formerly the province of religion, awareness training is now in the never-never borderland between religion and science. Whatever one thinks of religion, there is good reason to strengthen the science of awareness enhancement. Awareness enhancement is a part of the science of consciousness as well as an antidote to anxiety, stress, and tension. It is the purpose of this section to summarize the scientific basis of awareness training already touched upon in the section on transcendental meditation.

The first point to be made is that awareness training is not transcendental. In becoming aware of the mind's inner states, no recourse to mystical dualism is either required or justified. From a scientific point of view, awareness enhancement is purely secular. For those of you who seek a unified philosophy, this should be a welcome opportunity.

Muscular relaxation is a voluntary practice. In and of itself, relaxation constitutes proof of free will. When you let the muscular tension flow out of your fingers, your arms, your legs, your trunk, and your neck, you are simply reducing activation signals from your brain-mind to your body. Nothing really flows out of you. The flow idea is only a metaphor to help you instantiate a process now well understood by consciousness science. Muscle tone is regulated by the brain stem which, in waking, is responsive to instructions from the cortex as well as the cortical commands directing movement. This connection is automatically broken in REM sleep when muscle tone is abolished by brain stem inhibition. It should be clear to you that voluntary muscle relaxation moves you closer to REM in the AIM state space.

As a consequence of voluntary muscle relaxation, activation from the muscles to the reticular formation amplifies the relaxation response. The EMG can go toward zero amplitude by a self-enhancing negative feedback process. Physiologists refer to this process as *disfacilitation* (meaning decreased activation) to distinguish it from the muscle tone *inhibition* of REM sleep. Another negative feedback loop decreases reticular activation of the upper brain, again in the direction of sleep but rather to what adepts claim is heightened awareness. Scientists might argue that internal awareness is enhanced as external awareness is lessened. That relaxation is sleep-enhancing is recognized by insomniacs and their therapists.

The corticoreticular negative feedback loop is also enhanced by clearing cortical networks of the "noise" of thinking by instituting a mantra ("Om" in the Buddhist tradition) or by visualizing a peaceful scene (like a mountainside or a lake). The goal is to rid the brain-mind of its wake-state concerns like lists of things to do. This thought-anxiety laundromat cleanses the cortex and redirects attention from the busy outside world to the tranquil inside world. A deliberate focus on elemental physiology may choose breathing and reduce its rate while increasing its depth. These respiratory variables are still other brain stem reticular formation functions to which the volitional brain-mind has access.

Needed now are detailed subjective reports of awareness enhancement. To achieve this goal, it would seem desirable to remove awareness enhancement from its traditional mystical context. Is the mind ever really entirely concept and emotion free? We think not. Is the stream of consciousness ever an entirely immobile pool? Not until after death. Is meditation really superior to a good night of sleep? We wonder. We hope that these questions can be asked and answered without challenging the scientific validity of the relaxation response.

We cannot yet say exactly how the physical brain becomes the physical mind, but we are rapidly closing in on even this aspect of the problem. It seems to us that recognizing that both mind and brain are physical and that they are inextricably linked to each other reduces the problem to at least tolerable uncertainty. In other words, building upon the solid philosophical foundation of dual aspect monism and the scientific humanist position laid out in the previous chapter, the mind-body problem is at least (and at last) free of the quagmire of Cartesian dualism. This last part of this chapter is devoted to an exploration of the most appropriate philosophical response to the challenge issued by David Chalmers' famous "hard problem" argument.

Is There Really a Problem? Is the Problem Really Hard?

If we agree that there really is a problem, the next question is whether the problem is really hard. By hard, Chalmers implies that it is scientifically obdurate and possibly insoluble. To the many colleagues and critics of consciousness science, we say "No, we do not regard the problem as unsolvable or as even hard." For us, the hard problem grows softer every day and, far from being an obstacle to progress, it seems to us best left aside for the time being as we get on with the science. If the progress of the last seventy years continues at its current pace, — or speeds up, as we expect — the problem will not only be soft but it will be solved or shown to be irrelevant.

Our reasons for optimism stem, in part, from the progress made so far in understanding dreaming. Until 1949, it was not clear that the brain self-activated. Now we are sure that it does self-activate and we can specify how. We know, furthermore, that in sleep the brain self-activates and that the brain activation of sleep is correlated with dreaming. We know that reflexes are concomitantly modified by specific physiological mechanisms associated with self-activation of the brain. The way that information is processed is chemically modulated in such a way as to distinguish dreaming consciousness from that of waking.

Insofar as consciousness is information processing, we can be confident that consciousness is physical. The reader must ask himself, is the brain-dream problem not already solved? It may have been hard to solve it, but the problem itself is no longer hard. Indeed we ask, is it not already solved?

If the answer is yes (or even maybe), we should keep doing what we have been doing and take advantage of novel technological innovation knowing that our power to complete the job of solving this and other aspects of the mind-body problem will increase exponentially and that convincing success is only a matter of time. The technical assault on the problem must proceed, hand-in-hand, with subjective analysis, as we have pointed out earlier.

In concluding this section we would like to reiterate the philosophical concept of "virtual reality" which we have discussed at length in Chapter 24 and consider to be an indispensable tool for future research. It seems to us obvious that any and all perception is virtual in the sense that there is no real world within the brain, only copies of that world or predictive models of it. Thus the information of which consciousness is composed is just information. That information bears a tight relation to the world but it is an abstract (or virtual) reality, a transposition, as it were, of both external and internal reality.

Patricia Churchland (1943–) is the mother of Neurophilosophy. Neurophilosophy is a hybrid discipline which draws its strength from a careful reading of the data of modern neuroscience. Churchland might say that wake, sleep, and dream consciousness cannot be properly understood without knowing what the material brain does when persons experience their subjective mental states. She would adduce similar arguments to the understanding of visual perception and declarative memory. Churchland's impetus is rejected by many traditional philosophers who claim that their discipline starts where neuroscience leaves off. The philosopher's job, they assert, is to examine and criticize the logical rigor of neuroscience. Often resistance to Churchland's integrative message is not only territorial but antimaterialistic (or even mystical, spiritualistic, and philosophically idealistic).

Interdisciplinary Science and Consciousness

In order to succeed, consciousness science must be interdisciplinary. There is no one approach that can compete with all the others when they are appropriately integrated. The major scientific advantage of multidisciplinary work is the capacity to check one source of data against others. The best example is the check between the subjective mind and the objective brain. Even though we do not yet understand how the one becomes the other and the other becomes the one, we assume that they are causally linked continuous realities and that we can look for, and find, coherence across the interdisciplinary divide. For example, it is now a well established fact that the shared cognitive activation of the waking and dreaming mind is related to the shared physiological activation of the brain in both waking and sleep.

A major obstacle to interdisciplinary work is the divisive effect of scientific specialization. There are few psychologists who are also physiologists and few physiologists who are also psychologists. Likewise there are few clinical psychologists who are also experimentalists, and so on. Scientific specialization and subspecialization is therefore an enemy of progress in consciousness research.

Young students are told, in no uncertain terms, to decide upon one area of expertise at the expense of all others. This book is dedicated to encouraging students to look and move across field boundaries. We fear being ignored or opposed in counseling this sort of integration. Analysis and reductionism are the traditions of organized science and it is difficult to argue with their success. But we are confident that genetics, for example, will not solve the mind-body problem any more than Emmanuel Swedenborg saw the God he sought at the highest power of his microscope. REM sleep deprivation did the trick for Swedenborg, however. God's angels appeared to him and instructed him in the creation of the Church of the New Jerusalem.

A balance between reductionistic analysis and synthetic holism is a fundamental theme of the approach which we espouse. We ourselves have found it personally and intellectually rewarding to connect our day jobs to our night jobs (see again the Choral Dream Reports at the beginning of this chapter). By day, we work in laboratories guided by the analytic paradigm of traditional science. By night, and on weekends, we explore our own minds, recording our dream and fantasy recollections in search of testable hypotheses.

Besides making our lives more unified, we aim to render consciousness science more effective via interdisciplinary integration. To achieve this goal, we advocate improved communication between experts and between students and their far-flung teachers. This may now be achieved more easily via the internet, email, and Wikipedia than via departments, granting agencies, and books like this one. Most students, being born in the digital era, already know this but older experts may need their help to be helpful to them. To this end, we recommend the medium of consultation. Frame a question for yourself and ask it of someone who might know the answer. We practice what we preach, and are open to your inquiry at: allan_hobson@hms.harvard.edu.

Combining Top-down and Bottom-up Approaches

By top-down is meant both the figurative mind-to-brain paradigm and the more concrete cortex-to-subcortex chain of command. Bottom-up is the inverse. As with other interdisciplinary integrations, solution of the mind-body problem requires multiple perspectives. It is difficult to keep even one in focus. A good example of a top-down process is the effect of pre-sleep autosuggestion on dream lucidity and the mechanism by which this is accomplished; so far, only the psychological analysis is advanced; its physiological underpinnings have just begun to be established. By contrast, the bottom-up analysis of REM sleep-dreaming is advanced in both physiological and psychological terms. We respect both approaches and point out what further steps need be taken to flesh out our picture, emphasizing work that we believe to be within reach of students.

The terms "top-down" and "bottom-up" are symmetrical because the top of the brain is the locus of many subjective features (including the probable seat of will in the frontal cortex) and the bottom is the subcortical brain stem and spinal cord. No one proposes that the brain stem and spinal cord are sentient, although the cerebellum seems very likely to play a role complementary to that of the cerebral cortex.

The top controls the bottom but the bottom also controls the top. The ratio of control forces is state-dependent: in waking, top-down forces are relatively strong; the balance of power shifts to bottom-up forces in sleep. An important point is that the balance of forces is continuously variable, always dynamic and never absolute. Lucid dreaming is of great scientific interest because it reveals the relativity of waking and dreaming consciousness through their coexistence in a radically hybrid state.

The interaction and the admixture of top-down and bottom-up physiological forces and psychological experiences of them are always relative and never absolute. This statistical reality is not acceptable to many scientists. David Hubel, who shared the Nobel Prize for his physiological work on vision (see cameo in Chapter 4), abandoned the science of sleep when he encountered the need for statistics to characterize spontaneous cortical neuronal discharge. As many cortical neurons increased their activity as decrease discharge rate when his cats went to sleep, and they did so in a chaotic fashion unpredictable from the Sherringtonian reflex paradigm into which Hubel wisely retreated. The cerebral cortex top is not the spinal cord bottom. We may not like statistics but we recognize the need for them.

New Technology

We are on the cusp of an exciting new probe, fMRI. f stands for *functional* but also means *fast*. MRI means Magnetic Resonance Imaging. Before explaining this technique, realize that we are about to make 3D movies of the human brain in action, in real time, without making a hole in the head. How is this possible? By rapidly alternating the magnetic field around the head, the fMRI scans the polarity of the fluid in the brain, measures it as a proxy for neuronal activation, and plots the data as spectral color intensity. Hot spots of neuronal activation are red, cool spots are blue, and the changing color of the spots may be continuously visualized since a shot is taken every 100 milliseconds (that's 10 per second, about the frequency of the EEG alpha rhythm). The really good news is high temporal resolution, and this feature will certainly improve.

Gerald Edelman (1929–2014) won the Nobel Prize for his work in immunology before turning his attention to the neurobiology of consciousness. He applied his concept of Darwinian group selection from immunological molecules to functionally connected groups of neurons. To simplify this paradigm, consider the obvious connection between the immune system's record of infection with the brain's memory of an experience. Another important idea is that of reentrant signaling, by which the brain updates its model of the world at least six times a second. Thus we are constantly reinventing ourselves, and REM sleep offers a candidate for this creative process in its internal signal, the PGO wave. It may be no coincidence that these timing pulses occur at a frequency of 6 per second, setting the stage for dream consciousness.

What's the bad news? Poor spatial resolution. 1 or 2 millimeters spatial resolution is about 1 or 2 thousand times cruder than the microelectrode sensitivity in experimental animal studies. That's a huge gap and it isn't apparent that it will close soon. Another disadvantage of fMRI for consciousness science is that subjects find it difficult to sleep in the magnet. They cannot move a muscle because fMRI imaging requires total immobility and they must ignore that clanking noise caused by the magnet changing its polarity. How can we be optimistic, faced with these problems? Sleep-deprived subjects can sleep anywhere, including inside fMRI magnets, and pilot data reveal hidden truth not otherwise obtainable. The head movement that disrupts fMRI recording can be restrained mechanically, much as was done in the early days of microelectrode exploration of the cat brainstem.

Two pioneer investigators have already been successful despite these problems. One is Martin Dresler (and his team) at the Max Planck Institute for Psychiatry in Munich, Germany. The other is Charles Chong-Hwa Hong (and his group) at the Johns Hopkins Medical School in Baltimore, Maryland. Dresler has found activation in REM sleep of a forebrain circuit known to be important to background cognitive processing in waking. Hong has shown that REM sleep brain activation in humans mirrors that found in experimental animals. Both findings confirm the predictions of protoconsciousness and virtual reality theory.

If we were young students today, we would certainly consider a day job in a sleep lab or a high tech imaging center. But we would not quit our night job of watching our own consciousness in the comfort and privacy of our bedroom and making observations about consciousness that can only be answered via systematic self-observation.

Jennifer Windt is a philosopher who has integrated the new science of dreaming with the traditional scholarship in her field. Her recent book *Dreaming: A Conceptual Framework for Philosophy of Mind and Empirical Research* (2015) shows clearly that rational speculation alone fails to recognize empirical fact. Subjectivity can be misleading. Only experimentation can answer important phenomenological questions about dreaming. Windt thus follows Kant in her own critique of pure reason. She has also taken seriously the need to work across the artificial lines of consciousness research.

The Next Steps

We hope to have made clear that progress to date has depended upon:

- The development of the philosophical stance of dual-aspect monism.
- The formal analysis of subjective experience.
- The neurobiology of states that are correlated with waking and sleeping.
- Bijective mapping between the psychological and physiological domains.
- Integration of data that is coherent across domains.

To aid students in understanding these basic principles, we will discuss each of them concisely using dreaming as an example.

The philosophical stance of dual-aspect monism assumes that mind and brain are two aspects of a unified physical system. Brain is material and mind is invisible but both are physical. They are causal one upon the other and the other upon the one. When a person dreams, brain activation is concomitant with mind activation. Other features of dream experience are likewise explainable in physical terms. The philosophy of dual-aspect monism is compatible with mathematical modeling and will profit from more detailed and critical application to the science of consciousness.

The formal approach to subjective experience emphasizes the universal at the expense of the particular because the help now offered by neurophysiology is very limited. This strategy is surprisingly successful and answers questions which psychology alone cannot explain. For example, dream forgetting is likely to be the result of aminergic demodulation rather than repression. Formal analysis of dreaming is far more advanced than that of waking, including fantasy and imagination, which are now ripe for study. Waking consciousness is more accessible and more easily manipulated than that of sleep.

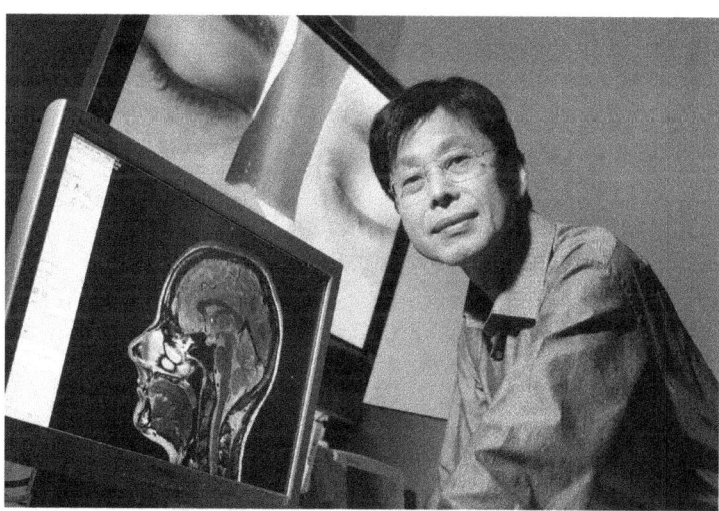

Charles Chong-Hwa Hong is a Korean-born scientist who worked at Johns Hopkins Medical School, where he did his heroic studies of human REM sleep using fMRI. Hong established the existence of PGO waves in man by using the REM sleep eye movements to trigger averaging of highly localized brain energy. Although he did not elicit dream reports, by demonstrating that the PGO wave activation of REM in the cat was present in the same brain state in which humans dream, a probable link between physiology and psychology was established. This is the cutting edge of conscious state research and invites further exploration.

The neurobiology of conscious states has, until recently, been pursued in animal models with inevitably limited application to man. Nevertheless, the fit is quite good. For example, PET studies of human REM sleep have provided data of brain stem activation confirming the predictions derived from animal work. New approaches to human neurobiology promise to further amplify our understanding of state-dependent cortical activation, which is still extremely limited. The neglect of the non-human animal brain is regrettable but, when it comes to the cortex, we can be quite sure that the basis for limitations of language and propositional thought is the absence of brain structural substrate.

Bijective mapping between the brain and mind domains has yielded correspondences of the general kind mentioned above, but it has also yielded unexpected findings such as the correlation of report word length with eye movement intensity in REM sleep and the related correlation between REM period quintile, neuronal discharge level in cats, and dream report word length in humans. The multiple measures that can now be made tend increasingly to support the dual-aspect monism model of consciousness.

There is much more to be done with readily available data. For example, it would be of interest to know if dream recall after awakening from NREM sleep stage IV early in the night was correlated with objective evidence of confusional awakening (persistent slow waves in the post-awakening EEG), slurred speech, supporting the hypothesis that early night awakenings fail to interrupt sleep, and elicited confabulation rather than reliable recall.

Integration of brain-mind correlations permits syndromic formulations. The several findings are not only each one scientifically impeccable, but taken together, they suggest new correlations. Dreaming is formally psychotic because of the undisputed presence of hallucinoid perceptual imagery and delusional belief in both states. Of the three major classes of major "mental" illness, dreaming most closely mimics organic psychosis. So faithful is the mimicry that, from a formal point of view, it is possible to state the hypothesis that dreaming is an organic psychosis. This is reductionism at its best: we seek to determine the fewest number of assumptions that explain the greatest amount of data:

The brain and its mind are united in the determination of human consciousness.

Dialogue 25. The Mind-Brain Problem

TH: Do you really think that the mind-brain problem is solved?

AH: More than solved. The so-called problem doesn't really exist.

TH: Please explain. I have wrestled with this problem all my life and expect to take it with me to my grave.

AH: You talk like a mysterian for whom subjectivity is forever irreducible.

TH: I would never lead with this kind of idea but I would feel obliged to share my disbelief with students.

AH: I would not hide my own skepticism but would argue for experimental dedication.

TH: Wouldn't you rather be rich and comfortable?

AH: I already have more money and comfort than I need. My motivation is to write a good book and put a positive spin on my story.

TH: You have led a privileged scientific life. I have been a teacher.

AH: You probably agree with our reviewers. Most students don't want to read my story.

TH: You write for the few. I write for the many.

AH: Many are called but few are chosen.

TH: That sounds biblical.

AH: Sorry for that because I think religion is a harder problem than consciousness itself.

TH: Atheism is attractive to me too. However, you do seem almost religious in your faithful adherence to neurobiology.

AH: I would hate to suppose that, like Freud, I have only created a new religion.

TH: Religious zeal may be as much a problem for psychiatrists as it is for psychologists. You have mentally disabled patients while I am confronted by ignorant, unmotivated students.

AH: It was precisely the unscientific nature of psychiatry that motivated my life work.

TH: Do you really think that psychiatry can be saved?

AH: Yes, and psychology, too. Even philosophy can be revolutionized.

TH: You will be long dead and gone before this happens.

AH: As Thomas Browning said: "A man's reach should exceed his grasp. Or what's a heaven for?"

Suggested Readings

Allison, T., & Cicchetti, D.V. (1976). Sleep in mammals: ecological and constitutional correlates. *Science*, 194(4266):732–4.

Antrobus, J.S. (1986). Dreaming: cortical activation and perceptual thresholds. *Journal of Mind and Behavior*, 7(2–3):193–212.

Ashby, W.R. (1952). *Design for a Brain: The Origin of Adaptive Behavior*. New York: J. Wiley and Sons.

Aserinsky, E., & Kleitman, N. (1953). Regularly occurring periods of ocular motility and concomitant phenomena during sleep. *Science*, 118:361–75.

Asimov, I. (1979). *In Memory Yet Green*. New York: Doubleday.

Baghdoyan, H.A., Rodrigo-Angulo, M.L., McCarley, R.W., & Hobson, J.A. (1987). A neuroanatomical gradient in the pontine tegmentum for the cholinoceptive induction of desynchronized sleep signs. *Brain Research,* 414:245–261.

Berger, H. (1929). Über das Elektrenkephalogramm des Menschen. *Arch Psychiatr Nervenkr*, 87:527–570.

Breton, A. (1928). *Nadja*. New York: Grove Press.

Brooks, J.E., & Vogelsong, J. (2000). *The conscious exploration of dreaming: discovering how we create and control our dreams*. Author House.

Cajal, Santiago Ramón y. (1966). *Recollections of My Life*. Cambridge, MA: MIT Press.

Cajal, Santiago Ramón y. (1995). *Histology of the Nervous System of Man and Vertebrates*. New York: Oxford University Press.

Cartwright, R.D., & Guilleminault, C. (2013). Defending sleepwalkers with science and an illustrative case. *Journal of Clinical Sleep Medicine*, 9(7):721–726.

Chalmers, D. (1997). *The Conscious Mind: In Search of a Fundamental Theory*. Oxford: Oxford University Press.

Chase, M.H. (2013). Motor control during sleep and wakefulness: Clarifying controversies and resolving paradoxes. *Sleep Medicine Reviews*, 17(4):299–312.

Coleridge, S.T. (1816). *Christabel, Kubla Khan, and the Pains of Sleep*. London: W. Bulmer and Co.

Darwin, C. (1872). *The Expression of Emotion in Man and the Animals*. London: John Murray.

Darwin. C. (1996). *The Origin of Species*. New York: Oxford University Press.

Descartes, R. (1649). *Passions of the Soul*. Voss, S.H., trans., 1989. Indianapolis: Hackett.

Descartes, R. (1641/1996). *Meditations on First Philosophy: With Selections from the Objections and Replies*, translated by John Cottingham. Cambridge: Cambridge University Press.

Chalmers, D.J. (1996). *The Conscious Mind: In Search of a Fundamental Theory*. New York: Oxford University Press.

Darwin, C. (1859). *On the Origin of Species*. London: John Murray.

Darwin, C. (1872). *The Expression of Emotion in Man and the Animals*. London: John Murray.

Dement, W. (1972). *Some must watch while some must sleep*. New York: W.H. Freeman.

Dennett, D.C. (1991). *Consciousness Explained*. Boston: Little, Brown.

De Quincey, T. (1821). *Confessions of an Opium Eater*. London: London Magazine.

D'Hervey de Saint-Denys, J.M.L., Marquis. (1982). In: *Dreams and the Means of Directing Them*, trans. van Schatzman, M., ed., Fry, N. London: Gerald Duckworth.

Dostoyevsky, F. (2005). *The Brothers Karamazov*. Trans. Constance Garnett. Minneapolis: Dover Publications, Incorporated.

Durrant, G. (1969). *William Wordsworth*. Cambridge: Cambridge University Press.

Edelman, G.M., & Mountcastle, V.B. (1982). *The Mindful Brain: Cortical Organization and the Group-Selective Theory of Higher Brain Function*. Cambridge, MA: MIT Press.

Edelman, G.M. (1992). *Bright Air, Brilliant Fire: On the Matter of the Mind*. New York: Basic Books.

Evarts, E.V. (1965). Neuronal activity in visual and motor cortex during sleep and waking. In: *Aspects anatomo-fonctionnels de la physiologie du sommeil*, pp. 189–212.

Fogel, S.M., Smith, C.T., & Cote, K.A (2007). Dissociable learning-dependent changes in REM and non-REM sleep in declarative and procedural memory systems. *Behavioural Brain Research*, 180(1): 48–61.

Freud, S. The Standard Edition of the Complete Psychological Works of Sigmund Freud. Volumes IV (1900) and V (1900–1901). *The Interpretation of Dreams*. London: Hogarth Press and the Institute of Psychoanalysis. Reprinted 1981.

Fuller, P.M., Saper, C.B., & Lu, J. (2007). The pontine REM switch: past and present. *Journal of Physiology-London*, 584(3):735–741.

Garrity, A.G., Botta, S., Lazar, S.B., Swor, E., Vanini, G., Baghdoyan, H.A., & Lydic, R. (2015). Dexmedetomidine-induced sedation does not mimic the neurobehavioral phenotypes of sleep in Sprague Dawley rat. *Sleep*, 38:73–84.

Geschwind, N. (2010). Disconnexion syndromes in animals and man: Part I. *Neuropsychology Review*, 20(2):128–157.

Gleick, J. (1987). *Chaos: Making a New Science*. New York: Vintage.

Greenblatt, S. (2011). *The Swerve: How the World Became Modern*. New York: W.W. Norton and Co.

Hanlon, E.C., & Van Cauter, E. (2011). Quantification of sleep behavior and of its impact on the cross-talk between the brain and peripheral metabolism. *Proceedings of The National Academy of Sciences of The United States of America*, 108 (Supplement 3):15609–15616.

Herculano-Houzel, S., Munk, M.H.J., Neuenschwander, S., & Singer, W. (1999). Precisely synchronized oscillatory firing patterns require electroencephalographic activation. *Journal of Neuroscience*, 19(10): 3992–4010.

Hobson, J.A. (2016). *Dreaming as Virtual Reality*. Self-published.

Hobson, J.A., & Friston, K.J. (2014). Consciousness, dreams, and inference: the Cartesian theatre revisited. *Journal of Consciousness Studies*, 21(1–2):6–32.

Hobson J.A., Hong C.C., & Friston, K.J. (2014). Virtual reality and consciousness inference in dreaming. *Front Psychol.*, 5:1133.

Hofstadter, D.R. (1999). *Gödel, Escher, Bach: An Eternal Golden Braid*. New York: Basic Books.

Hofstadter, D.R. (2007). *I Am a Strange Loop*. New York: Basic Books.

Hughlings Jackson, J. (1998). *Evolution and Dissolution of the Nervous System*. Thoemmes Press: Maruzen Co.

James, W. (1902). *The Varieties of Religious Experience*; a study in human nature; being the Gifford lectures on natural religion delivered at Edinburgh in 1901–1902, by William James. New York: Longmans, Green.

Jouvet, M. (1999). *The Paradox of Sleep: The Story of Dreaming*. Cambridge, MA: MIT Press.

Jung, C. (1961). *Memories, Dreams, Reflections*. New York: Random House.

Kahn, D. & Hobson J.A. (1994). Self organization theory of dreaming. *Dreaming*, 3:151–178.

Kant, I. (1998). *Critique of Pure Reason*. The Cambridge Edition of the Works of Immanuel Kant. New York, NY: Cambridge University Press.

Koch, C. (2004). *The Quest for Consciousness: a Neurobiological Approach*. New York: W.H. Freeman and Co.

Kraeplin, E. (1971). *Dementia praecox and paraphrenia*. Translated by R. Mary Barclay. Edited by George M. Robertson. Huntington, N.Y.: R. E. Krieger Pub. Co.

Krilowicz, B.L., Glotzbach, S.F., & Heller, H.C. (1988). Neuronal-activity during sleep and complete bouts of hibernation. *American Journal Of Physiology*, 255(6): R1008–R1019.

Laureys, S., & Tononi, G. (2009). *The Neurology of Consciousness: Cognitive Neuroscience and Neuropathology*. Academic Press.

Levy, R., & Goldman-Rakic, P.S. (2000). Segregation of working memory functions within the dorsolateral prefrontal cortex. *Experimental Brain Research,* 133:23–32.

Limb, C.J., & Braun, A.R. (2008). Neural substrates of spontaneous musical performance: An fMRI study of jazz improvisation. *PLoS ONE*, 3(2):e1679.

Livingstone, M.S., & Hubel, D.H. (1981). Effects of sleep and arousal on the processing of visual information in the cat. *Nature,* 291(5816):554–561.

Llewellyn, S. (2013). Such stuff as dreams are made on? Elaborative encoding, the ancient art of memory, and the hippocampus. *Behav Brain Sci,* 36(6):589–607.

Loewi, O. (1956). On the intraneural state of acetylcholine. *Experientia,* 12(9): 331–333.

London, J. (1986). *To Build a Fire*. New York: Bantam Classics.

Lucretius (1995). *On the Nature of Things (De rerum natura)* (edited and translated by Anthony Esolen). Baltimore: Johns Hopkins Press.

Lydic, R., McCarley, R.W., & Hobson J.A. (1987). Serotonin neurons and sleep: I. Long-term recordings of dorsal raphé discharge frequency and PGO waves. *Archives Italiennes de Biologie,* 125:317–343.

Lydic, R., McCarley, R.W., & Hobson J.A. (1987). Serotonin neurons and sleep: II. Time course of dorsal raphé discharge, PGO waves, and behavioral states. *Archives Italiennes de Biologie,* 126:1–28.

McGuire, W., ed. (1974). *The Freud/Jung Letters: Correspondence between Sigmund Freud and C.G. Jung*. Princeton, NJ: Princeton University Press.

Mesulam, M.M., Mufson, E.J., Wainer, B.H., et al. (1983). Central cholinergic pathways in the rat — an overview based on an alternative nomenclature (Ch. 1–Ch. 6). *Neuroscience,* 4:1185–1201.

Mesulam, M.M. (1990). Human brain cholinergic pathways. *Progress In Brain Research,* 84:231–241.

Metzinger, T. (2003). *Being No One: The Self-Model Theory of Subjectivity*. Cambridge, MA: MIT Press.

Metzinger, T. (2009). *The Ego Tunnel*. New York: Basic Books.

Miles, F.A., & Evarts, E.V. (1979). Concepts of motor organization. *Annual Review of Psychology,* 30: 327–362.

Minsky, M. & Papert, S. (1969). *Perceptrons: An Introduction to Computational Geometry*. Cambridge, MA: MIT Press.

Moruzzi, G. (1963). Active processes in the brain stem during sleep. *Harvey Lectures*, 58:233–297.

Moruzzi, G. (1966). The functional significance of sleep with particular regard to the brain mechanisms underlying consciousness. In: *Brain and Conscious Experience*, pp. 345–388. Ed. J.C. Eccles. New York: Springer.

Moruzzi, G. & Magoun, H.W. (1949). Brainstem reticular formation and activation of the EEG. *Electroencephalography and Clinical Neurophysiology,* 1:455–473.

Nabokov, V. (1951). *Speak, Memory*. London: Victor Gollancz.

Nauta, W.J.H. (1986). *Fundamental Neuroanatomy*. New York: W.H. Freeman and Co.

Nielsen, T.A. (2000). A review of mentation in REM and NREM sleep: "Covert" REM sleep as a possible reconciliation of two opposing models. *Behavioral and Brain Sciences,* 23(6):851–866.

Pace-Schott, E. (2003). *Sleep and Dreaming: Scientific Advances and Reconsiderations*. Cambridge: Cambridge University Press.

Pompeiano, O., & Morrison, A.R. (1966). Vestibular origin of rapid eye movements during desynchronized sleep. *Experientia*, 22(1):60.

Pompeiano, O. (1967). The neurophysiological mechanisms of the postural and motor events during desynchronized sleep. *Res. Publ. Assoc. Res. Nerv. Ment. Dis.*, 45:351–423.

Pompeiano, O., Manzoni, D., & Barnes, C.D. (1991). Responses of locus-coeruleus neurons to labyrinth and neck stimulation. *Progress in Brain Research*, 88:411–434.

Posner, M.I. (2012). Attentional networks and consciousness. *Frontiers In Psychology*, 3:64.

Rechtschaffen, A., Bergmann, B., Everson, C., Kushida, C., & Gilliland, M. (1989). Sleep deprivation in the rat. X. Integration and discussion of the findings. *Sleep*, 12(1):68–87.

Redfield-Jamison, K. (1996). *Touched with Fire: Manic Depressive Illness and the Artistic Temperament*. New York: Free Press.

Sacks, O. (1985). *The Man Who Mistook his Wife for a Hat*. London: G. Duckworth.

Sacks, O. (2012). *Hallucinations*. New York: Vintage.

Shepard, J.W., Buysse, D.J., Chesson, A.L., et al. (2005). History of the development of sleep medicine in the United States. *Journal of Clinical Sleep Medicine*, 1(1):61–82.

Sherrington, C.S. (1951). *Man on his Nature*. Cambridge: Cambridge University Press.

Singer, J.L., & Antrobus, J.S. (1972). *Dimensions of daydreaming: A factor analysis of imaginal processes and personality scales*. In P. Sheehan (ed.), *The function and nature of imagery*. New York: Academic Press.

Skinner, B.F. (1976). *Walden Two*. New York: Macmillan.

Snyder, F. (1966). Toward an Evolutionary Theory of Dreaming. *American Journal of Psychiatry*, 123 (2):121–142.

Solms, M. (1997). *The Neuropsychology of Dreams: A Clinico-Anatomical Study*. Mahwah, NJ: Lawrence Erlbaum Associates.

Spagna, T., Hobson, J.A., Bonfilio, D., Eldridge, R., & Mark, E. (2013) *Sleep*. New York: Universe.

Steriade, M., & Hobson J.A. (1976). Neuronal activity during the sleep-waking cycle. In: Kerkut G.A., Phillis, J.W., eds. *Progress in Neurobiology*, 6(3,4):155–376.

Thurber, J. (2006). *Secret Lives of Walter Mitty and of James Thurber*. New York: Harper Collins.

Tranquillo, N., Ed. (2014). *Dream Consciousness: Allan Hobson's New Approach to the Brain and Its Mind*. Cham, Switzerland: Springer.

Tranquillo, N., Ed. (2016). *London Bridges*. Self-published.

Twain, M. (1867). *The Celebrated Jumping Frog of Calaveras County, and Other Sketches*. New York: C.H. Webb.

Van Dongen, H.P.A., & Dinges, D.F. (2003). Investigating the interaction between the homeostatic and circadian processes of sleep-wake regulation for the prediction of waking neurobehavioural performance. *Journal of Sleep Research*, 12(3):181–187.

von Helmholtz, H. (1866). Concerning the perceptions in general. In: *Treatise on physiological optics* (3rd ed., Vol. III). (J. P. C. Southall, Trans., 1925 Opt. Soc. Am. Section 26, reprinted New York: Dover, 1962).

Voss, U., Holzmann, R., Tuin, I., & Hobson, J.A. (2009). Lucid dreaming: a state of consciousness with features of both waking and non-lucid dreaming. *Sleep*, 32(9):1191–2000.

Watson, J.D. & Crick, F.H.C. (1953). A structure for deoxyribose nucleic acid. *Nature*, 171:737–738.

Wiesel, T.N., & Hubel, D.H. (1963). Effects of visual deprivation on morphology and physiology of cells in the cats lateral geniculate body. *J. Neurophysiol.*, 26:978–93.

Windt, J. (2015). *Dreaming: A Conceptual Framework for Philosophy of Mind and Empirical Research*. Cambridge, MA: MIT Press.